The Ecology and Economic Impact
of Poisonous Plants
on Livestock Production

Sponsored by
the USDA-ARS Poisonous Plant Research Laboratory,
the Society for Range Management,
and the Utah Agricultural Experiment Station

The Ecology and Economic Impact of Poisonous Plants on Livestock Production

edited by

Lynn F. James, Michael H. Ralphs, and Darwin B. Nielsen

Routledge
Taylor & Francis Group

LONDON AND NEW YORK

First published 1988 by Westview Press, Inc.

Published 2019 by Routledge
52 Vanderbilt Avenue, New York, NY 10017
2 Park Square, Milton Park, Abingdon, Oxon OX14 4RN

Routledge is an imprint of the Taylor & Francis Group, an informa business

Copyright © 1988 Taylor & Francis

```
The Ecology and economic impact of poisonous plants on livestock
   production/edited by Lynn F. James, Michael H. Ralphs, Darwin B.
   Nielsen.
   p.  cm.--(Westview special studies in agriculture science
and policy)
   ISBN 0-8133-7453-7
   1. Livestock poisoning plants--Ecology.  2. Livestock poisoning
plants--Economic aspects.  I. James, Lynn F.  II. Ralphs, Michael.
III. Nielsen, Darwin B.  IV. Series.
SB617.E26 1988
636.089'5652--dc19                                      87-22444
                                                            CIP
```

ISBN 13: 978-0-367-29147-1 (hbk)
ISBN 13: 978-0-367-30693-9 (pbk)

Contents

v

vi

vii

Preface

Livestock poisoning by plants has been a problem to the livestock producers of the United States since our pioneer forefathers first grazed their herds of cattle and sheep on the vast rangelands and pastures of this country. It has long been recognized that poisonous plants are not only disruptive to the harvesting of the forage produced by our grazing lands, but are also one of the serious economic impediments to profitable livestock production.

In the past, most discussions about poisonous plants have centered around the effects of plant toxins on the grazing animal and less about the plant itself or the economic aspects of livestock poisoning. This is perhaps due to the fact that the first, most obvious, and emotionally striking aspect of an intoxication is the sick, dying, or dead animal. Thus the poisoned animal received attention first in the efforts to delineate and prevent livestock poisoning by plants. Yet to fully understand the problem of livestock poisoning by plants, the plant aspects must be defined and quantified. Many environmental and plant factors influence plant availability, toxin concentration, palatability, etc.

The economic aspects of livestock poisoning by plants were recognized early. However, they have been difficult to quantify. Many authors have attempted to give appropriate emphasis to the economics by statements such as, "Poisonous plants are one of the most important causes of economic loss to the livestock industry." But the scope and the elusive, hard-to-define, almost impossible-to-count, nature of the problem has made it difficult, if not impossible, to get a handle on. Yet in today's economic environment, we must have some way of putting a dollar value on the problem. How can we evaluate risk, management programs needed to accommodate poisonous plants, plant control programs, etc., without having some notion of the costs and benefits derived from our actions?

The content of this publication is based upon information given in a day-long symposium on poisonous plants held at the 40th Annual Meeting of the Society for Range Management at Boise, Idaho, February 8-13, 1987. The principal thrust of the symposium was to discuss the ecology of some of the important poisonous plants and the plant-animal interactions and to evaluate the economic impact of these plants on livestock production.

Lynn F. James
Michael H. Ralphs
Darwin B. Nielsen

Acknowledgments

We express our appreciation to those who contributed to the symposium and to the publication of these proceedings. We also acknowledge the support of the Program Committee for the 40th Annual Meeting of the Society for Range Management in allotting us time and facilities to hold the symposium.

Special thanks go to Betty Sigler and Terrie Wierenga for their outstanding efforts in formatting, proofreading, and preparing manuscripts for publication as well as their attention to the many other details involved in completing these proceedings.

Many of the program participants were staff members of various Universities and State Agricultural Experiment Stations. We express our appreciation to the following:

University of Arizona and
    Agricultural Experiment Station
Brigham Young University
Colorado State University and
    Agricultural Experiment Station
Montana State University and
    Agricultural Experiment Station
New Mexico State University and
    Agricultural Experiment Station
Oregon State University and
    Agricultural Experiment Station
South Dakota State University and
    Agricultural Experiment Station
Texas A & M University and
    Agricultural Experiment Station
Texas Tech University
University of Idaho and
    Forest, Wildlife, and Range
    Experiment Station
Utah State University and
    Agricultural Experiment Station
Wyoming State University and
    Agricultural Experiment Station

L.J., M.R., and D.N.

1

Introduction

Lynn F. James

Livestock poisoning by plants has been a problem to livestock producers since earliest times. Rangelands and pastures provide most of the forage for livestock and are of prime economic importance. Since poisonous plants can be found in most plant communities, it is little wonder that problems relating to them are a principal cause of economic loss to the livestock industry. Losses occur from deaths, abortions, birth defects, and pathologic effects on various organs that result in reproductive alterations, decreased performance, weight losses, wasting, and photosensitization. Some plants may interfere with estrus, and lengthened calving intervals may occur in cycling cows poisoned during the breeding season as well as in those that have had plant-induced abortions. Research and field observations have shown that plant-induced abortions and birth defects are more common than once thought. They are difficult to distinguish from viral, bacterial, and nutritional causes of abortion, and thus may go undiagnosed or be looked at with a "no cause" attitude.

Wasting (general deterioration of an animal due to the effect of the plant toxins) of livestock is a common occurrence on rangelands, but difficult to trace to a cause. It may result from sheep grazing sneezeweed (Helenium hoopesii); liver damage associated with cattle poisoned on Senecio spp.; photosensitization resulting in weight losses, abortion, and death from Tetradymia spp.; or emaciation and loss of performance associated with locoweed (Astragalus and Oxytropis spp). All are representative losses that are hard, if not impossible, to evaluate. Yet all available evidence indicates that these losses are extensive. Livestockmen, veterinarians, and land managers worry about these problems, yet solutions are few.

1

Toxins in various species of plants belong to a wide variety of chemical classes. Some affect a single organ; others affect multiple organs. Some affect an organ in one way; others may affect the same organ in a different way. The response of an animal to a specific toxin may also be mollified or intensified by the environmental conditions in which the animal finds itself, i.e. heat, cold, intense sunlight, wind, and moisture. Thus the methods that must be applied in researching, managing, or preventing poisonous plant problems must encompass many disciplines.

In the past, much of the emphasis on the problem of livestock poisoning by plants has been placed on determining the toxic effects of the plants on livestock and less on the plant aspect of the interaction. It is now recognized that the level of the toxin in some poisonous plants varies with environmental factors, such as moisture, temperature, site, stage of growth, plant part, etc., and it is probable that this is true for most poisonous plants. To develop intelligent strategies to prevent livestock poisoning by plants, methods to quantify and monitor plant toxins must be developed. The offending toxin in many plants has not yet been identified.

Livestock poisoning by plants can usually be traced to problems of management, overall range condition, or both, rather than the mere presence of poisonous plants. Many range management practices are conducive to plant poisoning of livestock. For example, lupine is more toxic to cattle than to sheep, yet thousands of sheep have died of lupine poisoning while few cattle have died. Halogeton (Halogeton glomeratus) is about equally toxic to sheep and cattle yet thousands of sheep have died from this plant, but very few cattle. This difference can be attributed to differences in management. Range sheep are kept in large flocks and moved from place to place as the shepherd deems necessary. In contrast, cattle are left to move freely about a range limited only by fences and natural barriers. Sheep are given access to feed and water only when allowed by the shepherd, while cattle search out the necessities as they feel the need.

Plant palatability, though often an unmeasurable entity, is a factor to be considered in livestock management. Some poisonous plants are sometimes palatable and at other times unpalatable. Palatability may vary with stage of plant growth, environmental conditions, and the relative palatability of other plants in the plant community where the animal is grazing. However, many poisonous plants are unpalatable and are consumed only when animals are hungry or when some other situation causes them to graze abnormally. Overgrazed ranges, trailing, trucking, or introducing animals onto a range that is new to them tends to induce hunger. Unfortunately, not all poisonous plants are unpalatable nor are they restricted to overgrazed pastures and ranges. Larkspur (Delphinium spp.) and locoweed, for example, are climax plants and often are readily grazed by livestock. Palatable plants that happen to have poisonous properties do not always kill or otherwise harm animals that consume them. For example, lupine and greasewood (Sarcobatus vermiculatus) may form a beneficial part of an animal's diet. The animal dies only when it consumes excessive amounts of such plants over a short period of time.

Direct effects of poisonous plants on livestock and factors influencing these effects, though difficult to assess, are most obvious. Less obvious, and perhaps even more difficult to assess or quantify, are the indirect effects. These include cost of increased fencing and altered management programs, increased labor cost, supplemental feed, increased veterinary fees, and others. These all must be considered, but their costs vary immensely depending on each individual ranching situation.

We give consideration to large death losses that may often occur in relation to events such as halogeton poisoning while paying little attention to the one or two cow losses per incident associated with waterhemlock (Cicuta douglasii) poisoning. Many animals die in numerous locations, and their deaths are accepted as normal occurrences and no attempt is made to determine the cause.

One thing is very evident: the cost of poisonous plants to the industry is high--very high. This cost almost defies definition because of the multitude of variables and contributing factors. For now, we are left with the next best thing, an educated guess by those most qualified to make it.

Despite the fact that cost figures are difficult to obtain and somewhat tenuous, if we are to deal with the problem of livestock poisoning by plants, we must at least have a best estimate that will allow us to say something about management, risk, evaluating improvement, range programs, determining land value, and other factors involved in livestock product programs. Management decisions require a thorough understanding of costs and benefits associated with shifts in programs to avoid problems like poisonous plants.

It becomes increasingly clear that to deal with the poisonous plant problem in its entirety, one must have a complete knowledge of the plant toxin involved and its effects on the animal and also understand the plant and how it functions in its environment. In other words, one must understand both the animal and plant sides of the equation to solve plant poisoning problems. In this equation, describing the interactions between plant and animal that leads to harm of the animal, one can only describe the plant as poisonous in terms of its effect on the animal. The undisturbed plant in and of itself does not harm an animal. The harm is done only when the animal consumes the plant at a rate and at an amount that causes intoxication in the animal. The conditions under which this occurs are of paramount interest to those dealing with livestock at risk from grazing plants. In this symposium we are addressing some of the more important aspects of this plant-animal interaction.

An additional purpose of the symposium was an attempt to put dollar values on the adverse effects of poisonous plants on livestock production. It is our hope that this effort will stimulate those directly and indirectly associated with poisonous plant problems to expand their interests in this direction and perhaps provide knowledge that will assist in meaningful research to solve some of the problems.

2

Economic Considerations of
Poisonous Plants on Livestock

Darwin B. Nielsen, Neil R. Rimbey,
and Lynn F. James

INTRODUCTION

Man's interest in poisonous plants probably predates
recorded time. His early interest must have been to learn
and avoid those plants that caused problems when used for
human consumption. Slow learners were probably soon
eliminated from the population. In some early civiliza-
tions, poisons derived from plants were used to make weapons
more deadly, thus increasing their effectiveness in hunting
and as a weapon of war. Ancient Greek and Roman herbalists
knew of plant toxicity and recognized the medicinal value of
some plants. For example, a tincture of larkspur was used
to destroy lice and mites in the hair. Other compounds
extracted from larkspurs were used as an insecticide to
control locusts, grasshoppers, maggots, and ticks (Kingsbury
1964, Crawford 1907). Poisonous plants probably have played
a part in man's efforts to provide himself with food and
shelter (the economy of survival) throughout history.

In modern times, man still has problems dealing
directly with the consumption of poisonous plants.
According to the 1975 records of the National Clearinghouse
for Poison Control Centers, plants were the second most
frequent substance ingested by children under 5 years of
age, and plants were the fifth most common substance
ingested by persons of all ages (Oehme 1978, p. 67).

There are other plants that would be classified as
poisonous if some people did not derive "pleasure" from the
use of them. Plant substances such as marijuana, cocaine,
opium, and tobacco have economic consequences associated
with their use that far exceed any associated with poisonous
plants. Man is more ingenious than livestock in continuing
to gain access to these plant substances. However,

5

participants in this symposium have enough challenges dealing with livestock and poisonous plants, so we will leave man's consumption of poisonous plants to the "drug experts."

Poisonous plants cause substantial economic losses and hardships to livestock operators. One result of these losses is that rangeland management becomes more challenging as one attempts to get a reasonable amount of use without excessive livestock losses. Grazing western rangelands and pastures while avoiding losses to poisonous plants is not new, since such losses have occurred since the livestock industry was in its infancy. Dwyer (1978) found evidence in the literature that significant numbers of livestock were being lost to poisonous plants as early as 1850 when ranchers were becoming suspicious of some plants. One author reported the loss of 70 of 104 merino bucks that had eaten "some kind of poison wild pea or bean vines" (Dwyer 1978, p. 13). Livestock losses were reported to occur repeatedly in the same area by these early livestock operators (Dwyer 1978). A Colorado publication (Durrell and Newsom 1936) reported that "each year these plants exact a costly toll of the livestock of this state, often as much as eight percent," although they did not specify whether the percentage applied to all livestock in the state, a particular herd, or a particular class or kind of livestock. One of the first range management textbooks published warned "poisonous plants cause great losses on the western range in death of animals and in decreased value, essentially through lesser gains" (Stoddart and Smith 1943). Many early researchers have noted that poisonous plants cause serious economic problems for the livestock industry. However, they had the same problem we face today: it is impossible, or at least impractical, to get definitive answers to make empirical estimates of the total value of economic losses to the livestock industry caused by poisonous plants.

Poisonous plants can be a threat when found on rangelands, pastures, crop aftermath, and even in some corrals. Hay or grain can be contaminated with poisonous plants. For example, overcrowded livestock may consume every plant in a corral, even those that are poisonous. Another example would be when the wind blows ponderosa pine limbs down or the power company trims them and they fall into the corral where they are consumed by the cows, causing an abortion problem.

Although poisoning of livestock occurs throughout the U.S., James (1978) notes that it is more of a problem in the western states:

"The economic losses of livestock due to poisonous plants are serious. Many of these losses are preventable. Approximately 75% of the land area of the western United States is rangeland. The forage on range areas can best be used by ruminant animals; i.e., sheep, cattle, and some wildlife. The least energy-dependent means of producing red meat is through foraging native range plants that require little fossil fuel production costs, and the forage resource is renewed each year without tillage, fertilization, or mechanical harvesting. Because grazing rangelands provides a most efficient means of converting solar energy into chemical energy (food), we must use this resource wisely." (p. 4)

Poisonous plants are a serious impediment to the wise use of grazing lands in the United States and the world.

STATE OF THE ART IN POISONOUS PLANT KNOWLEDGE

Information has been gathered over the last 85 years concerning the various aspects of poisonous plants and their effects on domestic livestock. Local areas where various poisonous plants are a problem have been identified, but similar information is often not readily available in published form at the state or regional level. As a result, valuable information may be lost if those with personal knowledge about these areas do not publish their results. This would be especially important concerning poisonous plants that rarely cause problems but can occasionally inflict devastating losses under certain conditions.

In general, most poisonous plants do not appear to be spreading into new areas. A few years ago, however, there was concern that halogeton would spread across the Great Basin, but this has not happened. Broom snakeweed is increasing and spreading into new areas. This poisonous plant out-competes the more palatable range plants and seriously reduces the carrying capacity of the rangeland.

Researchers have made substantial progress in establishing the relationship between various compounds in the plant and intoxication or poisoning of the animal. For example, it is known that poisonous plants produce such toxins as hydrocyanic acid, toxic alkaloids, oxalates, coumarin, and swainsonine. However, much more must be learned about the specific effects of the toxin on the animal. For example, larkspur contains many alkaloids, and we need to know which one or which combination causes the problem. If these specific relationships can be determined, one will be in a better position to devise methods of solving the problem or at least dealing with the situation on a different level than we have attempted in the past.

We also must learn to more accurately diagnose intoxication in animals. Animals only have a limited number of ways to show that they have problems. For example, if most poisonous plants caused animals to have a stomach ache, this symptom would not be of much value in determining which plant was causing the problem. Thus, new, accurate, and fast techniques for diagnoses of intoxications would be valuable.

So far, we are unable to predict the conditions under which intoxication or poisoning will occur. Poisonous plants may be abundant, yet no animals are poisoned. Yet, catastrophic loss may occur under similar circumstances. Was poisoning due to the particular animals, something inside the plant, the temperature, the moisture, the moon, or all of the above? Patterns of plant growth also vary. A pasture with few locoweed plants one year may be dominated by locoweed plants the following year. On some ranges, poisonous plants cause problems about every 15 years. What set of circumstances causes this? Can the problems be predicted? Obviously, there is still much to learn about poisonous plants and their effect on livestock.

Yet, as more knowledge is gained, we often find new and unexpected ways that these plants impact livestockmen who are trying to manage their animals within the environment where poisonous plants are a problem.

ECONOMIC IMPACT OF POISONOUS PLANTS ON THE
LIVESTOCK INDUSTRY

Ranchers are concerned because poisonous plants cost them money in increased death losses, reduced weight gains and reproductive efficiency, and decreased livestock values due to unthrifty animals. Added management costs include fencing, riding, veterinary fees, plant control, loss of usable forage due to direct competition from poisonous plants, and restricted land and livestock management options. When reductions in land values are added to these costs, the increased financial risk could put the ranch in jeopardy. Probably the most difficult cost to define is the increased stress on the owner/manager.

Death loss of sheep, cattle, goats, and horses is a very serious problem for some ranchers. These losses may occur in the same areas year after year and may be viewed as a necessary cost of doing business. In other areas, sporadic poisonings can generate considerable publicity and get ranchers very excited because they were unexpected. Currently, there is no system to report and record these losses nor is there likely to be one. Many ranchers do not recognize losses caused by poisonous plants, particularly when few livestock are lost, and they do not occur regularly. Nor are most poisonings brought to the attention of local veterinarians because losses cannot be prevented or the animals have died before treatment was possible. Thus, death losses due to poisonous plants are included in the average annual death loss reported for livestock ranching.

Losses also are caused by reduction in the reproductive efficiency of livestock. Several poisonous plants cause females to abort, to be infertile, to give birth to weak offspring that cannot survive, or to give birth to deformed offspring that are worthless. They also may cause infertility in males. This can be permanent or temporary, but it still impacts the reproduction cycle. The end result is that the calf, lamb, or kid crop is less than it would have been had the poisonous plants not been present.

About ten years ago, an estimate was made of the economic losses caused by poisonous plants, and this was presented in a paper at the annual meeting of the Society for Range Management (Nielsen 1978). Since then, a few comments were made about this estimate, for example: "it's too low," "it's too high," "it's arbitrary," and "who cares." It still appears that no one knows the extent of losses caused by poisonous plants, and obtaining reliable statistical estimates would be too expensive. Even if losses were measured, the estimate would be valid only for the period studied. Those involved in poisonous plant research need the loss data figures to justify and/or increase their programs.

Changes in the way USDA reports livestock numbers has made it difficult to compare livestock numbers over time. The livestock numbers used here are based on Agricultural Statistics for 1984. The estimated number of cattle and sheep that would likely be on rangelands or pastures in the 17 western states is found in Table 2.1.

An estimate of total annual death loss is 3% for cattle and 8% to 10% for sheep (Nielsen 1978). If about one-third of the death loss in these states is attributed to poisonous plants, then the losses would be:

    cattle: 280,560 head (1% of 28,056,000)
    sheep: 264,215 head (3.5% of 7,549,000).

    Current livestock values in the West are based on
information from Market News (Utah Dept. of Agr. 1987).
Average middle-aged stock cows are valued at $500 per head.
The number of replacement heifers reported (3,701,000) is
about 17 percent of the beef cow herd. Therefore, one can
assume that they would be moving into the cow herd within
one year. Given this age estimate, they are valued at $475
per head. Yearling steers are valued at $475 per head. The
average value for bulls, including breeding bulls, is $1,000
per head. Thus, the weighted average value for cattle based
on these values is $518 per head. Sheep are worth $90 per
head. Estimated death losses attributable to poisonous
plants would be those shown in Table 2.2.

TABLE 2.1
Estimated numbers of livestock on rangelands or pastures
in 17 western states

Cattle	
Beef cows	22,043,000 head
Replacement heifers	3,701,000 head
Steers 227 kg & over	1,046,000 head*
Bulls 227 kg & over	1,266,000 head**
TOTAL	28,056,000 head
Sheep	
Ewes, rams, & replacements	7,549,000 head

 *This figure represents 10% of the total number of steers
  227 kg and over (assumed to be on range pasture).
**This figure represents 90% of the total number of bulls
  227 kg and over (assumed to be on range pasture).

TABLE 2.2
Estimated death losses to poisonous plants in the 17
western states

Death Loss:		
Cattle	280,560 head x $518/hd =	$145,330,080
Sheep	264,215 head x $90/hd =	23,779,350
TOTAL		$169,109,430

The reduction in reproduction efficiency mentioned earlier also affects the lamb and calf crop percentages. The most meaningful lamb and calf crop percentage is the ratio of the number of off-spring weaned to the total number of potential mothers. Death loss of calves and lambs prior to weaning is taken care of in the lamb and calf crops. If it is assumed that poisonous plants decrease lamb and calf crops by 1%, then rancher income could be increased by the amounts shown in Table 2.3 if poisonous plants were not a problem.

TABLE 2.3
Estimated losses due to a 1% reduction in lamb and calf crop due to poisonous plants in the 17 western states

Calves	220,430 head x $275/head =	$60,618,250
Lambs	75,490 head x $60/head =	4,529,400
TOTAL		$65,147,650

In summary, an estimate of the total direct losses caused by poisonous plants in the 17 western states is given in Table 2.4.

TABLE 2.4
Estimated total cattle and sheep losses to poisonous plants in the 17 western states

Cattle	1% death loss	$145,330,080
Calves	1% increase calf crop	60,618,250
Sheep	3.5% death loss	23,779,350
Lambs	1% increase lamb crop	4,529,400
TOTAL		$234,257,080

There are additional losses not included in the $234,000,000. There are poisonous plants in the Eastern states, but even less seems to be known about the losses they cause; nor does the above estimate include losses to goats, horses, swine, and big-game animals. In some instances and under certain circumstances, they could be very high. For example, it is known that elk and antelope are intoxicated by locoweed consumption. There also have been losses reported from race horses that have been killed by poisonous plants in hay.

Poisonous plants also reduce animal performance. Smaller calves and/or lambs at weaning or sale time result in lower sales value, which reduces the revenue to ranchers. Calves that recover from poisoning usually require longer feeding periods to reach marketable weights. If poisonous plant consumption results in animals that have reduced performance potential through the finishing phase of production, the ranch or area where these animals originated can get a reputation that causes their calves and lambs to be discounted in the feeder markets.

Consumption of some poisonous plants may predispose animals to other diseases. Research in Utah has shown that calves that consume locoweed are predisposed to high altitude disease. Any time the consumption of poisonous plants reduces an animal's resistance to disease or causes the disease to be more severe than normal, there is a cost that should be assessed against these plants. This problem has never been given serious consideration in evaluating the economic significance of poisonous plants to the livestock industry.

Ranchers have significant increases in the cost of managing their livestock enterprise in areas where poisonous plants are common. These costs include: labor to find, treat, and/or remove affected animals; the cost of building and maintaining fences to keep livestock away from the plants; and underutilization of forage resources (ranges might be grazed at lighter stocking rates or livestock be kept off the ranges until late in the season in an attempt to reduce the consumption of the problem plants). Additional costs are incurred in controlling the plants, which include spraying with herbicides, grubbing or plowing, pulling the plants, burning, mowing, or any other control programs. There are areas in the West that are almost forced to produce one type of animal because of poisonous plants. Larkspur-infested ranges may have been limited to sheep production for years because ranchers do not put cattle on them for fear of incurring heavy death losses. This is an economic loss to the rancher in that his potential income is reduced because he cannot change to cattle or to a combination of sheep and cattle. This is an opportunity cost associated with failure to make the optimum use of resources.

Ranchers who operate where poisonous plants are a serious problem have added pressures which are impossible to measure, but which take their toll in the quality of life that they experience. Knowing that one mistake could seriously affect the future of the ranch would be a constant burden on the manager.

The increased uncertainty and risk caused by poisonous plants can have direct economic impacts on ranchers. Uncertainty makes it more difficult for ranchers to borrow money. If the potential losses from poisonous plants make ranch loans too risky, lenders are likely to refuse to make a loan or to increase interest rates and collateral requirements and shorten the loan repayment period.

Another area where poisonous plants can have a bearing on the economics of ranching is a reduction in real estate values caused by poisonous plant problems. This is a result of the other costs discussed above. There is an element of double counting if both losses are added together. That is, one may be prone to count the costs of lower net ranch income and also include the loss in value of the ranch/real estate. Many ranchers are reluctant to permit research on their ranches concerning poisonous plants because they fear calling attention to problems can reduce real estate values.

Many poisonous plant situations have the potential of putting ranchers out of business when a given set of circumstances occur to cause catastrophic losses. In some cases, ranchers may not be aware of the potential problem because they have just purchased a ranch, added new property to the ranch, or changed the livestock on their rangeland. A different herd of cattle or sheep grazing a range can result in catastrophic losses because new animals may graze differently than do animals that previously used these lands. Changing from sheep to cattle might cause an economic disaster on a larkspur range if the owner is not aware of the potential danger. A unique set of weather and vegetative growth circumstances could cause poisonous plant problems not usually encountered. Changing a herder without proper instructions or training concerning grazing management can set up the conditions for a problem. If the manager becomes lackadaisical, even though he knows he has poisonous plant problems, he can be setting himself up for failure.

Most ranchers want to hold on to their ranches at nearly any cost. Therefore, the potential for economic disaster is a very important consideration when dealing with the economics of poisonous plants. Even though the average loss over time might be something they can live with, a single loss so severe as to put a rancher out of business is possible. Several such losses have been reported to the Poisonous Plant Laboratory in Logan, Utah. A windstorm blows pine trees down and a rancher's cattle consume enough needles to cause 80 to 90 percent of his cows to abort. Sheep losses due to halogeton have the potential of putting sheepmen out of business. Sheep herds grazing near the Utah/Nevada border had losses of 450 head, 600 head, and 800 head. A rancher in Antelope Valley, Utah, lost 1,200 sheep at one time, a Nevada rancher lost 1,300 head at one time, while another lost 2,000 head over several years. In 1958, over 6,000 head of sheep died from locoweed poisoning in the Uintah Basin of Eastern Utah. In 1964, one sheep rancher lost $125,000, another rancher lost $65,000, and a third lost $55,000 due to locoweed poisoning. Locoweed consumption can cause abortion rates of up to 100% in cattle and sheep.

Ranchers are always interested in the market value of their livestock. There is much talk around "cow country" each fall about how and where to market for the best price. Expressing losses due to poisonous plants in terms of dollars per hundred weight might increase ranchers' interest in the problem. For example, the Manti Cattle Association grazes 837 cows on an allotment in the Manti-La Sal National Forest, and 36 head die from larkspur poisoning annually. Research has shown that 33 head could be saved annually by controlling larkspur. Assuming an 85% calf crop and a 20% replacement rate, 545 calves could be sold from these cows each year. If each cow saved is valued at $500, the annual savings would be $16,500. This $16,500 savings is equivalent to a $6.73 per cwt increase in the price of 545 calves averaging 450 lb per head.

At the present time, ranchers in the U.S. are under severe financial stress. Agricultural land values are declining or static or not increasing enough to offset the effects of the cost/price squeeze. Many financial institutions are reevaluating their ranch loans and often have given notice that credit has been overextended and that future credit will be much harder to secure. Some ranchers face foreclosure unless they can pay off a considerable portion of their debt. These financial pressures mean that a relatively small loss due to poisonous plants can cause a more serious financial problem than five or ten years ago. Also, ranchers will find it harder today to get financial help to control poisonous plant losses than in the past.

In response to these pressures, financial institutions are encouraging producers to consider cutting back their operations. The area of the operation selected for liquidation or cutbacks is often the range enterprise, particularly public range enterprises. This reflects the view of many bankers that the range enterprise is an area that generates only costs to the ranch operation with very little contribution to revenue.

LITERATURE CITED

Crawford, A.C. 1907. The larkspurs as poisonous plants. USDA Bureau of Plant Industries, Bull. 111.

Durrell, L.W., and I.E. Newsom. 1936. Poisonous and injurious plants of Colorado. Colo. Agr. Expt. Sta. Bull. 429.

Dwyer, D.D. 1978. Impact of poisonous plants on Western U.S. grazing systems and livestock operations, p. 13. In: R.F. Keeler, K.R. Van Kampen, and L.F. James (eds), Effects of poisonous plants on livestock. Academic Press, New York.

James, L.F. 1978. Overview of poisonous plant problems in the United States, p. 4. In: R.F. Keeler, K.R. Van Kampen, and L.F. James (eds), Effects of poisonous plants on livestock. Academic Press, New York.

Kingsbury, J.M. 1964. Poisonous plants of the United States and Canada. Prentice-Hall Inc., Englewood Cliffs, NJ.

Nielsen, D.B. 1978. The economic impact of poisonous plants on the range livestock industry in the 17 western states. J. Range Manage. 31:325-328.

Oehme, Frederick W. 1978. The hazard of plant toxicities to the human population. In: R.F. Keeler, K.R. Van Kampen, and L.F. James (eds), Effects of poisonous plants on livestock. Academic Press, New York.

Stoddart, L.A. and A.D. Smith. 1955. Range management. McGraw-Hill Book Company Inc., New York.

USDA. 1984. Agricultural statistics. US Govt. Printing Off., Washington, DC.

Utah Department of Agriculture. 1987. Market News 65(1&2). Div. of Agr. Devel. and Marketing, Salt Lake City, UT.

3

The Economic Impact
of Poisonous Plants on Land Values
and Grazing Privileges

E. Bruce Godfrey, Darwin B. Nielsen,
and Neil R. Rimbey

INTRODUCTION

Much will be said about the economic impact of specific
poisonous plants in the chapters written by others in this
book. This chapter and the preceding chapter by Nielsen et
al. (1987) outline the general principles needed to evaluate
the economic impact of poisonous plants on ranching
operations.   Nielsen et al. (1987) outlined the general
magnitude of the problem in the preceding chapter--a macro
orientation. This chapter will outline how the existence of
poisonous plants affect the value of land and grazing
privileges used by an operator--a micro orientation.

One general caveat should be understood at the outset
of this chapter. The numbers reported are fictitious because as far as we are able to determine, no empirical
estimates have been made and reported in the literature
concerning the firm or rancher level impacts outlined below.
The estimates are based on considerable professional
experience but will undoubtedly be altered by empirically
based estimates.

VALUATION OF ASSETS

There are basically three approaches to placing a value
on assets as outlined in appraisal literature.   These
include: (1) the sales or market; (2) income; and (3) cost
approaches.   The cost approach is of little use in
approaching the valuation of land and grazing privileges.
As a result, no further comment on this approach will be
given herein. However, both the income and sales approaches
can be used in evaluating these impacts.

Income Approach

The income approach uses the principles associated with capitol budgeting. The income approach basically can be summarized by the following formula:

$$\sum_{j=1}^{N} NR_j/(1+i)^j$$

where NR_j = net returns in period j, i = the discount rate, and N = length of the planning horizon or period over which the returns are to be received.

For example, if a rancher could lease a piece of property to his neighbor for 30 years with a payment of $8 per acre, the value of that stream of income would be $122.98

$$\sum_{j=1}^{30} \frac{8}{(1+i)^j}$$

if a discount rate of five percent was used. One could suggest that this is a fairly short planning horizon. If an infinite planning horizon was assumed (n=∞), this would only increase the value of this stream of income to $160 ($8/.05). Thus, the capitalized value of the land would be only $160 if a discount rate of five percent was used to discount a perpetual stream of income of $8 per year. This ($160) represents the income value of this land. If a higher discount rate and/or shorter planning horizon was assumed, this would lower the value of this stream of income to less than $160. If a lower discount rate (i < 5%) was assumed, this would increase the present value of this stream of income or value of the land. This basic approach has been used to value various types of resources including federal grazing privileges or permit values (Roberts 1963, Nielsen and Workman 1971). Many believe that this is the most valid approach to use when valuing resources because it reflects what the owner of the resources could afford to pay for the resource from the income to be received. However, the sales approach is also used to value resources such as land and grazing privileges.

Sales or Market Approach

The sales or market approach assumes that "a prudent person will not pay more for property than the amount for

which a comparable substitute property can be bought"
(American Institute of Real Estate Appraisers 1983). This
method relies on market sales to estimate value. The amount
willing buyers pay for comparable lands is used to estimate
the value of lands that may be sold. If a sale is made for
land that is not comparable to the subject property,
adjustments may be made in the sales value to make the two
pieces of property comparable. For example, if one parcel
of land has poisonous plants that detrimentally affect the
returns obtained from using the area compared to another
parcel that did not have poisonous plants, some downward
adjustment would be made. This adjustment would be
necessary because one would expect the parcel having
poisonous plants to sell for less than a similar parcel not
having poisonous plants as long as the buyer and seller were
aware of the existence and impact of these plants.

It should be noted that there is a basic difference in
what the value represents when the sales and income
approaches are used. The sales approach to value reflects
the amount a purchaser of a resource is "willing to pay"
while the income approach reflects what the purchaser is
"able to pay" from expected revenues. In general, the sales
and income approaches to value should yield essentially the
same value. When these estimates differ significantly at
some point in time, an adjustment will usually occur. For
example, it was common for estimates of land values using
the sales approach to be two or more times as great as the
value using the income approach in the late 1970s and early
1980s. One of the primary reasons for this difference was
the expectation that land values would continue to increase.
A common reason given for this expectation was that the
amount of land was fixed and that land could only increase
in value. However, this expectation was subsequently proven
to be false because land prices have fallen nearly 50
percent in many areas during the last three or four years.
This decline in the sales value has essentially made the
market and income value equal for lands in many areas. In
areas where they are not approaching equality, further
declines in the market value of land are expected.

IMPACT OF POISONOUS PLANTS ON ASSET VALUES

The preceding section outlines the general approaches
used to estimate the value of assets through appraisal
techniques. Once a value has been estimated for an asset
such as grazing land, we can turn our attention to the

impact any deleterious or detrimental agent or event may
have on asset values. The basic approach was summarized in
an article by Davis and James (1971): "anything that de-
tracts from or limits the ability of rangeland to serve this
function or increases costs of operation and management
decreases its productive value." In other words, any factor
that reduces the net returns that might be obtained from
using an area will diminish its value. Poisonous plants and
other noxious weeds are some of the factors that can reduce
net returns associated with the use of a piece of property.
These impacts can be estimated by using a "with" versus
"without" approach in the basic discounting or capitol
budgeting procedure discussed above. This is illustrated in
the following formula:

$$\sum_{j=1}^{N} \frac{NR(wo)_j}{(1+i)^j} - \sum_{j=1}^{N} \frac{NR(w)_j}{(1+i)^j}$$

where $NR(wo)_j$ = net returns without poisonous plants,
$NR(w)_j$ = net returns with poisonous plants, i = the discount
rate, and N = the planning horizon.

An analysis of the results of this formulation would
reveal the impact on asset values from the existence of
poisonous plants on an area of land. In general, if some
factor, such as the existence of poisonous plants, yielded
lower net returns "with" the factor than the net returns
"without" the factor, this would yield a corresponding
reduction in asset values. For example, if poisonous plants
reduced the net returns from $8 per acre to $6 per acre,
this would reduce the value of the land from $122.98 to
$92.23 if a five percent discount rate and a 30-year plan-
ning horizon were assumed. In theory, if poisonous plants
have a negative impact on net returns, the difference should
be reflected in decreased asset values (land values and/or
grazing permit values).

It must be emphasized that if differences in asset
values associated with poisonous plants are correctly
estimated, they should be approximately equal to the general
losses estimated by Nielsen et al. (1987). Thus, the
estimates in Chapter 2 cannot be added to the values
estimated using the procedures outlined in this chapter
because this would effectively be double-counting the
economic impact of poisonous plants. The procedures
outlined above may appear to be fairly straightforward but
numerous problems are encountered in estimating the net

returns suggested above. The basic equation for calculating net returns associated with an enterprise is:

net returns = gross returns or revenue less total expenses

In general, poisonous plants can impact net returns by changing revenues, costs, or both. These possible impacts are outlined below.

Returns

Perhaps the most obvious impact associated with poisonous plants is when an animal dies. However, this is not the only factor that can affect the revenue or returns side of the net returns equation. For example, animals that ingest poisonous plants may not die but their performance (weight gain, reproduction) may be impaired. Some operators may also be forced to graze a different type of animal (e.g., sheep versus cattle) if poisonous plants exist in that area. Any combination of these as well as other possible factors could affect the returns obtained from using an area where poisonous plants exist. These impacts, especially death losses, tend to be quite obvious.

Not all impacts associated with poisonous plants result in reduced revenues. For example, some operators may find it necessary to breed livestock so they calve or lamb earlier in an effort to reduce the chance of offspring consuming some plant(s) because the toxicity of some poisonous plants varies greatly over time. This will often yield higher weaning and/or sale weights. Obviously, there are additional costs and risks associated with this change that have to be considered. These changes may positively affect gross returns, but the existence of poisonous plants will generally yield decreased net returns because costs increase more than returns.

Additional Costs

In most cases, the costs incurred by an operator to avoid the effects caused by the existence of poisonous plants are not as obvious as are losses in revenues outlined above. A number of extra costs are often incurred when an operator tries to use an area having poisonous plants. Some of the most obvious costs include: (1) elimination of the plant(s); (2) prevention of animals from using an area; and

(3) taking the risks associated with the use of an area infested with poisonous plants.

One of the first alternatives that some operators consider is the elimination or reduction of poisonous plants in an area. Some operators invest considerable time, effort, and/or money to eradicate poisonous plants. For example, a large but unknown portion of the rangeland that was plowed, sprayed, or burned and seeded primarily to crested wheatgrass in the southern part of Idaho in the 1950s and 1960s was treated in an effort to eliminate the effects of halogeton (Godfrey 1972). Smaller areas are often treated intensively (e.g., hand grubbing) to reduce the potential for the spread of a poisonous plant.

Several alternatives exist that can be used to prevent animals from using an area where the risks of losses from poisonous plants are high. For example, some operators will spend considerable effort to herd animals in an attempt to keep them away from areas where losses are likely to occur. They may also fence an area so animals cannot use that area. Fencing, herding, and other means of excluding animals from using an area increase the costs of operating in areas infested with poisonous plants.

Sometimes efforts to prevent the possible use of an area having poisonous plants can be very expensive. One example will illustrate this problem. Some operators near Challis, Idaho, were not willing to risk the chance of cattle losses to larkspur (Delphinium occidentale) on a Forest Service allotment and were unable to remain on a bordering Bureau of Land Management (BLM) allotment until the larkspur matured and its toxicity declined. This required one of three alternatives to be used. First, the animals could be gathered and taken back to private sources of forage. This, however, was a very expensive alternative. Secondly, the animals could use BLM land longer. This was the preferred alternative for these operators. However, this would put additional grazing pressure on BLM lands which would not be tolerated by BLM administrators. The third choice was to use the Forest Service allotment and either risk losses or eradicate or control the larkspur. Each of these alternatives was expensive. This case, however, does illustrate two factors that must be considered in evaluating the effects of poisonous plants. First, use (or nonuse) of an area where poisonous plants exist often has an impact on the use of other areas or resources. An evaluation must, therefore, consider use of the total system and not just the effects that may be associated with a particular piece of property. Secondly, the effects of using an area having poisonous plants are rarely known with certainty. Risk is always involved. As a result, these risks must be considered when the economic effect of poisonous plants is being estimated.

Risks

Any evaluation of the losses associated with poisonous plants must either explicitly consider how risks vary over time or assume that the losses measured in any single year are typical or average. Risk can be evaluated in several ways. First, one could try to obtain some estimate(s) of the probability of losses (reduced returns and/or additional costs) that might be incurred if lands having poisonous plants are used. This would provide a basis for estimating the expected value of these returns. This type of evaluation is illustrated in Table 3.1. These data would allow one to estimate the expected value (the sum of the estimated net returns times their respective probability of occurrence) and the standard deviation of these expected values. For example, the hypothetical data illustrated in Table 3.1 indicate not only that the expected value of the returns are lower with poisonous plants ($7.80 without versus $5.30 with poisonous plants), but the risks are higher because the standard deviation of the returns is higher ($1.25 without versus $2.54 with poisonous plants). These data suggest that the value of lands having poisonous plants should not only be less than those not having poisonous plants but that the variability in their value should also be greater.

Another method that could be used to account for the uncertainty associated with using areas infested with poisonous plants would be to increase the discount rate used (a risky versus a relatively riskless rate) to derive land or permit values. This would have the same general effect (the amount of magnitude may be different) as finding the expected values of the returns.

The data in Table 3.1 also can be used to illustrate how poisonous plants would affect land values. If a 30-year planning horizon and a five percent discount rate were assumed, the data in Table 3.1 would suggest that the value of the land without poisonous plants ($111.91 per acre) would be one-third larger than it would be if these lands had poisonous plants ($81.47). The difference in net returns ($2.50 = $7.80-$5.30) illustrated in Table 3.1 that may result from reduced production and/or increased costs

TABLE 3.1
Hypothetical net returns per hectare from using an area with versus without poisonous plants

Returns without poisonous plants:

Returns	Net Probability	Expected Value
$10.00	.2	$2.00
8.00	.3	2.40
7.00	.4	2.80
6.00	.1	.60
5.00	.0	.00
4.00	.0	.00
0.00	.0	.00
Total		$7.80
Standard deviation of returns		$1.25

Returns with poisonous plants:

Net Returns	Probability	Expected Value
$10.00	.1	$1.00
8.00	.1	.80
7.00	.2	1.40
6.00	.2	1.20
5.00	.1	.50
4.00	.1	.40
0.00	.2	.00
Total		$5.30
Standard deviation of returns		$2.54

could make the use of some areas unprofitable. It should be emphasized that these differences will vary by area, and generalizations are difficult at best. It also indicates that lawsuits may be successfully tried in court when lands having poisonous plants are misrepresented by sellers-- buyers must consider the possibility of poisonous plant problems whenever lands are sold.

No data are currently available that can be used to perform the analysis suggested above. However, Rimbey (1982) surveyed 26 livestock operators to determine the cost of their grazing on federal lands. He interviewed two operators who had significant losses associated with poisonous plants. Sixty percent of the ewes reported lost during a three-week period were due to poisonous plants--an estimated death loss of $900. A cattle operator had death losses associated with grazing an area due to poisonous plants that amounted to $7.32 per AUM, which is nearly four times the current federal grazing fee per AUM. This suggests that the costs associated with using an area having poisonous plants can be substantial for some operators. These costs would result in decreased value of the lands and/or grazing privileges owned by these operators.

LITERATURE CITED

American Institute of Real Estate Appraisers. 1983. The Appraisal of Rural Property. American Institute of Real Estate Appraisers, 430 North Michigan Ave., Chicago, IL. p. 123.

Davis, L.H., and L.F. James. 1972. Impact of poisonous plants on rangeland appraisal and management. J. Amer. Soc. of Farm Mgrs. and Rural Appraisers 36:49- 54.

Godfrey, E.B. 1972. Rangeland improvement practices in Idaho. Forest, Wildlife, and Range Experiment Station. Information series No.1, Univ. of Idaho, Moscow.

Nielsen, D.B., N.R. Rimbey, and L.F. James. 1987. Economic considerations of poisonous plants on livestock. In: L.F. James, M.H. Ralphs, and D.B. Nielsen (eds), The ecology and economic impact of poisonous plants on livestock production. Westview Press, Boulder, CO.

Nielsen, D.B., and J.P. Workman. 1971. The importance of renewable grazing resources on federal lands in the 11 western states. Utah Agr. Exp. Station Circular 155.

Rimbey, N.R. 1984. Economics and poisonous plants. Presented at the annual meetings of the Idaho Weed Control Association held at Boise, ID, March 21, 1984.

Roberts, N.K. 1963. Economic foundations for grazing fees on public lands. J. Farm Economics 45(4):721-732.

4

Ecological Status of
Poisonous Plants on Rangelands

W.A. Laycock, J.A. Young, and D.N. Ueckert

INTRODUCTION

Poisonous plants often are the least understood and most misrepresented class of plants on rangelands. Most poisonous plants are forbs, and Cook (1983) indicated that most range managers treat forbs as weeds or least desirable plants. The usual characterization of poisonous plants is that they are unpalatable and do not occur in abundance on properly grazed rangelands. One of the leading text books on range management (Stoddart et al. 1975) suggested that overgrazing is the major cause of plant poisoning because poisonous plants increase with overuse as the desired forage species decrease. Animals are then forced to eat poisonous plants. Some poisonous plants fit this characterization, but many others are a natural part of high condition range communities and should be recognized as such. We hope that this symposium will lead to a better appreciation of the ecology of poisonous plants.

COMMUNITY ECOLOGY OF POISONOUS PLANTS

Because of the great variation in poisonous plants, they are represented in a host of plant communities from early seral to climax. We intended to develop a comprehensive list of poisonous plants by occurrence in different seral stages or with differential responses to grazing. To do this, we examined range site descriptions and condition criteria of the major land management agencies. We found that most lists either ignored poisonous plants, gave them low species composition values, or classified them as increasers, invaders, or undesirable, often without regard to their true ecological status.

27

Native poisonous plants should be classified the same as all other plants when considering the ecological status of plant communities. The fact that these plants contain chemical compounds that may be toxic to animals has no bearing on the relationship between plants. The toxic properties are important only when considering animal/plant relationships. The animal attacks the plant, not the reverse. The presence of poisonous plants usually is a problem only when management, weather, or other conditions are such that animals graze these plants excessively.

It seems anachronistic that range condition criteria that are supposed to be based on ecological principles ignore, downplay, or simply misinterpret the role of poisonous plants in pristine communities. What agencies advocate as a strictly ecological approach to determine range condition actually is something between an ecological rating and a resource value rating. The latter approach allows management goals to be set which include lesser amounts of poisonous plants than might have been present in the original community.

Table 4.1 contains a list of some of the species of poisonous plants that occur in climax or high-seral plant communities in the western United States. The list is not intended to be comprehensive or complete. Of the species listed, the following may decrease with grazing pressure: white snakeroot, deathcamas, some locoweeds, some lupines, some larkspurs, and rayless goldenrod. Grazing has little effect on many of the woody poisonous species such as greasewood and chokecherry. Other species listed in Table 4.1 are natural components of pristine rangelands but increase with grazing pressure. Table 4.2 lists some of the species of poisonous plants that are aliens, annuals, invade overgrazed rangelands, or are characteristic of low-seral situations. The seasonal and year-to-year abundance of many of the species listed in Tables 4.1 and 4.2 is largely regulated by precipitation as it affects above-ground biomass, seed germination, and seedling establishment.

Except for the compounds that are digestion inhibitors (the tannins in the oaks and the terpenoids in sagebrush), no group of compounds appears to be confined to just climax poisonous plants (Table 4.1) or low-seral plants (Table 4.2). According to Cates and Orians (1975), digestion-inhibiting compounds are most prevalent in perennial shrubs and trees that are available to herbivores during the entire year.

TABLE 4.1
Examples of species of poisonous plants that are present in "climax" or high-seral plant communities.

Common Name	Scientific Name	Poisonous Compound	Response to Grazing
WEST-WIDE			
Locoweeds	Astragalus mollissimus, A. lentiginosus, A. wootonii, Oxytropis lambertii	Alkaloids	Increasers, decreasers, or not affected
Waterhemlock	Cicuta spp.	Cicutoxin	Not affected
Poison-hemlock	Conium maculatum	Alkaloids	Not affected or increaser
White snakeroot	Eupatorium rugosum	Tremetol	Decreaser
Oaks	Quercus spp.	Tannins	Shrubs or trees--increasers or not affected
Greasewood	Sarcobatus vermiculatus	Oxalates	Shrub--not affected
Groundsels	Senecio longilobus and S. riddellii	Pyrrolizidine alkaloids	Sub-shrubs--increasers
Snakeweeds	Xanthocephalum sarothrae and X. microcephala	Saponins	Sub-shrubs--increasers but cyclic
Deathcamas	Zigadenus spp.	Alkaloids	Decreaser or not affected
NORTHERN AND CENTRAL ROCKIES, GREAT BASIN			
Sagebrush	Artemisia spp.	Terpenoids	Increasers or decreasers
Larkspurs	Delphinium spp.	Alkaloids	Some species are decreasers
Orange sneezeweed	Helenium hoopesii	Glycoside	Increaser
Lupines	Lupinus spp.	Alkaloids	Some species are increasers, others are decreasers
Chokecherry	Prunus virginiana	Cyanogen	Shrub--not affected by grazing
False hellebore	Veratrum californicum	Alkaloids	Probably an increaser
SOUTHWEST AND SOUTHERN PLAINS			
White thorn	Acacia constricta	Cyanogenic glycoside	Increaser on some sites
Lecheguilla	Agave lecheguilla	Saponin	Increaser
Twoleaf senna	Cassia roemeriana	Glycoside	Increaser
Rayless goldenrod	Isocoma wrightii	Tremetol	Sub-shrub--decreaser
Sacahuista	Nolina texana	Saponin	Increaser

TABLE 4.2
Examples of species of poisonous plants that are aliens, invaders or characteristic of early-seral plant communities.

Common Name	Scientific Name	Poisonous Compound	Response to Grazing
WEST-WIDE			
Milkweeds	Asclepias spp.	Cardiac glycoside	Perennial--increaser or invader
Jimmyweed	Datura stramonium	Alkaloids	Early seral
Tansymustard	Descurainia pinnata	Unknown	Annual--early seral
Bitterweed	Hymenoxys odorata	Sesquiterpene lactone	Increaser or invader
Summercypress	Kochia scoparia	Saponins, Alkaloids, Oxalates, Nitrates	Alien annual invader
Nightshades	Solanum spp.	Alkaloids	Increaser, invader, or early seral
Cocklebur	Xanthium strumarium	Glycoside	Annual
GREAT BASIN			
Halogeton	Halogeton glomeratus	Oxalates	Alien annual invader
SOUTHWEST AND SOUTHERN GREAT PLAINS			
Milkvetch	Astragalus emoryanus	Miserotoxin (a nitro-compound)	Winter annual, early seral
Coffeesenna	Cassia occidentalis	Glycoside	Early seral
Alfombrilla	Drymaria arenarioides	Saponins	Invader or increaser
Burroweed	Happlopappus tenuisectus	Tremetol	Sub-shrub, invader or increaser
Sneezeweeds	Helenium amarum and H. microcephalum	Sesquiterpene lactones	Invader or increaser-- early seral
Lobelia	Lobelia berlandieri	Alkaloids	Annual increaser or invader
Paperflower	Psilostrophe tagetina and P. gnaphalodes	Psilotropin (sesquiterpene lactone)	Increasers and invaders
Johnson grass	Sorghum halepense	Cyanogen	Alien perennial grass--invader

EVOLUTION OF POISONOUS PLANTS

Some understanding of how the poisonous properties of plants may have evolved is important to understanding the role of poisonous plants in natural communities. This section draws heavily on information summarized by Laycock (1978) and Cronin et al. (1978).

What is a poisonous plant? A poisonous plant contains some specific substance which, when consumed by herbivores in sufficient quantities and under specific circumstances, causes injury to susceptible animals. Have toxins in plants evolved as defense mechanisms against herbivores? Not much information exists to answer this question, and the subject has been almost completely ignored in range management literature until recently. However, in entomology and plant biochemistry, the role of toxic secondary substances in plants has been a subject of debate for many years. Much of the early literature considered these compounds as waste products or as having unknown functions (Muenscher 1958).

Stahl (1888) may have been the first to suggest that poisonous compounds are defense mechanisms against herbivores. This hypothesis has been strongly defended by Janzen (1973), Freeland and Janzen (1974). Rhoades and Cates (1976), Swain (1977), and others. Literature concerning insect herbivores contains many theories and discussions of the possible ways some toxic compounds may function as anti-herbivore defense systems (Laycock 1978). The effectiveness of these compounds as defense mechanisms is relatively clear. However, proof that they evolved for that purpose is impossible.

The reasons advanced to support the defense mechanism theories include:

1. In some plants, concentration of poisonous compounds is so great or the compounds are so highly complex structurally (Swain 1977) that the energy cost of producing and storing them would appear to be too high unless the compounds had some function to increase fitness, such as a defense mechanism against herbivores (Rhoades and Cates 1976).
2. The great number of different kinds of poisonous compounds and species of plants containing these poisons appears to be too high for their evolution to be accidental.
3. Few poisonous compounds can be classified either as by-products or as compounds

essential for plant metabolism. This may imply that these compounds serve other functions, such as defense mechanisms against herbivores.

4. Insects and some large herbivores have developed resistance to or ways to detoxify, sequester, or otherwise render ineffective specific plant poisons. This suggests coevolution of poisonous plants and herbivores.

DOES TOXICITY CONVEY AN ECOLOGICAL ADVANTAGE?

The potential ecological or evolutionary advantage of being poisonous is difficult to assess. Assuming that the production of toxic substances by plants is always a defense against grazing predation probably is erroneous. For example, oxalates in halogeton may help the plant overcome osmotic stress and, by doing so, reduce moisture stress. The succulence of halogeton is proportional to the concentration of sodium or potassium chloride in the growth medium; the chloride ion is thought to be the element responsible for increased succulence (Williams 1960). If so, then the absorption of potassium or sodium chloride must result in an excess of potassium or sodium cations. Production of oxalic acid provides a means of tieing up the excess cations to produce an acceptable sodium balance in halogeton (Waisel 1972). This explanation appears logical in view of the increasing concentration of oxalates in halogeton (Williams 1960) as the soil moisture is depleted over the growing season (Cronin 1965).

If toxic secondary compounds in plants function as defense mechanisms, there are various ways that they might operate, including: (1) extreme toxicity; (2) poisonous properties associated with palatability; and (3) aversive conditioning. All proposed mechanisms assume that being poisonous reduces photosynthetic tissue removed by herbivores which, in turn, maintains the vigor and enhances survival of the plant.

Extreme toxicity

Poisonous species might gain a competitive advantage if they were so toxic that consumption by an herbivore resulted in death or lowered fitness of the animal in terms of growth rate or fecundity (Rhoades and Cates 1976). However, for large herbivores such as cattle, horses, and sheep, this assumption may be false. Being poisonous cannot reduce consumption of a plant's photosynthetic tissue unless the plant is so acutely toxic that animals drop dead at the first taste (a rare occurrence) or unless animals can sense the toxicity of the plant and do not graze the herbage.

Most large herbivores are generalists, i.e., they eat a great variety of plants. If a plant species eaten by a generalist kills or makes the animal ill, thus reducing food intake, the "advantage" to the poisonous species may or may not accrue. Affected herbivores would not only stop feeding on the poisonous plants but also on all other palatable plant species. Thus, the poisonous plants might not gain any competitive advantage in the community or any long-term evolutionary advantage.

Reduction in consumption because of extreme toxicity would be most effective if the herbivores involved were specialists, i.e., those that feed only on one (or a few closely related) plant species. Many insects are specialists and it seems possible that, for many plant species, poisonous properties have evolved as defense mechanisms against insects (Feeny 1975) or against disease-causing organisms (Swain 1977). The fact that they are poisonous to large herbivores may be accidental. Examples of this may be some species of Lupinus and Astragalus. Both are subject to heavy seed predation by insect larvae. It is often impossible to find more than an occasional intact seed. This seed predation would seem to exert tremendous selective pressure on populations for the evolution of metabolic chemicals to provide resistance against the insects (Breedlove and Ehrlich 1968). This resistance to insects may be at least partially responsible for these plants being poisonous to livestock.

Poisonous Properties Correlated With Palatability

Unpalatable plants may gain a competitive advantage in a community that is subjected to heavy grazing pressure. However, poisonous plants occur in a wide spectrum of palatability (Cronin et al. 1978). In addition, palatability is relative because it depends upon the species of animal, the choice of other plants available, and many other factors. Linking poisonous properties to palatability may not explain how poisonous properties might work as defense mechanisms because an unpalatable plant does not need to be poisonous to have a competitive advantage over palatable plants (Bate-Smith 1972). For unpalatability to have operated as an evolutionary factor, it probably has to be linked to aversive conditioning.

Aversive Conditioning

This phenomenon is also called conditioned taste aversion or learned food aversion. The basic premise is that, if an animal consumes a flavored food and subsequently becomes ill, the animal will avoid or drastically reduce consumption of that flavor (Gustavson 1977). The strength of the resulting aversion is related to the length of time between consumption and the onset of the illness, the discriminability or novelty of the flavor, and the intensity of the illness.

Taste aversions have been experimentally induced in a wide variety of animals (Gustavson 1977). The sampling of small quantities of novel foods by rats behaviorally enhances an ability to distinguish tastes of and to subsequently avoid food flavors that have made them ill. Aversive stimuli [e.g., X-ray irradiation or injections of lithium chloride (LiCl)] administered as much as 18 hours later can produce conditioned avoidance of the last food consumed by rats (Revunsky and Garcia 1970).

One of the most controversial uses of aversive conditioning has been the attempt to prevent coyotes from attacking sheep. Gustavson et al. (1974) reported that coyotes can be prevented from attacking sheep under controlled conditions by feeding them baits of lamb meat, hide, and wool laced with LiCl, a fast acting, powerful gastrointestinal poison. However, another study with mice and chickens has raised questions about the effectiveness of LiCl in averting coyotes (Conover et al. 1977). Field trials using lamb baits laced with LiCl have been conducted (Ellens et al. 1977; Gustavson 1977) with some reported success.

HAVE ANIMALS EVOLVED TO COPE WITH POISONOUS PLANTS?

If plants have developed poisonous properties as defense mechanisms against herbivores, animals should have coevolved either to avoid eating the plants or to detoxify the poisonous compounds to prevent being poisoned. Much of the literature dealing with the coevolution theory related to poisonous plants has been confined to insects (Ehrlich and Raven 1965; Fraenkel 1969; Whittaker and Feeny 1971; Jones 1973; Feeny 1975). An understanding of possible coevolution of mammals and poisonous plants is essential to this discussion. Some of the mechanisms by which large herbivores could be demonstrating evolutionary adaptations to poisonous plants are discussed.

Generalized Diet

Most large herbivores, including livestock, are generalists. One advantage of such a diet would be to reduce the probability of eating a toxic amount of any one poisonous species. Provenza et al. (1987) point out that recent research has shown that in feedlot situations livestock are neophobic, i.e., they ingest small quantities of novel foods initially and gradually increase ingestion, provided the gastrointestinal consequences are not adverse. It is not known whether livestock foraging on rangeland with a diverse array of plant species also display neophobia, but it is logical to assume that they do.

Ability to Detect and Avoid Poisonous Plants

The ability to detect poisonous plants and avoid consumption is one aspect of the complex behavior pattern of animals about which we have little information. Native animals that coevolved with the vegetation should be able to avoid native poisonous plants better than domestic animals that have been moved from their native habitat. Arnold and Hill (1972) observed that sheep in western Australia preferentially eat plants of Gastrolobium and Oxylobium, whose monofluoroacetic acid content is lethal, while native kangaroos avoid these plants. Dixon (1934) found that, although plants poisonous to livestock (such as azalea and larkspur) were abundant on summer range in California, mule deer did not graze these species. Laycock (1978) summarized studies which indicate that domestic livestock also have the ability to detect and avoid eating poisonous plants, at least in some circumstances. The fact that large numbers of livestock die after eating poisonous plants indicates that this ability only occurs for certain species or certain compounds, that management of livestock may alter the ability to detect poisonous plants, or that learning may have some role. Provenza and Balph (in press) indicated that animals moved to new environments may be more vulnerable to unfamiliar poisonous plants than animals familiar with such plants.

Generally, when animals avoid eating poisonous plants, it is assumed that the poisonous compounds are detectable to animals and thus are related to palatability. While all the senses play some role in food selection (Arnold 1964; Krueger et al. 1974), taste is one of the most important aspects of palatability (Arnold and Hill 1972).

Many studies have been made on the reasons for the preferences shown by large herbivores, but no universal relationships between chemical properties and palatability have been found (Arnold and Hill 1972). Sugar or sweet flavors have been shown to be preferred by many animals. Astringency apparently deters consumption of some herbage containing tannins (Swain 1977). Bitterness is generally repellent to most animals (Bate-Smith 1972). Bitterness is a property of many substances, but alkaloids and cyanogenic glycosides are universally bitter, at least to humans. Bate-Smith (1972) stated: "that these happen to be extremely toxic is incidental--it is their bitterness which is a repellent rather than the fact that they are deadly." In contrast, Garcia and Hankins (1974) contended that: "natural aversions to bitter substances have been acquired by a wide variety of species through natural selection." Thus, in their opinion, rejection of bitter substances is a species behavioral trait. Sometime in evolutionary history, survival was enhanced for genotypes of herbivores that rejected bitter-tasting plants.

Leopold (1948) stated that mammals have highly developed senses of taste and smell and thus avoid poisonous plants through the educational effect of unpleasant individual experiences, i.e., through what is now called aversive conditioning.

Taste aversion to toxic substances has been demonstrated in a wide variety of animals, but few studies have been conducted with large, generalist herbivores until recently. Zahorik and Houpt (1977) reported that an aversion to a single food was created by introducing LiCl into the rumen of cattle immediately after feeding. Olsen and Ralphs (1986) determined that cattle can form strong feed aversions. Research with Hereford heifers indicated that infusion with larkspur extract or LiCl induced a persistent aversion to an otherwise palatable feed (alfalfa pellets). They concluded that the larkspur extract might be used to teach cattle to avoid grazing larkspur on the range.

Thorhallsdottir et al. (in press) induced feed aversion in sheep using rolled barley treated with LiCl. Sheep thus exposed significantly reduced consumption of the treated grain but did not stop sampling or eating it entirely. They readily consumed grain that had not been treated with LiCl. The sheep remembered the experience caused by the LiCl for at least three months after the initial treatment. However, the repeated sampling of the feed that occurred would present problems in the training of animals to avoid poisonous plants on the range.

Sly and Bell (1981) reported that sodium-depleted steers developed an aversion to LiCl solution after three to five consecutive days of oral consumption. Some animals still had some aversion after three months.

These recent studies show that aversions can be formed in domestic livestock and, in our opinion, this is the first real breakthrough in poisonous plant research in a rather long time. Further research is needed to determine if sheep and cattle can be conditioned to completely avoid a particular poisonous plant under rangeland conditions and the implications for management if such aversions can be induced.

Ability to Detoxify Plant Poisons

The ability to detoxify plant poisons is another adaptation of herbivores to plants containing toxic secondary compounds. For a given poisonous compound, this ability varies with different animal species. Sheep can eat a larger percentage of their body weight of larkspur without damage than can cattle (James and Johnson 1976). Cattle and horses are poisoned by tansy ragwort (Senecio jacobaea), but sheep and goats (Cheeke 1977) and blacktailed deer (Dean and Winward 1974) are immune.

Much detoxification takes place in the digestive system, but microsomal enzymes in the liver or kidneys also may play a role (Freeland and Janzen 1974). In the rumen, detoxification is accomplished by the rumen microflora. Various nonruminant animals such as the kangaroo, hippopotamus, and sloth have forestomachs in which microbial fermentation also takes place (McBee 1971). According to Moir (1968), "a bacterial population in the stomach offers possibilities of colonizing other plant environments normally dangerous to mammals. If movement into these environments is gradual, some portion of the bacteria population quickly adapts to, and metabolizes, toxic substances to harmless substances..."

Some research has been conducted on transplanting rumen flora from a species of animal that is not poisoned by a specific plant to another susceptible species. Goats in Hawaii have rumen bacteria that degrade the toxic amino acid, mimosine, in Leucaena leucocephala and are not poisoned. Goats and steers in Australia are poisoned by Leucaena. Transfer of cultures of bacteria from stomachs of goats in Hawaii to a goat and a steer in Australia resulted in both animals being able to eat Leucaena without adverse effects (Jones 1985). Transplanting rumen flora is another area of research that could lead to a reduction of livestock losses due to poisonous plants.

Native animals that coevolved with the plant community should be more able to detoxify poisonous compounds than domestic livestock. In general, this seems to be true. Although native big game animals avoid some poisonous species, they apparently can eat others without serious effects. Stoddart and Rasmussen (1945) stated that, in Utah, "a number of plants poisonous to livestock are eaten in large quantities by (mule) deer without harm. These include deathcamas, low larkspur, chokecherry, and others." Similar statements have been made about other wild animals (Laycock 1978).

In Australia, Oliver et al. (1977) studied susceptibility of native mammals to Compound 1080, which is a fluoroacetate compound similar to compounds found in native plant species. Western Australian populations of animals had a substantially higher tolerance to this poison than was found in eastern populations of the same species. In South Africa, Basson et al. (1984) determined that the native eland and kudu were much more resistant than domestic sheep or cattle to poisoning from gifblaar (Dichapetalum cymosum), which contains monofluoroacetate, and to poisoning from slangkop (Urginea sanguinea), which contains cardiac glycosides.

While it is generally true that native wild animals are not killed by poisonous plants, examples of losses do exist for almost all native big game animals in North America. This indicates that the abilities to avoid or to detoxify poisons are not infallible (Laycock 1978; Fowler 1983).

CONCLUSIONS

Poisonous plants have a wide range of palatability to herbivores, occur in a wide range of successional communities from early seral to climax, and differ greatly in response to grazing. Range condition criteria based on ecological status should recognize that many climax or pristine communities do contain substantial amounts of poisonous plants, and these plants need to be included in reference lists in the proper amounts. An alternative may be to make more widespread use of resource value ratings on ranges which contain poisonous plants. Resource value ratings allow the manager to specify that the management goal is a community that has lower amounts of specific poisonous plants than occurred in undisturbed communities.

The lack of overwhelming evidence that being poisonous is always advantageous to plants or that herbivores have developed mechanisms that always operate to prevent poisoning by these plants indicates that coevolution probably has taken place. Thus, neither the poisonous plant nor the animal has achieved any clear-cut selective advantage. Jones (1973), in discussing coevolution of plants and insects, proposed a "leap-frogging" of evolutionary development of toxic compounds in plants and the adaptations of insects to cope with these toxins. The same should hold true in the evolution of large herbivores.

Research on the role of aversive conditioning in livestock and its relationship to the poisonous plant problem needs to be continued and strengthened. If effective and long-lasting aversions to specific poisonous plants can be formed, management of ranges containing poisonous plants might become easier.

Other research is needed on the feasibility of transplanting rumen flora from animals resistant to poisoning from a specific plant to other species of animals which are susceptible to that plant.

LITERATURE CITED

Arnold, G.W. 1964. Some principles in the investigation of selective grazing. Proc. Austr. Soc. Anim. Prod. 5:258-271.

Arnold, G.W., and J.L. Hill. 1972. Chemical factors affecting selection of food plants by ruminants, pp. 71-101. In: J.B. Harborne (ed), Phytochemical ecology. Academic Press, New York.

Basson, P.A., A.G. Norval, J.M. Hofmeyr, H. Ebedes, R.A.Schultz, T.S. Kellerman, and J.A. Minne. 1984. Antelopes and poisonous plants, pp. 695-700. In: Proc. XIII World Congress on Diseases of Cattle. Hoechst, Durban, Republic of South Africa.

Bate-Smith, E.C. 1972. Attractants and repellents in higher animals, pp. 45-56. In: J.B. Harborne (ed), Phytochemical ecology, Academic Press, New York.

Breedlove, D.E. and P.R. Ehrlich. 1968. Plant-herbivore co-evolution: Lupines and lycaenids. Sci. 162:671-672. Cates, R.G., and G.H. Orians. 1975. Successional status and the palatability of plants to generalized herbivores. Ecology 56:410-418.

Cheeke, Peter R. 1977. What makes tansy tick? Anim. Nutr. and Health 32:15-16.

Conover, M.R., J.G. Francik, and D.E. Miller. 1977. An experimental evaluation of aversive conditioning for controlling coyote predation. J. Wildl. Manage. 41:775-779.

Cook, C.W. 1983. "Forbs" need proper ecological recognition. Rangelands 5:217-220.

Cronin, E.H. 1965. Ecological and physiological factors influencing chemical control of Halogeton glomeratus. U.S. Dep. Agr. Tech. Bull. 1325.

Cronin, E.H., P.R. Ogden, J.A. Young, and W.A. Laycock. 1978. The ecological niches of poisonous plants in range communities. J. Range Manage. 31:328-334.

Dean, R.E., and A.H. Winward. 1974. An investigation into the possibility of tansy ragwort poisoning of black-tailed deer. J. Wildl. Dis. 10:166-169.

Dixon, J.S. 1934. A study of the life history and food habits of mule deer in California, Part 2. Food habits. Cal. Fish and Game 20:315-354.

Ehrlich, P.R., and P.H. Raven. 1965. Butterflies and plants: a study in coevolution. Evol. 18:586-608.

Ellens, S.R.. S.M. Catalano, and S.A. Schechinger. 1977. Conditioned taste aversion: a field application to coyote predation on sheep. Behav. Biol. 20:91-95.

Feeny, P. 1975. Biochemical coevolution between plants and their insect herbivores, pp. 3-19. In: L.E. Gilbert, and P.H. Raven (eds), Coevolution of animals and plants. University Texas Press, Austin.

Fowler, M.E. 1983. Plant poisoning in free-living wild animals: a review. J. Wildl. Dis. 19:34-43.

Fraenkel, G.S. 1969. The raison d'etre of secondary plant substances. Sci. 129:1466-1470.

Freeland, W.J., and D.H. Janzen. 1974. Strategies in herbivory by mammals. Amer. Natur. 108:269-289.

Garcia, J., and W.G. Hankins. 1974. The evolution of bitter and the acquisition of toxiphobia, pp. 1-12. In: D. Denton (ed), Fifth International Symposium on Olfaction and Taste. Melbourne, Australia.

Gustavson, C.R. 1977. Comparative and field aspects of learned food aversions, pp. 23-43. In: L.M. Barker, M.R. Best, and M. Domjan (eds), Learning mechanisms in food selection, Baylor University Press, Texas.

Gustavson, C.R., J. Garcia, W.G. Hankins, and K.W. Rusinak. 1974. Coyote predation control by aversive conditioning. Sci. 184:581-583.

James, L.F., and A.E. Johnson. 1976. Some major plant toxicities of the Western United States. J. Range Manage. 29:356-363.

Janzen, D.H. 1973. Community structure of secondary compounds in plants. Pure and Applied Chem. 34:529-538.

Jones, D.A. 1973. Co-evolution and cyanogenesis, pp. 213-242. In: V.H. Heywood (ed), Taxonomy and ecology. Academic Press, New York.

Jones, R.J. 1985. Leucaena toxicity and the ruminal degradation of mimosine, pp. 111-119. In: A.A. Seawright, M.P. Hegarty, L.F. James, and R.F. Keeler (eds), Plant toxicology, Proc. Australia-U.S.A. Poisonous Plant Symposium, Brisbane, Queensland Dept. Prim. Ind., Yeerongpilly.

Krueger, W.C., W.A. Laycock, and D.A. Price. 1974. Relationships of taste, smell, sight and touch to forage selection. J. Range Manage. 27:258-262.

Laycock, W.A. 1978. Coevolution of poisonous plants and large herbivores on rangelands. J. Range Manage. 31:335-342.

Leopold, A. 1948. Game management. Charles Scribner's and Sons, New York.

McBee, R.H. 1971. Significance of intestinal microflora in herbivory. Ann. Rev. Ecol. and System. 2:165-176.

Moir, R.J. 1968. Ruminant digestion and evolution, pp. 2673-2694. In: Handbook of physiology, alimentary canal, Vol. 5. Amer. Physiol. Soc., Washington D.C.

Muenscher, W.C. 1958. Poisonous plants of the United States. The Macmillan Co., New York.

Oliver, A.J., D.R. King, and R.J. Mead. 1977. The evolution of resistance to fluoroacetate intoxication in mammals. Search 8:130-132.

Olsen, J.D., and M.H. Ralphs. 1986. Feed aversion induced by intraruminal infusion with larkspur extract in cattle. Amer. J. Vet. Res. 47:1829-1833.

Provenza, F.D., and D.F. Balph. (In Press.) Diet learning by domestic ruminants: theory, evidence, and practical implications. Appl. Anim. Behav. Sci.

42

Provenza, F.D., D.F. Balph, J.D. Olsen, D.D. Dwyer, and M.H. Ralphs. 1987. Toward understanding the behavioral responses of livestock to poisonous plants. In: Ecology and the economic impact of poisonous plants on livestock production. Westview Press, Boulder.

Revunsky, S.H., and J. Garcia. 1970. Learned associations over long delays, pp. 1-84. In: G.H. Bower and J.T. Spence (eds), The psychology of learning and motivation. Academic Press, New York.

Rhoades, D.F., and R.G. Cates. 1976. Toward a general theory of plant antiherbivore chemistry. In: J. Wallace and R. Mansell (eds), Biochemical interaction between plants and insects. Recent Advances in Phytochem. 10:168-213.

Sly, J., and F.R. Bell. 1981. Effect of lithium intake on sodium and lithium appetite in sodium-deficient cattle. Physiol. Behav. 27:147-152.

Stahl, E. 1888. Pflanzen and Schnecken. Eine biologische Studie uber die Schutzmittel der Pflanzen gegen Schneckenfrass. Jena. Z. Med. Natur. 22:557-684. (cited by Jones 1973).

Stoddart, L.A., and D.I. Rasmussen. 1945. Deer management and range livestock production. Utah Agr. Exp. Sta. Circ. 121.

Stoddart, L.A., A.D. Smith, and T.W. Box. 1975. Range management. 3rd Ed. McGraw-Hill Book Co., New York. Swain, T. 1977. Secondary compounds as protective agents. Ann. Rev. Plant Physiol. 28:479-500.

Thorhallsdottir, A.G., F.D. Provenza, and D.F. Balph. (In Press.) Food aversion learning in lambs with or without a mother: discrimination, novelty, and persistence. Appl. Anim. Behav. Sci.

Waisel, Y. 1972. Biology of halophytes. Academic Press, New York.

Whittaker, R.H., and P. Feeny. 1971. Allelochemics: chemical interactions between species. Sci. 171:757-770.

Williams, M.C. 1960. Effect of sodium and potassium salts on growth and oxalate content of halogeton. Plant Physiol. 35:500-505.

Zahorik, D.M., and K.A. Houpt. 1977. The concept of nutritional wisdom: Applicability of laboratory learning models to large herbivores, pp. 45-46. In: L.M. Barker, M.R. Best, and M. Domjan (eds), Learning mechanisms in food selection. Baylor University Press, Texas.

5

Ecology, Mangagement and Poisonous Properties Associated with Perennial Snakeweeds

Kirk C. McDaniel and Ron E. Sosebee

INTRODUCTION

Broom snakeweed (Gutierrezia sarothrae) and threadleaf snakeweed (G. microcephala) are aggressive weeds common on western rangelands from Canada to Mexico. The two species, which are similar in appearance, are distinguished primarily by the number of ray flowers per flower head. Broom snakeweed usually has three or more ray flowers per head whereas threadleaf snakeweed has only one or two. In the southwest, they are often collectively referred to as perennial snakeweeds.

Platt (1959) estimated that these perennial plants occur on more than 57 million ha of rangelands in the United States. Broom and threadleaf snakeweed are found on about 60% of New Mexico's rangeland and about 22% of the rangeland of west Texas. Plant numbers periodically reach high densities and cause large economic losses to ranchers in the plains, prairies, and desert areas of the southwest. Dense broom snakeweed stands can reduce grassland forage production by 70% or more (Ueckert 1979; McDaniel et al. 1982. Herbage reduction may be a more serious problem than poisonous properties associated with the species. Broom and threadleaf snakeweed have low forage value, and livestock seldom graze them unless other forage supplies are low. Both are toxic and cause abortion or death in livestock, particularly in winter or when short forage supplies force livestock to graze the plant. More subtle losses include lightweight, sickly calves that never perform as well as those from snakeweed-free areas and retained placentas that may result in a decrease in reproductivity.

Broom and threadleaf snakeweed are extremely variable in morphology, phenology, population density, and presumably physiology (Solbrig 1960; Lane 1980b; DePuit and Caldwell 1975a,b). This variability allows them to successfully inhabit a broad range of environments. The variability in growth and occurrence of these species has prompted numerous investigations into the taxonomy and ecological relationships of the plants on western rangelands (Solbrig 1960, 1964, 1970, 1971; Ruffin 1971, 1974, 1977; Lane 1980a,b, 1982, 1985).

TAXONOMY AND DISTRIBUTION

The taxonomic history of Gutierrezia has been reviewed in detail elsewhere (Solbrig 1960; Ruffin 1974; Lane 1980 a or b, 1982, 1985). According to Lane (1985), Gutierrezia is a genus of seven woody shrubs, three herbaceous perennials, and six suffrutescent annuals in North America (Table 5.1). Solbrig (1960) described 11 species in South America.

Broom snakeweed (Figure 5.1) is the most widely distributed North American species of the genus Gutierrezia and is the only perennial species common throughout the central plains and prairie regions. Broom snakeweed grows in the cold temperate climates of southern British Columbia and Saskatchewan, Canada, to subtropical areas in Nuevo Leon and Sinola, Mexico. The range of broom snakeweed extends from the subhumid Great Plains across the arid and semiarid Rocky Mountain and Intermountain regions to the Sierra

TABLE 5.1
Gutierrezia species in North America as described by Lane (1985).

Gutierrezia Nomenclature		
Annual Herbs	**Perennial Herbs**	**Perennial Shrubs**
G. wrightii	G. alamanii var.	G. argyrocarpa
conoidea	alamanii	californica
arizonica	alamanii var.	grandis
sphaerocephala	megalocephala	microcephala
texana var.	sericocarpa	ramulosa
texana	petradoria	sarothrae
texana var.		serptoma
glutinosa		
triflora		

FIGURE 5.1
Gutierrezia sarothrae growing in eastern New Mexico.

Cascade axis (Solbrig 1960, 1964; Lane 1980a). Threadleaf
snakeweed often grows intermixed with broom snakeweed but is
restricted primarily to the Mohave, Sonoran, and Chihuahuan
deserts of southwestern United States and Northern Mexico.

While flower number can be used to easily distinguish
broom and threadleaf snakeweed, they are often confused
because both species are shallow-rooted, low-growing with
rounded canopies, and when in bloom, densely covered with
yellow flowers. Threadleaf snakeweed is usually taller (> 1
m) than broom snakeweed (10 cm to 1 m tall) where the two
species occur together. Narrow, linear leaves grow
alternately on unbranched erect stems which originate from a
woody base. Leaves are often shed in times of drought or at
maturity. According to Lane (1985), a high degree of
environmentally induced phenotypic variation occurs within
individual broom and presumably threadleaf snakeweed plants
on a season-to-season basis or among plants growing from
different populations. For example, during relatively wet
years, well-developed leaves are borne singly. However,
during dry years or even drier seasons of the same year, the
same plant may bear filiform leaves in fascicles. Peduncle
length, number of heads, and florets per head are also
affected by soil moisture levels. Plant taxonomists of the

late nineteenth and early twentieth centuries described a
number of species to account for the variability in the taxa
(Green 1899), but contemporary plant taxonomists have
related most of these taxa to the synonymy of broom
snakeweed (Solbrig 1960; Lane 1980a,b, 1985).

In New Mexico and west Texas, moderate to dense
infestations of broom, and occasionally threadleaf snakeweed
grow on open grassland regions found in these states. Soils
where very dense broom snakeweed stands grow (greater than
20% canopy cover or 20 plants/m^2) are characteristically
shallow sandy loam to loam and underlain by caliche or
limestone bedrock. Dense broom snakeweed stands are
commonly associated with Bouteloua spp. (B. gracilis, B.
curtipendula, and B. eriopoda) dominated grasslands.
Numbers of broom snakeweed are not, as a rule, as high in
plant communities with a woody overstory layer as compared
to shrub-free grasslands. However, broom snakeweed growing
intermixed among mesquite (Prosopis glandulosa),
creosotebush (Larrea tridentata), sagebrush (Artemesia
tridentata), salt desert shrubs, pinyon-juniper, and other
shrub-dominated communities offer significant competition
with desirable grasses for soil moisture.

GROWTH AND DEVELOPMENT

Snakeweed are short-lived perennial plants with a life
expectancy range of three to ten years once plants are
established. However, they depend on seed to maintain the
cyclic population. They are prolific seed producers. An
average broom snakeweed plant will produce 9,717 seeds with
a range of 1,263 to nearly 21,853 seeds per plant (Ragsdale
1969). As many as 35,000 yellow radiant flowers have been
counted during one season from a single threadleaf snakeweed
plant collected in southern New Mexico, with at least one
viable seed produced per involucre (McDaniel 1981).

The environmental mechanisms that trigger a mass
germination of seeds in the soil are not well understood. A
prolific seed source is made available to support future
reinvasions but seedlings are not usually produced each
year. Snakeweed populations have been described as being
cyclic (Jameson 1979), which could be partially due to
autotoxicity caused by allelopathic agents found in
snakeweed.

Germination requirements are rather broad. Seeds
require a short period of after ripening, after which they
will germinate under alternating temperatures ranging from

10 to 30 C (Mayeux and Leotta 1981). Observations suggest broom snakeweed can germinate any time during the year when soil moisture and temperature conditions are conducive to germination. Seedlings develop a taproot and a central top stem which are about equal in length after a month (Figure 5.2). An extensive root system develops rapidly, enabling broom snakeweed to become a strong competitor with other plants in the community (Dwyer and DeGarmo 1970). Further stem elongation and branching occurs until about midsummer when flower buds form.

Flowering and seed development follow formation of flowering buds at approximately 3-week intervals. Flowering usually occurs earlier in plants near the southern range of distribution compared to plants in northern regions. When plants are stressed because of inadequate soil moisture, leaves and lower branches are usually shed which is a useful strategy because snakeweed leaves have little stomatal control at low water potentials (Depuit and Caldwell 1975b). If soil moisture and air temperatures are not limiting, the snakeweeds remain evergreen year-round, particularly in the southwestern U.S.

Knowledge of the basic biology and physiology of snakeweed is beneficial in understanding its interaction with other species and its ability to be managed and controlled. DePuit and Caldwell (1975b) found that by use of the C_3 pathway of photosynthesis, photosynthetic rate of broom snakeweed in northern Utah was highest in spring and early summer when soil moisture was not limiting. Maximum assimilation rates were found between -0.8 MPa and -1.5 MPa water stress for plants growing between 20 and 30 C. Stem photosynthesis of snakeweeds might contribute greatly to the total plant carbon gain during a year. DePuit and Caldwell (1975a) estimate that during warm dry weather in August, 20% of the carbon gain was contributed by the stems. They concluded that lack of stomatal control and stem photosynthesis might be advantageous for a drought-deciduous species like snakeweed. Ehleringer et al. (1987) found in the Sonoran desert that threadleaf snakeweed stems have a photosynthetic rate four times higher than broom snakeweed. This could explain the different distributions of the two perennial plants. Threadleaf snakeweed having a higher stem photosynthesis should be able to survive in regions with higher temperatures and more erratic precipitation than broom snakeweed.

Total nonstructural carbohydrate (TNC) trends closely follow phenological development in broom snakeweed when soil moisture is not limiting (Sosebee 1985). TNC concentrations

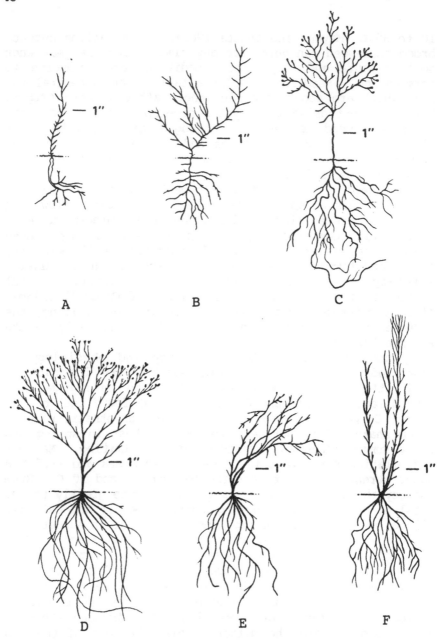

FIGURE 5.2
First year growth cycle of <u>Gutierrezia</u> <u>sarothrae</u> germinating
in April (adapted from photographs by Ragsdale 1969). (A)
seedling, 3 weeks; (B) vegetative growth, 13 weeks; (C)
flower bud development, 20 weeks; (D) bloom to post-bloom,
31 weeks; (E) perennating bud development and leaf shed, 38
weeks; (F) vegetative growth, 52 weeks.

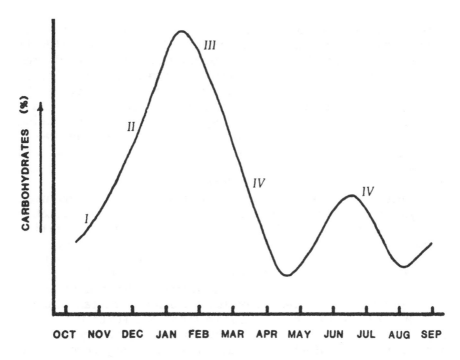

Generalized reserve carbohydrate curve for snakeweed in the southwestern U.S.

FIGURE 5.3
Generalized reserve carbohydrate curve for <u>Gutierrezia</u>
<u>sarothrae</u> in the southwestern U.S. Full bloom (I), post-
bloom (II), new terminal growth (III), and vegetative (IV).

in both stems and roots typically decline most rapidly
during vegetative development and increase during and after
flowering (Alliney 1982, Courtney 1984) (Figure 5.3). In
areas where soil moisture is not favorable for growth,
plants will become quiescent (quasidormant) during the dry
period. Under these conditions, a significant amount of
variation in phenological development and carbohydrate
accumulation trends can occur among plants growing on
different range sites.

ECOLOGY

Perennial snakeweeds compete with desirable forage for
moisture, space, nutrients, and light. Of these factors,
moisture competition is probably most important. Snakeweeds

were found to have a low water use efficiency (high water use) compared to other range plants (DePuit and Caldwell 1975a). The plant's competitive advantage is derived from its drought deciduous habit and the numerous adventitious (fibrous) roots produced in the upper 15 cm of soil depth, especially on shallow and loamy soils. Because the plant exhibits high rates of photosynthesis and transpiration under both optimal and suboptimal temperature and water availability conditions in the fall and winter (DePuit and Caldwell 1975a), it can be expected to use much of the soil water required for the spring "green up" of grasses.

Perennial snakeweeds are a component of the natural ecosystem, but under pristine conditions, they probably would contribute about 5 to 10% of the species composition (Campbell and Bomberger 1934). Under these conditions, the amount present would fluctuate over time according to climatic conditions. However, with the introduction of livestock, and in some cases overgrazing, the densities of these plants have increased on many southwestern rangelands.

Historically, broom snakeweed populations were reportedly found mainly on rocky ridges, gravelly slopes, and immature or infertile soils. Gradually they become more noticeable around watering holes, trails, and other areas where livestock tended to congregate (Parker 1939). The species aggressively invades disturbed ranges that are heavily grazed; some believe the plant is a valuable range condition indicator (Barnes 1913; Wooten 1915; Jardine and Forsling 1922; Talbot 1926; Dayton 1931; Costello and Turner 1941; Green 1951; Nichol 1951; Humphrey 1955; Schmutz et al. 1968). However, broom snakeweed populations have been observed to increase and decrease in a cyclic pattern according to climatic conditions, causing others to believe the plant is not necessarily an indicator of overgrazing (Jameson 1970; Vallentine 1971; Ueckert 1979; McDaniel et al. 1984).

POISONING CIRCUMSTANCES

Livestock do not normally eat perennial snakeweed when adequate forage is available. Grazing animals, however, will eat considerable amounts of snakeweed foliage when other herbage is in limited supply.

Most cases of snakeweed poisoning occur in late winter or early spring when other herbaceous forage is depleted by winter grazing. Warm temperatures during this period stimulate snakeweed growth, but warm-season grasses remain dormant. Snakeweed poisoning is more common during this period since grass growth is limited and forage supplies are sparse.

For 75 years, ranchers on the southern Great Plains suspected that snakeweeds were poisonous to cattle. Initial research conducted in the 1930s, in response to ranchers' concerns, failed to show toxic effects of broom and threadleaf snakeweed (Mathew 1936). Later testing demonstrated that cattle, sheep, and goats are susceptible to snakeweed poisoning (Dollahite et al. 1962). In sheep, death resulted from consuming 3.6 kg of green snakeweed foliage in 5 days, in a steer from consuming 10.9 kg in 3 days, and in a goat from consuming 5 kg in 14 days. Rabbits and guinea pigs also died as a result of consuming snakeweed foliage. Baby chicks that received an alcohol extract of snakeweed gained less weight than control chicks.

Broom snakeweed is poisonous to cattle and sheep. There is considerable individual variation in the amount of plant material required to cause intoxication. Signs of poisoning include listlessness, anorexia, rough coat, diarrhea, and/or constipation. In severe cases hematuria may occur. The muzzle may become crusted and peel. Death may occur. Lesions associated with intoxication include gastroenteritis and degenerative changes in the liver and kidneys.

Cows and occasionally sheep may abort following the grazing of broom snakeweed, with the rates reportedly higher when animals graze plants from sandy soils compared to those from loam or clay soils (Sperry et al. 1977). The abortions occur primarily in the early spring when the snakeweed is starting its spring growth and prior to the growth of the grasses and other desirable forage. Calves born near term may survive, although many such calves are small and weak and may die. A persistent retained placenta is common. Retained placentas are common in cows grazing broomweed even though the calf appears normal.

Cows may abort without showing many indications of the abortion. However, if a cow has been grazing the broomweed over a period of time, she may exhibit swelling of the external genital and filling of the udder.

The uterus is edematous. Death of the cow may follow the abortion. This is likely to be due to a peritonitis and toxemia associated with the abortion.

There is still some uncertainty as to the toxic compounds in the snakeweeds. J. W. Dollahite et al. (1962) isolated a saponin from threadleaf snakeweed. This saponin, when injected into rabbits, goats, and cattle, produced symptoms similar to those of snakeweed poisoning. Visible signs of poisoning after saponin injections were prolonged appetite loss, mucopurulent nasal discharges, diarrhea changing to constipation, thick white mucus in the feces, and mucus passing from the vagina. Lesions resulting from the saponin injections were pneumonia, cardiac hemorrhages, fat degeneration, gastroenteritis, and hemorrhaging into the abomasum, abdominal cavity, and uterus. Additional research showed that saponin isolated from Gutierrezia had abortifacient properties when orally administered.

Saponins are large polymers (with steroid configurations) of sapogenins and sugars (glucose, galactose, rhamnose, and arabinose). They form colloidal solutions and produce soapy foams when shaken in water. Saponins irritate the gastrointestinal tract and enter the bloodstream through the injuries in the tract. Once in the bloodstream, the triterpenoids lyse (hemolyze) red blood cells and disrupt cells and tissues by reacting with cholesterol and damaging cell membranes. The literature is vague about how saponins disrupt the hormone balance of pregnant animals to produce abortions; however, some saponins contain steroidal groups resembling estrogens and progesterone.

SNAKEWEED MANAGEMENT

Mechanical, biological, and burning methods have been used to control broom and threadleaf snakeweed but, in general, herbicides are the most common and practical method for yielding immediate range improvement results. Early studies conducted with foliar herbicides emphasized optimum application rates and the importance of soil moisture and soil temperature in the control of snakeweed. Foliarapplied herbicides have the advantage of selectively killing snakeweed but not harming beneficial grasses. Without snakeweed competition, tremendous increases in grass production usually occur.

Recommendations for control of the snakeweeds with foliar applied herbicides call for moist soils with soil temperatures from 15 to 25 C. The plants should be actively growing and, preferably, less than one to two years old. Older, mature plants are susceptible to the herbicides, but suppressing young snakeweed stands is likely to give a longer-term increase in grass production. With the introduction of picloram in the early 1970s, more consistent results over a wider time span could be obtained than with the phenoxy herbicides. Several studies have shown that picloram applied at 0.28 to 0.56 kg a.i./ha during September through December after the plant has bloomed is consistently effective for snakeweed control (Sosebee 1985, McDaniel 1984).

Pelleted formulations of picloram and tebuthiuron, applied aerially at 0.56 to 0.87 kg a.i./ha, are in widespread use today. Control is usually better on coarse soils, especially when applied before fall or winter precipitation. Generally more expensive than comparable liquid herbicides, the pellets are often used to control undesirable brush or weed species in a complex stand which includes snakeweed as a major plant component.

Grass response following snakeweed control with herbicides is usually positive, even without artificially reseeding the rangeland (this assumes an available seed source). If snakeweed dies in the late winter or early spring, grass will respond immediately with spring and summer rains. Unfortunately, if the snakeweed does not die until late summer because of lack of spring and summer rainfall, warm season grasses will not respond until the following spring. Grass production can be increased as much as tenfold (e.g. 80 to 800 kg/ha), depending upon the site and infestation of snakeweed (McDaniel 1984).

LITERATURE CITED

Alliney, J.E. 1982. Carbohydrate trends in broom snakeweed. M.S. Thesis, Texas Tech Univ., Lubbock, TX.
Barnes, W.C. 1913. Western grazing grounds and forest ranges. Sanders Publishing Co., Chicago.
Campbell, R.S., and E.H. Bomberger. 1934. The occurrence of Gutierrezia sarothrae on Bouteloua eriopoda ranges in southern New Mexico. Ecology 15:49-61.
Costello, D.F., and G.T. Turner. 1941. Vegetational changes following exclusion of livestock from grazed ranges. J. For. 39:310.
Dayton, W.A. 1931. Important western browse plants. USDA Misc. Pub. 101.
DePuit, E.J., and M.M. Caldwell. 1975a. Stem and leaf gas exchange of two arid land shrubs. Amer. J. of Bot. 62:954-961.
DePuit, E.J., and M.M. Caldwell. 1975b. Gas exchange of three cool semi-desert species in relation to temperature and water stress. J. Ecol. 63:835-858.

Dollahite, J.W., T. Shaver, and B.J. Camp. 1962. Injected saponins or abortifacients. J. Amer. Vet. Res. 23:1261-1263.

Dwyer, D.D., and H.C. DeGarmo. 1970. Greenhouse productivity and water use efficiency of selected desert shrubs and grasses under four soil moisture levels. New Mexico Agr. Exp. Sta. Bull. 570. NMSU, Las Cruces, NM.

Ehleringer, J.R., J.P. Comstock, and T.A. Cooper. 1987. Leaf-twig carbon ratio differences in photosynthetic-twig desert shrubs. Oecologia 71: 318-320.

Greene, R.L. 1951. Utilization of winter forage by sheep. J. Range Manage. 4:233.

Greene, E.L. 1899. A decade of new Gutierrezia's. Pitonia. 4:54-48.

Humphrey, R.R. 1955. Arizona range resources. II. Yavapui County, Ariz. Ariz. Agr. Exp. Sta. Bull. 229.

Jameson, A.D. 1970. Value of broom snakeweed as a range condition indicator. J. Range Manage. 23:302-304.

Jardine, J.T., and C.L. Forsling. 1922. Range and cattle management during drought. USDA Bull. 1031.

Lane, M.A. 1980a. Systematics of Amphiachyris, Greenella, Gutierrezia, Gymnosperma, Thurovia, and Xanthocephalum (Compositae: Astereae). Ph.D. Diss., University of Texas at Austin. (University Microfilms 81-00925).

Lane, M.A. 1980b. New and reinstituted combinations in Gutierrezia (Compositae: Astereae). SIDA 8:313-314.

Lane, M.A. 1982. Generic limits of Xanthocephalum, Gutierrezia, Amphiachyris, Gymnosperma, Greenella, and Thurovia (Compositae: Astereae). Systemtic Botany 7:405-416.

Lane, M.A. 1985. Taxonomy of Gutierrezia (Compositae: Astereae) in North America. Systm. Bot. 10:7-28.

McDaniel, K.C. 1981. Ecology, toxicity, and control of broom snakeweed, pp. 10-127. In: A. McGinty (ed), Proc.--Poison Plant Management in the Trans-Pecos. Texas A & M Univ., Ft. Stockton, TX.

McDaniel, K.C. 1984. Snakeweed control with herbicides. New Mexico State University Agri. Exp. Sta. Bull. 706.

McDaniel, K.C., R.D. Pieper, and G.B. Donart. 1982. Grass response following thinning of broom snakeweed. J. Range Manage. 35:219-222.

McDaniel, K.C., R.D. Pieper, L.E. Loomis, and A.A. Osman. 1984. Taxonomy and ecology of perennial snakeweeds in New Mexico. N. M. Agr. Exp. Sta. Bull. 711.

Mayeux, H.S. and L. Leotta. 1981. Germination of broom snakeweed (Gutierrezia sarothrae) and threadleaf snakeweed (G. microcephala) seed. Weed Sci. 29:530-534.

Nichol, A.A. 1952. The natural vegetation of Arizona. Arizona Agr. Exp. Sta. Tech. Bull. 127. pp. 187-230.

Parker, K.W. 1939. The control of snakeweed in the southwest. Southwest For. and Range Exp. Sta. Res. Note 76.

Platt, K.B. 1959. Plant control--some possibilities and limitations. J. Range Manage. 12:64-68.

Ragsdale, B.J. 1969. Ecological and phenological characteristics of perennial broomweed. Ph.D. diss., Texas A & M University, College Station.

Ruffin, J. 1971. Morphology and anatomy of the genera Amphiachyris, Greenella, Gutierrezia, Gymnosperma, Thurovia, and Xanthocephalum (Compositae). Ph.D. diss., Kansas State University, Manhattan.

Ruffin, J. 1974. A taxonomic re-evaluation of the genera Amphiachyris, Amphipappus, Greenella, Gutierrezia, Gymnosperma, Thurovia and Xanthocephalum (Compositae). SIDA 5:301-333.

Ruffin, J. 1977. Polyphyletic survey of the genera Amphiachyris, Amphipappus, Greenella, Gutierrezia, Gymnosperma and Xanthocephalum. Contrib. Gray Herb. 207:117-131.

Schmutz, E.M., B.N. Freeman, and R.E. Reed. 1968. Livestock poisoning plants of Arizona. Univ. of Arizona Press, Tucson.

Solbrig, O.T. 1960. Cytotaxonomic and evolutionary studies in the North American species of Gutierrezia (Compositae). Contrib. Gray Herb. 188:1-63.

Solbrig, O.T. 1964. Infraspecific variations in the Gutierrezia sarothrae complex (Compositae-Astereae). Contrib. Gray Herb. 193:67-115.

Solbrig, O.T. 1979. The phylogeny of Gutierrezia: An eclectic approach. Brittonia 22:217-229.

Solbrig, O.T. 1971. Polyphyletic origin of tetraploid populations of Gutierrezia sarothrae (Compositae). Madrono 21:21-25.

Sosebee, R.E. 1985. Timing--the key to herbicidal control of broom snakeweed. Texas Tech. Management Note 6. Texas Tech Univ., Lubbock.

Sperry, O.E., J.W. Dollahite, G.O. Hoffman and B.J. Camp. 1977. Texas plants poisonous to livestock. Texas A&M Ext. Ser. Rep. B-1028. College Station, TX.

56

Talbot, M.V.   1926.   Indicators of southwestern range
    conditions. USDA Farmers Bull. 1782.
Ueckert, D.N. 1979. Broom snakeweed: Effect on shortgrass
    forage production and soil water depletion.   J. Range
    Manage. 32:216-219.
Vallentine, J.F.   1971.   Range development and improvements.
    Brigham Young Univ. Press, Provo.
Wooten, E.O.   1915.   Factors affecting range management in
    New Mexico. USDA Bull. 211.

6

Economic Impacts of
Perennial Snakeweed Infestations

L. Allen Torell, Hal W. Gordon, Kirk C. McDaniel,
and Allan McGinty

INTRODUCTION

Significant economic losses from perennial snakeweed
(Gutierrezia spp.) infestations have occurred. These losses
include suppression of desirable forage production with
reduced rangeland carrying capacity; livestock poisoning
with increased abortions, poor conception, poor animal
health, reduced sale weights and increased death losses.

As discussed in the companion paper by McDaniel and
Sosebee (1987), broom snakeweed (Gutierrezia sarothrae) and
threadleaf snakeweed (G. microcephala) occur throughout the
central plains and prairie regions of the United States. The
greatest plant numbers and most severe management problems
exist on the southern High Plains and the Canadian-Pecos
Valleys of western Texas and eastern New Mexico. An
estimated 22% of the rangeland in Texas and 60% in New
Mexico are infested to some degree by snakeweed. In these
two states, an estimated 3.5 million ha of rangeland are
classified as having moderate to dense, or dense to very
dense, infestations of snakeweed (Figure 6.1). On these
acreages, rangeland is so densely covered with the weed that
forage production is greatly suppressed and livestock
poisoning is a problem (McDaniel et al. 1986; McGinty and
Welch, in press).

The dollar loss in reduced livestock production from
snakeweed infested rangeland is substantial, but not well
quantified. The objectives of this paper are to determine
economic losses to perennial snakeweed under various
livestock price situations and to investigate the economic
potential for implementing control programs. Only losses in
Texas and New Mexico are considered.

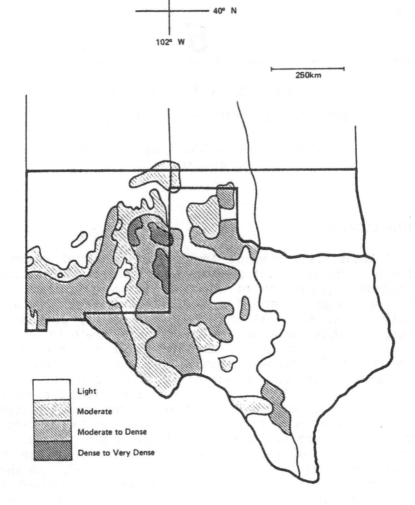

40° N

102° W

250km

Light

Moderate

Moderate to Dense

Dense to Very Dense

FIGURE 6.1
Density and distribution of snakeweed infestations in Texas
and New Mexico.

LIVESTOCK AND FORAGE LOSSES

Reduced Forage Production

Heavy stands of snakeweed can greatly reduce production
of desirable forage species. Removing a snakeweed stand

makes soil moisture and nutrients more available to preferred forage species. The results can be dramatic, as shown in Figure 6.2 for a control project near Wagon Mound, New Mexico. One growing season after spraying, forage production had increased 600 percent. Yield increases of from 100% to 800% are common (Ueckert 1979; McDaniel et al. 1982).

Forage response from controlling snakeweed depends on the amount of snakeweed present before treatment, the amount of grass in the understory, effectiveness of the control treatment, subsequent rainfall, and follow-up grazing management (McDaniel and Torell 1987). In general, recent control projects have been very successful, resulting largely from development and refinement of new chemical control techniques (see McDaniel and Sosebee 1987 for recommended control practices).

Research plots at two locations in New Mexico's central plains grassland region near Roswell and Vaughn, New Mexico, resulted in a 300% to 500% increase in grass production after dense snakeweed stands were controlled (McDaniel, in press). Before herbicide application, density of mature broom snakeweed averaged 37 plants/m^2 and 20% canopy cover at the Roswell site, and 47 plants/m^2 and 22% canopy cover at the Vaughn site. Recommended herbicide treatment resulted in snakeweed mortality rates exceeding 95%, with a dramatic increase in the production of perennial warm-season grasses the following year. McDaniel (in press) estimated that carrying capacity at the Roswell site increased from about 3 AUY (animal unit year-long)/section to 15 AUY/section. Similarly, carrying capacity at the Vaughn site increased from about 4 AUY/section to 13 AUY/section. This stocking rate increase was sustained over three growing seasons.

Livestock Poisoning and Production Losses

McGinty and Welch (in press) surveyed county agents in west Texas and estimated a 1% annual death loss in cattle and a 2.9% annual abortion rate which could be attributed to snakeweed poisoning. Both death and abortion losses were somewhat lower for sheep as compared to cattle, 0.7% and 1.3%, respectively. Goats, which are less susceptible to snakeweed poisoning, had a death loss of only 0.4% and abortion loss of 0.7%.

Livestock weaning and sale weights are also reduced on snakeweed infested rangeland, but the amount of loss is variable and not well quantified. In addition, production

FIGURE 6.2
Snakeweed is easily controlled and forage response is rapid
after spraying with picloram. Area in top photo was sprayed
in October 1985, and the bottom photo is an immediately
adjacent unsprayed area near Wagon Mound, New Mexico,
October 1986.

expenses, especially supplemental feed costs, are higher on
snakeweed infested rangeland. Some ranchers do not reduce
livestock numbers to levels consistent with forage
production and availability on these poor producing areas.
Although forage production may only support 3 to 5 AUY/
section, actual stocking rate may be double this amount. As
a result, supplemental feed costs usually must be increased
and the nutritional requirement of grazing animals may not
be met. In many cases, animal production is reduced, and
snakeweed poisoning is increased.

Table 6.1 summarizes the level of livestock production
considered to be representative for an eastern New Mexico or
western Texas cow/calf producer, both with and without heavy
snakeweed infestations. An average level is specified for
each parameter, based on livestock budgets prepared in both
the panhandle of Texas and the northeastern corner of New
Mexico, and input from knowledgeable researchers and
ranchers.

Livestock production parameters with or without
snakeweed would be expected to vary by area and for a
particular ranch. Because of this variability and uncertainty, a conservative estimate of forage and livestock
benefits was used in our economic evaluation. Substantially
higher forage and livestock responses have been reported
after heavy infestations of snakeweed were controlled,
especially on sandy soils (McDaniel and Sosebee 1987).

ECONOMIC LOSSES

A partial budgeting approach was used to evaluate
economic losses from snakeweed infestations. Livestock do
not normally eat perennial snakeweed when adequate
forage is available, and forage production is not greatly
reduced under light to moderate snakeweed densities
(McDaniel et al. 1982). Thus, only moderate to dense, and
dense to very dense stands of snakeweed were considered to
be of economic importance and valued in this analysis.

Losses from these heavier snakeweed infestations were
determined using a with snakeweed and without snakeweed
comparison of livestock production from a representative
section of rangeland, assuming year-long grazing capacity.
Potential benefits from controlling snakeweed, including
increased carrying capacity and improved production of
grazing animals, were considered.

Using the parameter specifications given in Table 6.1,
livestock numbers, livestock production levels and sales per

TABLE 6.1
Specified average livestock and forage production parameters
with and without heavy snakeweed infestations[1]

Description	Dense Snakeweed Infestation	Little or no Snakeweed	Percent Difference
Rangeland Carrying Capacity:			
AUY/Section	10.0	20.0	100%
Livestock Production:			
Calf crop	75%	85%	10%
Replacement rate	15%	15%	0%
Cow to bull ratio	20:1	20:1	0%
Cow death loss	2%	1%	1%
Calf death loss	4%	3%	1%
Steer Calf Weight	428	475	10%
Heifer Calf Weight	405	450	10%
Cull Cow Weight	855	950	10%
Cull Bull Weight	1,080	1,200	10%

[1]Literature used in defining production parameters:
Dollahite and Anthony 1956; Sperry et al. 1964; Sosebee et
al. 1979; Ueckert 1979; McDaniel et al. 1982; Texas A&M
1984a,b; McDaniel 1984; Torell et al. 1986a,b,c; McGinty
and Welch, in press.

section were calculated using a computerized spreadsheet
program. Economic losses were computed for three different
beef price situations: 1985-86 average prices, which
represent the current depressed cattle market; 1979 prices,
which reflect record high levels; and 1980 prices, which
were intermediate (Table 6.2).

Table 6.3 summarizes the estimated loss in annual
income to New Mexico and Texas ranchers under each of these
price situations. The average economic loss for a dense
snakeweed infestation ranged from $11.12/ha under 1985-86
beef prices to $14.65/ha under relatively high 1979 prices.
These loss estimates include reduced stocking rates from
forage suppression, livestock poisoning, and diminished
production rates on snakeweed infested rangeland.

Approximately 72% of the losses from heavy snakeweed
infestations result from suppressed forage production. This

TABLE 6.2
Beef prices used in the economic analysis

Livestock Class	Beef Price Situation ($/kg)		
	1985-86	1980	1979
Steer Calves (181-227 kg)	1.61	1.81	2.09
Heifer Calves (181-227 kg)	1.33	1.54	1.80
Cull Cows	0.85	1.01	1.12
Cull Bulls	1.07	1.18	1.32

Source: Annual average prices quoted for the Clovis, New Mexico, auction by the U.S. Department of Agriculture, Agricultural Marketing Service.

implies that the major economic loss from snakeweed is not livestock poisoning but the loss of desirable forage by competition with snakeweed.

An estimated 1.6 million ha were heavily infested with snakeweed in Texas during 1984 (McGinty and Welch, in press). Our model predicted losses of $18 million on these infested areas using relatively low 1985-86 beef prices. This estimate is slightly higher than the $16.9 million loss estimated by McGinty and Welch (in press) using 1984 average beef prices and Texas livestock statistics. The estimated economic loss was considerably higher ($23.7 million) when 1979 beef prices were considered (Table 6.3).

New Mexico rangeland densely infested with snakeweed was estimated to total 1.9 million ha during 1985 (McDaniel et al. 1986). The estimated economic losses to snakeweed in New Mexico ranged from $20.7 million with 1985-86 beef prices to $27.3 million with 1979 beef prices (Table 6.3). Losses for both states totaled $38.7 million, $44.3 million, and $51.0 million under 1985-86, 1980 and 1979 beef price situations, respectively.

ECONOMIC POTENTIAL FOR SNAKEWEED CONTROL

Economic losses from snakeweed to Texas and New Mexico ranchers have been substantial, but control measures are also costly and relatively short-lived. The net economic benefit from controlling snakeweed will vary depending on livestock prices, production, costs, treatment costs, treatment life, and treatment success.

TABLE 6.3
Economic losses from heavy snakeweed infestations in Texas
and New Mexico under alternative beef price situations

	Beef Price Situation		
	1985-86	1980	1979
Livestock Sales ($/Section):			
With Snakeweed	1,888	2,163	2,484
Without Snakeweed[1]	4,769	5,460	6,277
Increased Sales Without Snakeweed	2,881	3,297	3,793
Percent of Increase Resulting From Increased Forage[2]	72%	72%	72%
Average Losses Due to Snakeweed Infestations:			
($/ha)	11.12	12.72	14.65
Total Loss in Texas[3] (million)	18.01	20.61	23.71
Total Loss in New Mexico[4] (million)	20.71	23.70	27.26
Total Loss both States (million)	38.72	44.31	50.97

[1]Includes stocking rate increase of 10 AUY/section and
livestock production increases as given in Table 6.1.
[2]Estimated by computing increased sales per section when the
rangeland carrying capacity in the computer model was
increased while holding livestock production parameters
constant at pretreatment levels (Table 6.1).
[3]An estimated 1.6 million ha of rangeland were heavily
infested with snakeweed in Texas during 1984 (McGinty and
Welch, in press).
[4]An estimated 1.9 million ha of rangeland were heavily
infested with snakeweed in New Mexico during 1985 (McDaniel
et al. 1986).

Potential benefits from controlling a heavy infestation of snakeweed have been identified, but more research is needed to quantify production relationships for detailed economic evaluation. At present, only estimates that grossly define the differences in livestock and forage production levels before and after snakeweed control are available from knowledgeable researchers and ranchers (Table 6.1). Site-specific evaluations of potential benefits and costs should be made when assessing the economic potential of specific control projects.

Three standard economic criteria commonly used to analyze range improvement investments, including net present value (NPV), the benefit/cost ratio, and the internal rate of return (IRR), are presented in summarizing the economic potential of snakeweed control (Table 6.4). Complete discussions of these three measures of investment profitability are given by Workman (1986) and in other financial textbooks. In general, an economically feasible investment is indicated by an IRR greater than the prevailing market interest rate, a positive NPV, or a benefit/cost ratio greater than 1.0.

Good control results have been obtained on large-scale spray programs, but treatment life is variable because of the cyclic nature of snakeweed populations (McDaniel 1984). A dense snakeweed stand may appear with favorable growing conditions, then completely die out in later years as a result of insect damage, old age, drought or for other reasons (McDaniel and Sosebee 1987).

A control program could yield little, if any, benefits if a natural die-off occurred shortly after treatment implementation. Further, forage response could be minimal if adequate rainfall does not occur after the treatment is implemented. Snakeweed control is not a riskless investment. Judgement about the economic potential for control can only be made based on realistic expectations of future events occurring.

For purposes of this analysis a five year treatment life was assumed with uniform annual forage response and livestock benefits, beginning after a one-year grazing deferment period. A 7% discount rate was used to calculate the NPV of the stream of earnings from snakeweed control. This interest rate is taken as an expected real rate, including a risk premium under the assumption of constant relative price levels. Control cost was estimated to average $23.22/ha, including herbicide and application costs (McDaniel et al. 1986).

TABLE 6.4
Benefit/cost analysis of controlling heavily snakeweed
infested rangeland

	($/Section) Beef Price Situation		
	1985-86	1980	1979

I. Gross Benefits:

Increased Livestock Sales (from Table 6.3)	2,881	3,297	3,793
Increased Annual Livestock Production Costs[1]	808	808	808
Net Annual Benefit	2,073	2,489	2,986
Present Value of Net Annual Benefits (5 year uniform benefit, discounted at 7%)[2]	7,947	9,541	11,441

II. Project Costs And Returns:

Spraying with .28 kg/ha of Picloram @ $23.22/ha (includes herbicide and application costs)	6,016	6,016	6,016
Net cost of Additional Breeding stock[3]	1,027	1,027	1,027
Total Costs	7,043	7,043	7,043
Net Present value	904	2,498	4,398
Benefit/Cost Ratio	1.07:1	1.20:1	1.36:1
IRR	9.40%	13.41%	17.87%

[1]Includes miscellaneous expenses for feed, veterinary and medicine, fuel and repairs, and others (see Torell et al. 1986a,b,c).
[2]All discounting starts at the beginning of year two because of the assumed one year of grazing deferment.
[3]Cow numbers are increased by about eight head with the additional carrying capacity. This represents a cost when the control treatment is initiated, but a salvage value is added at the end of the planning period to account for the additional value of livestock accumulated. The net cost of the additional breeding livestock is $1,027/section.

As shown in Table 6.4, control of heavily infested snakeweed acreages is an economical management option with the production responses assumed (Table 6.1). Even under the relatively low 1985-86 beef price situation, NPV is estimated to be positive. The benefit/cost ratio was estimated to be 1.07:1 with 1985 prices and 1.36:1 with 1979 prices.

Increased forage production is the major benefit of controlling snakeweed. An estimated 72% of foregone livestock sales result from decreased stocking rate under dense snakeweed stands (Table 6.3). However, it takes a combination of a positive forage response (increased stocking rate) and improved livestock production levels to pay for snakeweed control. If, after snakeweed control, no improvement in livestock production levels were realized (only the positive stocking rate increase considered), net annual benefits would be about 30% less. The NPV of the investment would then be negative, except under the most optimistic 1979 beef price situation. With 1979 prices, increased forage production alone would pay for the control treatment.

SUMMARY AND CONCLUSIONS

Economic losses from heavy snakeweed infestations have been substantial. An estimated 3.5 million ha in Texas and New Mexico are so densely covered with snakeweed that favorable forage production is greatly suppressed and livestock poisoning is a problem.

Forage production can be increased from 100% to 800% by controlling dense snakeweed stands. In addition, increased death losses and abortion rates, coupled with decreased livestock weight gains, result in substantial reductions in livestock production and performance.

Economic losses from snakeweed infestations in Texas and New Mexico were estimated to be $44.3 million using average 1980 beef prices. This includes economic losses from both reduced stocking rate and from livestock poisoning.

With current technology, snakeweed can be satisfactorily controlled, with kill rates exceeding 95%. However, treatment life is relatively short and variable because of the cyclic nature of snakeweed populations. Although the treatment is not generally long lasting, herbicidal control of snakeweed is an economical range improvement practice.

At current low beef prices, and with only a 5-year treatment life, positive economic benefits from snakeweed control are still realized, with discounted benefits exceeding costs by an estimated 7%. If record high 1979 beef price levels are considered, benefits of control exceed costs by an estimated 36%. With 1980 beef price levels, benefits of snakeweed control exceed costs by 20%. In this case, the NPV of the range improvement investment was estimated to be about $9.64/ha, and the investment return was estimated to be about 13%, as indicated by the IRR. Investment returns are considerably less under the current relatively low beef price situation.

LITERATURE CITED

Dollahite, J.W., and W.V. Anthony. 1956. Experimental production of abortion, premature calves and retained placentas by feeding a species of perennial broomweed. Texas Agr. Exp. Prog. Rep. 1884.

McDaniel, K.C. 1984. Snakeweed control with herbicides. New Mexico State University Agri. Exp. Sta. Bull. 706.

McDaniel, K.C. (In Press). Broom snakeweed (Gutierrezia sarothrae) control with spring and fall application of picloram and metsulfuron. Weed Science.

McDaniel, K.C., R.D. Pieper, and G.B. Donart. 1982. Grass response following thinning of broom snakeweed. J. Range Manage. 35:219-222.

McDaniel, K.C. and R.E. Sosebee. 1987. Taxonomy, ecology, and poisonous properties associated with perennial snakeweeds. In: L.F. James, M.H. Ralphs, and D.B. Nielsen (eds), The ecology and economic impact of poisonous plants on livestock production, Westview Press, Boulder, CO.

McDaniel, K.C., and L.A. Torell. 1987. Ecology and management of broom snakeweed. In: J.L. Capinera (ed), Integrated pest management on rangeland: A shortgrass prairie perspective. Westview Press, Boulder, CO.

McDaniel, K.C., L.A. Torell, J.M. Fowler, and K.W. Duncan. 1986. Brush control on New Mexico rangeland. New Mexico State University, Coop. Ext. Serv. Bull. 400B-18.

McGinty, A., and T. Welch. (In Press). Perennial broomweed and its impact on Texas ranching. Rangelands.

Sosebee, R.E., W.E. Boyd, and C.S. Brumley. 1979. Broom snakeweed control with tebuthiuron. J. Range Manage. 32:179.

Sperry, O.E., J.W. Dollahite, G.O. Hoffman, and B.J. Camp. 1964. Texas plants poisonous to livestock. Texas Agri. Exp. Sta. Bull. 1028.

Texas A&M University. 1984a. Texas livestock enterprise budgets: Texas High Plains Region II. Texas Agri. Ext. Serv. Report B-1241(L02).

Texas A&M University. 1984b. Texas livestock enterprise budgets: Texas High Plains Region III. Texas Agri. Ext. Serv. Report B-1241(L03).

Torell, L.A., B.A. Brockman, K.M. Garrett, and H.W. Gordon. 1986a. 1985 costs and returns for a large cow-calf enterprise, northeastern New Mexico. New Mexico State University Coop. Ext. Serv. Report CRE85-LA-NE1.

Torell, L.A., B.A. Brockman, K.M. Garrett, and H.W. Gordon. 1986b. 1985 costs and returns for a medium cow-calf enterprise, northeastern New Mexico. New Mexico State University Coop. Ext. Serv. Report CRE85-LA-NE2.

Torell, L.A., B.A. Brockman, K.M. Garrett, and H.W. Gordon. 1986c. 1985 costs and returns for a small cow-calf enterprise, northeastern New Mexico. New Mexico State University Coop. Ext. Serv. Report CRE85-LA-NE3.

Ueckert, D.N. 1979. Broom snakeweed: effect on shortgrass forage production and soil water depletion. J. Range Manage. 32:216-219.

Workman, J.P. 1986. Range economics. Macmillan Publishing Company, New York.

7

Cattle Abortion from Ponderosa Pine Needles: Ecological and Range Management Considerations

F. Robert Gartner, Frederic D. Johnson,
and Penelope Morgan

Ponderosa pine (Pinus ponderosa) occurs on a great variety of landforms, soil types, and climatic regimes both in pure and mixed stands which dominate a vast array of understory plants. The tree is a valuable timber species in most areas. Ponderosa pine forests provide forage and cover for domestic livestock and wildlife, watershed protection, and recreational use. Estimates vary, but at least 16.8 million ha are occupied by ponderosa pine in western North America (Burns 1983). Much of this area is grazed by cattle, yet cows that ingest pine needles in the later stages of pregnancy frequently abort or have weak calves that often die (MacDonald 1952). The magnitude of losses has generated minimal support for a broad array of research, but the abortifacient compound or compounds in ponderosa pine needles remain undetected.

ECOLOGY OF PONDEROSA PINE

Distribution and Nomenclature

Ponderosa pine is the most widespread of the North American pines. It ranges from central British Columbia to central Mexico, from central Nebraska to the Pacific Coast (Figure 7.1). Throughout this vast range it is primarily a lower montane forest species, and yet it grows at elevations from near sea level in the Puget Trough of Washington to above 3,050 m in the southern Rockies and Mexico (Fowells 1965; Harlow et al. 1979).

Three varieties are recognized, although there is great morphological diversity within the varietal boundaries (Figure 7.1). Varieties are differentiated by the number of

71

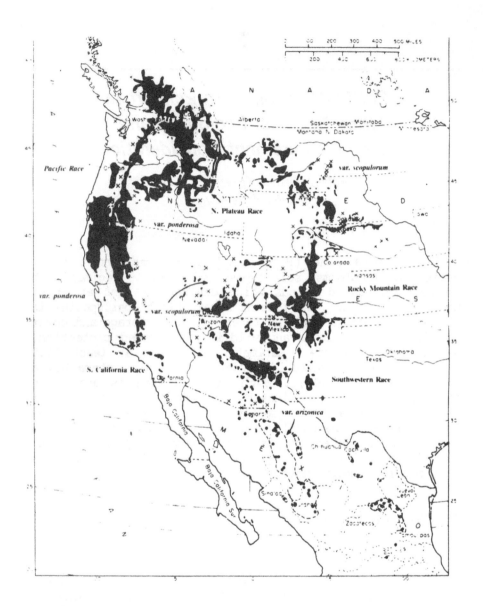

FIGURE 7.1
Distribution of <u>Pinus ponderosa</u>. General distribution from
Critchfield and Little (1966); races from Critchfield
(1984); varieties from Little (1953). Transition areas are
stippled.

needles per fascicle: two or three needles (scopulorum); three needles (ponderosa); four or five needles (arizonica). Critchfield (1984) recognized five races. Three races, Southern California, Pacific, and North Plateau, were combined in the variety ponderosa (Pacific ponderosa pine). This variety is geographically separated from the Rocky Mountain and Southwestern races which constitute the variety scopulorum (Rocky Mountain ponderosa pine). Other varieties occur in Mexico (Martinez 1948); the variety arizonica (Arizona ponderosa pine) extends northward into southern Arizona and New Mexico.

Needle lengths are extremely variable, from about 9 to 30 cm. Cones are 5 to 15 cm long, subsessile, the scale apex rhomboidal and armed with a small outward-pointing prickle. The bark of immature trees has dark-grey to black papery scales, while bark of mature trees is deeply black-furrowed. On old trunks, vertical furrows are more pronounced with cinnamon to yellow-brown flat scales on plates between furrows.

Synecology

The literature on ponderosa pine reflects its extensive range. Axelton (1967, 1974, 1977) listed numerous citations on the biology and management of this important species. Fowells (1965) summarized the silvical characteristics and included a brief listing of major community types in which ponderosa pine occurs.

In the Rocky Mountains, from southern Canada to northern Mexico, a lower elevation forest of climax ponderosa pine generally appears (Daubenmire 1943). An exception is the Snake River gap, centered in eastern Idaho, which separates two major varieties (Figure 7.1). The ponderosa forest often takes the form of an open stand with vegetation of the adjacent nonforested type beneath. Climax ponderosa pine is often just above juniper-pinyon, mountain shrub, or shrub/grassland; ecotones with these types may be quite complex and cover large areas. Where ponderosa pine forests grade to grassland with decreasing elevation, a pine savanna results which provides excellent grazing (Pearson 1942). Almost everywhere in the Rockies, Douglas-fir (Pseudotsuga menziesii) climax forests replace ponderosa pine as elevation increases. In these forests, ponderosa pine is a major seral species; with disturbance, these forests often provide excellent forage, but pine reproduction is generally more common than in the drier ponderosa pine climaxes.

The North Plateau race of ponderosa pine occurs in a climax formation in the lower elevations of forest development on the east slopes of the Cascade Mountains. These forests are generally similar to the Northern Rockies (Franklin and Dyrness 1973), with Douglas-fir forests above and bunchgrass or sagebrush (Artemisia spp.) steppe below. In southern Oregon, on both sides of the Cascades, the Sierran influence is manifested by a strong mixture of sclerophyllous shrubs in the understory of the lower elevation ponderosa pine forests and the presence of other tree species in the overstory.

In California, the Pacific race of ponderosa pine occurs on both sides of the Sierra Nevada as well as in the North Coast Ranges. There are nineteen species of pine native to California and, except for the three relict species along the coast and two high subalpine species, the others share some segment of their habitat with ponderosa pine. Ponderosa pine is an integral part of the lower montane forests of the Sierra Nevada (Barbour and Major 1977). At lower elevations various oaks (Quercus spp.), chaparral vegetation, and California incensecedar (Libocedrus decurrens) are common associates. Jeffrey pine (Pinus jeffreyi), a very close relative, replaces ponderosa pine as elevation increases. In the North Coast Ranges, ponderosa pine plays a relatively minor role in the Douglas-fir/mixed hardwood forests. As in southern Oregon, the predominant understory vegetation is largely chaparral-type shrubs. The Southern California race shares forest dominance with Jeffrey pine with which it intergrades; shrub vegetation dominates the understory.

One of the most widely used references on North American forest vegetation is the "Forest Cover Types of the United States and Canada" (Soc. of Amer. Foresters 1980). Three types feature ponderosa pine: interior ponderosa pine (Type 237) extends throughout the Rockies and the east slopes of the Cascades; Pacific ponderosa pine (Type 245) occurs in southern Oregon and California; the Sierra Nevada mixed conifer (Type 243) consists of seral ponderosa pine mixed with other conifers and some broadleaf trees. These three types are combined as a "ponderosa pine" type in Figure 7.1. Ponderosa pine also appears as a lesser component of many other western forest types. In this role it nearly always occurs as a seral species in the forest zone above the climax ponderosa pine forests of lower elevations. The ponderosa pine forests of the Black Hills of South Dakota and the nearby Bear Lodge Mountains of northeastern Wyoming are isolated from the rest of the interior ponderosa pine cover types by wide expanses of nonforested land. They are also sufficiently distinctive in their silvical characteristics to warrant separate description and specialized silvicultural treatment (Boldt et al. 1983).

Habitat types, first described by Daubenmire (1952), are increasingly used to delineate forest and nonforest habitats on a much more refined basis than cover types. A habitat type consists of all units of land that will support a specific plant association at climax. They are based on the premise that plants integrate the environment and that climax plants are superior indicators of specific habitats. Most of the forests and much of the nonforest vegetation of the western United States are described using this system. Where continued disturbance makes it impossible to find representative climax vegetation, community types have been used to describe vegetation (Volland 1976). Other authors prefer the more traditional term "association" (Williams and Lillybridge 1983), but the end results are virtually identical. These community descriptions include cover and constancy data for major understory plants. Some include figures for total herbage production and key forage species (Johnson and Simon 1985).

Alexander (1985) summarized forest habitat type classification in the Rocky Mountains. General location, successional status, major tree, and understory associates are presented in tabular form. Ponderosa pine appears as the principal climax species in 39 habitat types, and as a minor climax, seral, or accidental species in 79 other habitat types. Work now underway in the Pacific Northwest, central and southern Rockies, and in the Black Hills will add more habitat types supporting ponderosa pine in the near future.

The northern Rocky Mountains is the only large region where virtually all forests have been classified to habitat types (Pfister et al. 1977; Steele et al. 1981; Steele et al. 1983; Cooper et al., unpublished reference). A summary of habitat types which support climax ponderosa pine from this area will serve to illustrate habitat types and similar classifications (Table 7.1). Since habitat types, community types, and associations are named by dominant and/or characteristic overstory/understory plants, a general picture of each habitat type can be gained by being familiar with the plants or the class of plants. Thus, the forage production potential of the ponderosa pine/Idaho fescue (Festuca idahoensis) habitat type is clearly greater than that of the ponderosa pine/curlleaf mountain mahogany (Cercocarpus ledifolius) habitat type.

TABLE 7.1
Habitat types of the Northern Rockies supporting ponderosa pine at climax.

	Montana[1,2]	Central Idaho[3]	E. Idaho W. Wyoming[4]	Northern Idaho[5]
PONDEROSA PINE SOLE CLIMAX DOMINANT				
ponderosa pine/bluestem	X	·		·
ponderosa pine/bluebunch wheatgrass	X	X		X
ponderosa pine/Idaho fescue	X	X		X
ponderosa pine/western needlegrass	·	X	--ponderosa pine absent - see Figure 7.1--	·
ponderosa pine/antelope bitterbrush	X	X		X
ponderosa pine/common snowberry	X	X		·
ponderosa pine/mountain snowberry	·	·		X
ponderosa pine/ninebark	·	·		·
ponderosa pine/common chokecherry	X	·		·
PONDEROSA PINE SHARES CLIMAX DOMINANCE				
limber pine/bluebunch wheatgrass	X	·		·
Douglas-fir/bluebunch wheatgrass	X	X		X
Douglas-fir/Idaho fescue	X	X		X
Douglas-fir/elk sedge	·	·		X
Douglas-fir/Oregon grape	·	X		·
Douglas-fir/mountain snowberry	·	X		·
Douglas-fir/curlleaf mountain mahogany	·	X		·

[1] Pfister et al. 1977
[2] Canopy cover must exceed 25%
[3] Steele et al. 1981
[4] Steele et al. 1983
[5] Cooper et al. 1985

Once climax communities are described, successional studies can proceed with greater assurance of understanding total seral sequences. The successional models developed by Steele (1984) are coupled with management and include time-amplitudes of successional plants, forage values per successional unit, and plant autecological information. Management implications for cattle and other uses are described (Steele and Geier-Hayes 1982, 1983, 1984, 1985, unpublished reference).

Autecology and Population Dynamics

Fowells (1965) summarized the vast array of soils, landforms, and climates which determine the habitats occupied by ponderosa pine. As expected, the extremes are great: mean annual temperatures from 5.5 to 11.1 C; soil pH from 4.9 to 9.1; and soil parent material of igneous, metamorphic, or sedimentary origin. Ponderosa pine is mainly a slope species that is favored on sites with deep, well-drained soils; it is largely absent from soils with impeded drainage, although it is often present in riparian zones if the substrate is sufficiently rocky.

Individual ponderosa pine trees may live over 250 years. Major causes of mortality of mature trees are, in order of importance, diseases, insects, fire, and windthrow (Schubert 1974). Diseases that result in slower growth and mortality of both mature and young trees include several dwarf mistletoes and a variety of root rots. Periodic outbreaks of bark beetles cause patches of mortality in mature individuals and young trees in overstocked stands. Growth of young trees may be slowed by pine tip moth, shoot borer, and other insects.

Large seed crops occur at intervals of one to five years or more (Fowells and Schubert 1956; Schubert 1974; Sundahl 1971). Many insects cause seed loss (Hedlin et al. 1981), as do birds and mammals. Seeds are wind disseminated, but 75% fall within 37 m of the source (Barrett 1979).

Ponderosa pine most frequently regenerates following disturbance by fire, logging, insect or disease outbreaks, or heavy grazing. Successful establishment often depends on the combination of adequate rainfall, openings in overstory canopy, absence of fire until seedlings are large enough to survive, presence of bare mineral soil, and low rodent populations.

Seeds may germinate in the spring or fall, with best survival on loose mineral soil. Seedlings grow rapidly, especially on productive sites. High seedling mortality, due to frost heaving (especially on heavy soils), drought, rodent damage, competition from grass and shrubs, and episodic establishment result in uneven recruitment of trees, particularly on drier, low elevation sites.

Pre-settlement fires occurred every 2 to 25 years in climax ponderosa pine communities (Weaver 1951; Cooper 1960; Biswell 1973; Dietrich 1980). Ponderosa pine is well adapted to survive low-intensity fires and will be favored by surface fires that expose some mineral soil. The thick bark and high open crowns of large trees protect meristematic tissue, while small trees are readily killed by fire. Aggressive suppression of fires since the early 1900s has resulted in increased fuel accumulations, stagnated nutrient cycles (Covington and Sackett 1984), denser stands, and greater fire intensities when fire occurs. Decreases in forage quantity and quality are associated with absence of periodic fire.

Timber harvest and prescribed burning can increase forage for cattle and wildlife in both seral and climax ponderosa pine communities, depending on understory plant composition, range condition, fire effects, and site. Herbage production may increase following fires that open tree canopies and encourage organic decomposition (Oswald and Covington 1984). Understories vary from sparse bunchgrasses to dense chaparral in ponderosa climax forests. Under continual grazing regimes, forests where ponderosa pine is seral are often more productive for livestock forage than adjacent climax stands of ponderosa pine.

There can be little disagreement that ponderosa pine is one of the most majestic conifers in western North America. Designated the state tree of Montana, this tree has been used by landscape architects in urban areas and planted in farm and ranch shelterbelts. It has been a valuable source of lumber throughout its range. Ponderosa pine trees also provide summer shade and protection from winter storms for domestic livestock. Paradoxically, the ingestion of ponderosa pine needles is suspected of causing abortion among pregnant cows. Reports from livestock producers, county agents, veterinarians, and research scientists indicate that economic losses attributed to abortion due to pine needle ingestion are of major proportions. Therefore, ponderosa pine has been included in most, if not all, poisonous plant publications, and several million dollars have been spent on researching the causative agent.

CATTLE ABORTION FROM PINE NEEDLE INGESTION

Several investigators in the western United States and southwestern Canada have reported cattle abortions resulting from ingestion of pine needles (MacDonald 1952; Stevenson et al. 1972; Cogswell 1974; Kamstra 1975; James et al. 1977). Indeed, pine needle abortion is one of the most important poisonous plant problems in many areas because of the magnitude of economic consequences (Panter and James 1985). Where ponderosa pine is the predominant tree, many cattle producers are annually confronted with potential economic losses when pregnant cows have the opportunity to consume pine needles.

The Black Hills of South Dakota and Wyoming have been identified as one of the areas where pine needle abortion in cattle is a serious livestock health and economic problem (Figure 7.2). Calf losses attributed to abortion following pine needle consumption by pregnant cows were reported by nearly 75% of 173 livestock producers surveyed in seven Black Hills counties (Cogswell 1974). Cattle management aimed at minimizing losses from abortion also affects cow/calf operators in California, Colorado, Idaho, Montana, Oregon, Washington, and in some areas of western Canada (DeHaan 1979). It has not been established that abortions occur everywhere ponderosa pine grows. Yet, publications on poisonous range plants of widely separated geographic areas of the West, such as New Mexico (Norris and Valentine 1954) and Montana (Leininger et al. 1977), allude to the potential danger of pine needle consumption. Thus, there appears to be ample evidence that consumption of ponderosa pine needles by pregnant cows may lead to abortions.

Causes and Symptoms of Pine Needle Abortions

Hill (1917) concluded that cattle do not browse pine reproduction in the Southwest by preference but because of insufficient palatable forage. Another major factor causing cattle in the Southwest to browse pine seedlings was thirst, according to Cassidy (1937). Call and James (1978) noted that undefined causative factors such as environmental stress, body condition, disease, and biological differences may influence the effect of ingested pine needles on the cow. Shue (1983) listed several factors that cause cattle to consume pine needles: (1) sudden weather changes such as cold winds or snow storms which cause cattle to seek the protection of pine trees; (2) extreme hunger due to scarcity

FIGURE 7.2
Keeping pregnant cows from consuming pine needles is difficult and costly in the ponderosa pine grasslands of the Black Hills. Ranch headquarters are typically located in the foothills with calving pastures nearby.

of forage near the end of the grazing period; (3) changes in feed, especially to unfamiliar or poor quality forage; (4) sudden access to pine needles, especially when cattle are moved from nonforested to forested range or pasture; (5) boredom; and (6) accidental ingestion with forage.

Abortions in cattle have been reported following grazing of green ponderosa pine needles directly from trees, slash, and windfalls, and from dried needles that have fallen to the ground (Stevenson et al. 1972; James et al. 1977). Abortions generally occur when cows are in the last trimester of pregnancy, although there have been some reported abortions during earlier stages. Stevenson et al. (1972) reported that pine needle-induced abortions often occur one to three days after pregnant cows have consumed needles or buds, but may be delayed two to three weeks. Even after cattle have been removed from available pine needles, abortion may continue for up to two weeks. Cattle abortion from pine needle ingestion may be a misnomer because the fetus is viable (James et al. 1986, unpublished reference). They suggested that "premature induced parturition" may more correctly describe abortion caused by pine needle consumption since near-term calves may survive with proper care and nursing. Abortion, or premature calving, apparently occurs suddenly. If calves live, they appear weak (Olson 1976).

Cows that graze pine needles may show signs of preparing for parturition before aborting, i.e., swelling of genital organs and filling of the udder. Milk production will be initiated, but if the cow continues to consume pine needles, she may abort (James et al. 1986, unpublished reference). If needles are removed from the diet, the swelling disappears and the cow will have a calf at the normal time. Most field observations reveal few advance signs of an impending abortion following ingestion of pine needles. Cows will usually appear depressed and dull prior to aborting (Call and James 1978). Shue (1983) noted that some ranchers have reported a bloody discharge from the reproductive tract of cows eating pine needles. Some abortions are characterized by "weak parturition contractions, excessive uterine hemorrhage, incomplete dilation of the cervix, and often a characteristic nauseating odor" (James et al. 1986, unpublished reference). Other aborting cows follow a "more normal sequence through parturition." Frequently cows will need obstetrical assistance at parturition because of the condition of the genital tract. If a calf is born rapidly, or assisted from the uterus in early parturition, its survival chances are improved. Calves may be born weak or die shortly after birth due to placental damage or delay in delivery.

A persistent retained placenta was a consistent characteristic of abortions induced by ponderosa pine needles, regardless of the stage of gestation (Stevenson et al. 1972). Manual attempts to remove the placenta were mostly unsuccessful and often associated with excessive hemorrhage. Mortality of cows that aborted was high. Cows that survive abortion may breed back but subsequently have an increased calving interval (James et al. 1986, unpublished reference).

At the time of abortion, some cows appear to have a toxemia. Death may occur before or from several hours to several days after the abortion. The death of the cow may be so rapid as to suggest a toxic effect from the ingested pine needles (James et al. 1986, unpublished reference).

Composition and Toxic Properties of Pine Needles

Cogswell (1974) reported only slight seasonal variability of physical and chemical characteristics of ponderosa pine needles collected from the foothills of the Black Hills. The nutritive value of pine needles was comparable to low quality hay, and the agent which causes cows to abort was assumed to be present in pine needles year around (Kamstra and Cogswell 1975).

Researchers at several locations in the West have been trying for years to isolate and characterize causative agents in ponderosa pine needles which trigger abortion in pregnant cows. A number of toxic compounds in pine needles have been suggested as the causative agent (Shue 1983), but no specific agent has been isolated or identified. Yet, there is justification, both from field observations and experimental feedings, to conclude that pine needles can be toxic under certain conditions (Short et al. 1987). In addition to possible toxins in pine needles, several other factors may be associated with and/or predispose animals to abort following pine needle consumption, including: (1) stage of gestation when pine needles are consumed; (2) environmental stresses; (3) animal condition; and (4) general physiology of the animal (Shue 1983). Because of the complexities these factors impose on solutions, researchers have been unable to consistently induce abortions in cattle under experimental conditions.

Laboratory Investigations with Small Animals

Controlled research with large animals having long gestation periods is not only relatively slow but also expensive. For these reasons, most exploratory research into specific causes of pine needle-induced abortion has been done with small laboratory animals. This approach assumes that the anatomy of cattle as well as their responses to treatments would be similar to those of mice and rats.

Laboratory research has resulted in the isolation of several agents in pine needles that are toxic or otherwise adversely affect pregnant mice and rats (Cook 1960; Allen and Kitts 1961; Tucker 1961; Cook and Kitts 1964; Allison and Kitts 1964; Chow et al. 1972, 1974; Cogswell 1974; Anderson and Lozano 1977, 1979; Kubik and Jackson 1981; Manners et al. 1982; Neff et al. 1982; Shue 1983). Mice served as a successful biological assay since liquid extracts of pine needles disrupted pregnancy through death and resorption of the fetuses. Early embryo mortality in pregnant mice fed aqueous and acetone extracts of ponderosa pine needles was reported by Cogswell and Kamstra (1980). That laboratory method was said to have value as a quick and economical bioassay for the toxicity of ponderosa pine needles.

DeHaan (1979) conducted a series of experiments to determine the value of chick embryo bioassays for measuring the detrimental effect of pine needle extracts and their subfractions. This approach proved to be a good screening technique, but he concluded that "the search for the abortive factor in pine needles is still clouded by contradictions."

The research of Cook (1960), Manners et al. (1982), and Shue (1983) suggests that certain luteolytic agents contained in pine needles affect pregnant laboratory animals. Shue (1983) studied the effects of various forms of prostaglandins on pregnant mice and concluded that the evidence is unclear that prostaglandins are active agents in pine needle-induced abortions of mice and cattle. Toner (1971) found that plasma estrogen levels decreased when bred heifers were drenched with an aqueous solution of ground ponderosa pine needles.

Other researchers have noted that the potential causative agents in pine needles may be either fungal mycotoxins (Chow et al. 1974; Anderson and Lozano 1977) or infectious microorganisms such as Listeria monocytogenes (Adams et al. 1979). Several laboratory attempts to isolate specific causal agents were reported by Panter and James (1985). Even though a large number of agents in or on ponderosa pine needles have been suggested as the cause of abortions, it is evident that the exact cause is yet unknown (James et al. 1986, unpublished reference).

Investigations with Livestock

Bruce (1927) may have been the first to label ponderosa pine a poisonous plant and suggest that abortions may occur when cows eat needles and buds of ponderosa pine. In 1950, the British Columbia Beef Growers Association requested research to determine whether pine needles cause abortion in range beef cattle. MacDonald (1952), reporting on that research, concluded that pine needles and buds caused abortion and weak calves. He also found that pine needles and buds were palatable to wintering stock and would be ingested by pregnant cows even though adequately fed. These studies were repeated in the winter of 1953 with inconsistent results: one cow in the control group of six aborted, while only two cows of the twelve consuming pine needles aborted (Majak et al. 1977). Later trials with cows in British Columbia were also inconclusive. Early accounts of abortion due to pine needle ingestion were discounted because there were several other causes of abortion in cattle.

Some studies suggest that abortions are more frequent in cows in their last trimester of gestation (MacDonald 1952; Faulkner 1969; Stevenson et al. 1972; James et al. 1977). None of the cows involved in a three-year study at Colorado State University aborted in less than 170 days after breeding, regardless of the length of time they had been eating needles (Anonymous 1960). In the last trimester of gestation, though, abortion occurred as soon as two weeks after the animals were placed on a daily ration of pine needles.

Short et al. (1987) reported variable results in several pine needle feeding trials. Incidence of cattle abortions varied from 0% to 100%. Stevenson et al. (1972) noted that the effects of pine needle consumption on reproduction in sheep and deer had not been determined. Pregnancy and lambing rates of 30 mature ewes were not affected when fed pine needles during days 60 to 90 of gestation (Call and James 1976). In recent studies, neither sheep nor goats aborted after being fed pine needles (Short et al. 1987). Information is lacking on the effects of pine needles on pregnant horses and elk. Bison reproduction in Wind Cave National Park is apparently not affected by pine needle consumption (Richard W. Klukas, Research Biologist, Wind Cave National Park, personal communication).

Some authors referred to ponderosa pine needles as harmful when consumed in large quantities, though specific amounts were not quantified (Muenscher 1947; Kingsbury 1964; James et al. 1980). In a recent study, Short et al. (1987) reported that daily feeding of small quantities (0.68 kg and 1.36 kg) of pine needles caused most cows to abort. Cows fed 2.72 kg of pine needles daily had the highest abortion percentage. Decreasing the number of days needles were fed "drastically reduced the abortion response," but some abortions occurred after only one and three days of feeding pine needles. Schmutz and co-workers (1968) reported that cows will consume 2.27 to 2.72 kg of fresh pine needles daily. Others noted that range cattle seldom consume pine needles because of the resinous taste (Muenscher 1947), and may do so only when the native forage is covered with snow or in short supply (Bracken 1968). Uresk and Lowrey (1984) and Uresk and Paintner (1985) reported that pine needles were commonly part of cattle diets from June through October in the Black Hills. Diet composition was not reported for other months of the year. Among livestock producers and practicing veterinarians there are mixed reports about the quantity of pine needles that a cow may consume whether or not adequate forage is available.

Scientists have been unable to determine with certainty if cows in early gestational stages are less susceptible to toxicity of pine needles than those in later stages. Neither has it been established that fewer needles are consumed in earlier stages of gestation because forage is more abundant and/or more available than in later stages. Pine needle consumption by cattle in the Black Hills increased from 4.5% of the diet in June to 9.4% in September (Uresk and Paintner 1985). Regardless of the gestation stage when needles are eaten, aborted cows retain placentas which may lead to death or cause them to be barren the following year.

The inconsistent results reported when pregnant cows in controlled studies were fed pine needles are in concert with experiences of livestock producers. A northern Black Hills rancher, whose cattle graze ponderosa pine grassland year around, stated that cattle abortions are not an annual occurrence, and abortions in wild animals are also inconsistent. He noted that the extent and severity of pine needle abortion varies from the northern to the southern Black Hills within the same year (Wesley W. Thompson, Spearfish, South Dakota, personal communication).

RANGE MANAGEMENT CONSIDERATIONS

Because the causative agent or toxic principal of ponderosa pine needles is unknown, no therapeutic measures or alleviators are available. MacDonald (1952) listed several management recommendations for minimizing losses:

(1) keep pregnant cows from ranges where ponderosa pine occurs;
(2) if ponderosa pine ranges must be used with bred cows, herbaceous vegetation should be maintained in a vigorous, high condition;
(3) do not move pregnant cows under stress into areas with pine trees;

> (4) if pastures having pine must be used by pregnant cows, trim lower branches of trees within reach of cattle and remove slash;
>
> (5) suspend or postpone ponderosa pine logging operations where bred cows are grazed.

Range managers have been able to expand very little on these recommendations because the poisonous principal is unknown. Some producers have successfully avoided the potential abortion problem by changing from a cow/calf to a yearling business.

Ponderosa pine stands at higher elevations in the Black Hills and in the foothills have greatly increased in density since the beginning of settlement due to the lack of recurrent wildfires and improved fire suppression techniques (Thompson and Gartner 1971, Progulske and Sowell 1974). Pine encroachment into adjacent grasslands has been a continuing problem to livestock producers on private foothill ranges (Gartner and Thompson 1973, Gartner and White 1986).

Prescribed burning of private rangelands in the foothills of the Black Hills has been quite common during the past decade. Objectives include reduction of ponderosa pine seedlings and saplings in grasslands in order to reduce competition for native forage plants. With improved firing and fire control techniques in recent years, some burn prescriptions have aimed at thinning post- and pole-size doghair pine. Burning, especially on a repeated basis, will prune lower branches of trees, effectively raising crowns out of the reach of cattle. Needles on the ground are also consumed by fires. Thus, the total quantity of needles available to cattle is effectively reduced, and forage quality may be increased. Black Hills area rangelands should probably be burned on a five to ten year rotation to minimize availability of pine needles, but grazing management must also aim for high range condition. Since ponderosa pine growth and reproduction, associated plant species, and climate are different in other regions, prescribed burning methods and objectives may differ from those developed for the Black Hills. Careful cattle management (grazing use) has been cited as the watchword on ponderosa pine rangelands to minimize losses from abortion caused by pine needle ingestion. Paradoxically, careful grazing use was suggested as the best method to minimize cattle damage to natural and artificial pine regeneration on ponderosa pine-bunchgrass range in central Colorado (Currie et al. 1978).

SUMMARY

Reports from cattle producers and field observations by researchers and veterinarians confirm that abortions induced by pine needles are a major economic problem of the cattle industry in many areas. Economic losses result not only from calves lost by abortion but also from death losses of cows. The prolonged calving interval associated with placental retention further escalates the cost of losses. Presently, there is no clinical diagnosis to positively identify abortion due to pine needle ingestion. When the causative agent or agents become known, prevention will become possible with benefits accruing directly to the producers and indirectly to the overall economy of many geographic areas.

LITERATURE CITED

Adams, C.J., T.E. Neff, and L.L. Jackson. 1979. Induction of Listeria monocytogenes infection by the consumption of ponderosa pine needles. Infect. Immun. 25:117-120.

Alexander, R.R. 1985. Major habitat types and plant communities in the Rocky Mountains. USDA Rocky Mountain Forest and Range Exp. Sta. Gen. Tech. Rep. RM-123.

Allen, M.R., and W.D. Kitts. 1961. The effects of yellow pine (Pinus ponderosa Laws.) needles on the reproductivity of the laboratory female mouse. Can. J. Anim. Sci. 41:1-8.

Allison, C.A., and W.D. Kitts. 1964. Further studies on the anti-estrogenic activity of yellow pine needles. J. Anim. Sci. 23:1155-1159.

Anderson, C.K., and E.A. Lozano. 1977. Pine needle toxicity in pregnant mice. Cornell Vet. 67:229-235.

Anderson, C.K., and E.A. Lozano. 1979. Embryotoxic effects of pine needles and pine needle extracts. Cornell Vet. 69:169-175.

Anonymous. 1932. Ponderosa pine now official. J. Forest. 30:510.

Anonymous. 1960. Scientists advise on pine needle abortion. Colorado State Univ. Agri. Exp. Sta., Colo. Farm & Home Res. 10(1):5.

Axelton, E.A. 1967. Ponderosa pine bibliography through 1967. USDA Intermountain Forest and Range Exp. Sta. Res. Paper INT-40.

Axelton, E.A. 1974. Ponderosa pine bibliography II: 1966-1970. USDA Intermountain Forest and Range Exp. Sta. Gen. Tech. Rep. INT-12.

Axelton, E.A. 1977. Ponderosa pine bibliography III: 1971 through 1975. USDA Intermountain Forest and Range Exp. Sta. Gen. Tech. Rep. INT-33.

Barbour, M.G., and J. Major, (eds). 1977. Terrestrial vegetation of California. John Wiley & Sons, Inc., New York.

Barrett, J.W. 1979. Silviculture of ponderosa pine in the Pacific Northwest: the state of our knowledge. USDA Pacific Northwest Forest and Range Exp. Sta. Gen. Tech. Rep. PNW-97.

Biswell, H.H. 1973. Fire ecology in ponderosa pine-grassland. Proc. Tall Timbers Fire Ecology Conf. 12:69-96.

Boldt, C.E., R.R. Alexander, and M.J. Larson. 1983. Interior ponderosa pine in the Black Hills, pp. 80-83. In: R.M. Burns (tech. compiler), Silvicultural systems for the major forest types of the United States. USDA Forest Serv. Agri. Handbook 445.

Bracken, F.K. 1968. Losses due to disease, p. 86. In: Prenatal and postnatal mortality in cattle. Nat. Acad. Sci., Washington, DC.

Bruce, E.A. 1927. Astragalus serotinus and other stock poisoning plants of British Columbia. Dominion of Canada Dept. Agri. Bull. No. 88.

Burns, R.M., (tech. compiler). 1983. Silvicultural systems for the major forest types of the United States. USDA Forest Serv. Agri. Handbook 445.

Call, J.W., and L.F. James. 1978. Pine needle abortion in cattle, pp. 587-590. In: R.F. Keeler, K.R. Van Kampen, and L.F. James (eds), Effects of poisonous plants on livestock. Academic Press, New York.

Cassidy, H.O. 1937. How cattle may use cut-over ponderosa pine-bunchgrass ranges with minimum injury to reproduction. USDA Pac. Southwest Forest & Range Exp. Sta. Res. Note 15.

Chow, F.C., D.W. Hamar, and R.H. Udall. 1974. Mycotoxic effect on fetal development: pine needle abortion in mice. J. Repro. Fertil. 40:203-204.

Chow, F.C., K.J. Hanson, D.W. Hamar, and R.H. Udall. 1972. Reproductive failure of mice caused by pine needle ingestion. J. Repro. Fertil. 30:169-172.

Cogswell, C.A. 1974. Pine needle (Pinus ponderosa) abortive factor and its biological determination. Ph.D. Thesis, South Dakota State Univ., Brookings, SD.

Cogswell, C., and L.D. Kamstra. 1980. Toxic extracts in ponderosa pine needles that produce abortion in mice. J. Range Manage. 33:46-48.

Cook, H. 1960. The effects of certain extracts of birdsfoot trefoil and yellow pine needles on the reproductive processes of the laboratory mouse and rat. M.S. Thesis, Univ. of British Columbia, Vancouver, B.C., Canada.

Cook, H., and W.D. Kitts. 1964. Anti-estrogenic activity in yellow pine needles (Pinus ponderosa). Acta Endocrin. 45:33-39.

Cooper, C.F. 1960. Changes in vegetation, structure, and growth of southwestern pine forests since white settlement. Ecol. Monogr. 30:129-164.

Covington, W.W., and S.S. Sackett. 1984. The effect of a prescribed burn in southwestern ponderosa pine on organic matter and nutrients in woody debris and forest floor. Forest Sci. 30:183-192.

Critchfield, W.B. 1984. Crossability and relationships of Washoe pine. Madrono 31(3):144-170.

Critchfield, W.B., and E.L. Little, Jr. 1966. Geographic distribution of the pines of the world. USDA Forest Serv. Misc. Publ. 991.

Currie, P.O., C.B. Edminister, and F.W. Knott. 1978. Effects of cattle grazing on ponderosa pine regeneration in central Colorado. USDA Rocky Mountain Forest and Range Exp. Sta. Res. Paper RM-201.

Daubenmire, R.F. 1943. Vegetational zonation in the Rocky Mountains. Botan. Rev. 9:325-393.

Daubenmire, R.F. 1952. Forest vegetation of northern Idaho and adjacent Washington, and its bearing on concepts of vegetation classification. Ecol. Monogr. 22:301-330.

DeHaan, K.A. 1979. Embryonic screening and isolation of pine needle abortive factors. M.S. Thesis, South Dakota State Univ., Brookings, SD.

Dietrich, J.H. 1980. Chimney Spring forest fire history. USDA Rocky Mountain Forest & Range Exp. Sta. Res. Paper RM-220.

Faulkner, L.C. 1969. Pine needle abortion, pp. 75-79. In: L.C. Faulkner (ed), Abortion diseases of livestock. Charles C. Thomas, Springfield, IL.

Fowells, H.A., (compiler). 1965. Silvics of forest trees of the United States. USDA Forest Serv. Agri. Handbook 271.

Fowells, H.A., and G.H. Schubert. 1956. Seed crops of forest trees in the pine region of California. USDA Forest Serv. Tech. Bull. 1150.

Franklin, J.F., and C.T. Dyrness. 1973. Natural vegetation of Oregon and Washington. USDA Pacific Northwest Forest and Range Exp. Sta. Gen. Tech. Rep. PNW-8.

Gartner, F.R., and W.W. Thompson. 1973. Fire in the Black Hills forest-grass ecotone. Proc. Tall Timbers Fire Ecology Conf. 12:37-68.

Gartner, F.R., and E.M. White. 1986. Fire in the northern Great Plains and its use in management, pp. 13-21. In: E.V. Komarek, S.S. Coleman, C.E. Lewis, and G.W. Tanner (compilers), Proc. Symp. Prescribed fire and smoke management. Soc. for Range Manage., Denver, CO.

Harlow, W.M., E.S. Harrar, and F.M. White. 1979. Textbook of dendrology. McGraw-Hill Book Co., New York.

Hedlin, A.F, H.O. Yates III, D.T. Cibrian, B.H. Ebel, and others. 1981. Cone and seed insects of North American conifers. USDA Forest Serv., Washington, D.C.

Hill, R.R. 1917. Effects of grazing upon western yellow-pine reproduction in the national forests of Arizona and New Mexico. USDA Bull. 580.

James, L.F., J.W. Call, and A.H. Stevenson. 1977. Experimentally induced pine needle abortion in range cattle. Cornell Vet. 67:293-299.

James, L.F., R.F. Keeler, A.E. Johnson, M.C. Williams, E.H. Cronin, and J.D. Olsen. 1980. Plants poisonous to livestock in the western states. USDA Agri. Info. Bull. 415.

Johnson, C.G., Jr., and S.A. Simon. 1985. Plant associations of Hells Canyon National Recreation Area. USDA Forest Serv. Wallowa-Whitman National Forest, Burns, OR.

Kamstra, L.D. 1975. Pine needle abortions can happen all year. Crops and Soils Magazine 27(8):25-26.

Kamstra, L.D., and C. Cogswell. 1975. Pine needle abortion among cows grazing foothill ranges. A progress report. South Dakota State Univ., Dept. of Anim. Sci., Anim. Sci. Series 75-44:48-50.

Kingsbury, J.M. 1964. Poisonous plants of the United States and Canada. Prentice-Hall, Inc., Englewood Cliffs, NJ.

Kubik, Y.M., and L.L. Jackson. 1981. Embryo resorptions in mice induced by diterpene resin acids of Pinus ponderosa needles. Cornell Vet. 71: 34-42.

Leininger, W.C., J.E. Taylor, and C.L. Wambolt. 1977. Poisonous range plants in Montana. Montana State Univ. Coop. Ext. Serv. Bull. 348.

Little, E.L. 1953. Checklist of native and naturalized trees of the United States (including Alaska). USDA Forest Serv. Agri. Handbook 41.

Little, E.L. 1979. Checklist of United States trees, native and naturalized. USDA Forest Serv. Agri. Handbook 541.

MacDonald, M.A. 1952. Pine needle abortion in range beef cattle. J. Range Manage. 5:150-155.

Majak, W., D.E. Waldern, and A. McLean. 1977. Pine needle water extracts as potential abortive agents in gestating cow diets. J. Range Manage. 30:318-319.

Manners, G.D., D.D. Penn, L. Jurd, and L.F. James. 1982. Chemistry of toxic range plants. Water-soluble lignols of ponderosa pine needles. J. Agri. Food Chem. 30:401.

Martinez, M. 1948. Los Pinos Mexicanos, 2nd Ed. Mexico City, Mexico.

Muenscher, W.C. 1947. Poisonous plants of the United States. The MacMillan Co., New York.

Neff, T.E., C.J. Adams, and L.L. Jackson. 1982. Pathological effects of pine needle ingestion in pregnant mice. Cornell Vet. 72:128.

Norris, J.J., and K.A. Valentine. 1954. Principal livestock-poisoning plants of New Mexico ranges. New Mexico Coll. of Agri. and Mech. Arts Agri. Exp. Sta. Bull. 390.

Olson, D.P. 1976. Pine needle abortion disease in cattle. Univ. of Idaho Coop. Ext. Serv. Current Info. Series 354.

Oswald, B.P., and W.W. Covington. 1984. Effect of a prescribed fire on herbage production in southwestern ponderosa pine on sedimentary soils. Forest Sci. 30:22-25.

Panter, K.E., and L.F. James. 1985. A review of pine needle and broom snakeweed abortion in cattle, pp. 125-130. In: F.D. Provenza, J.T. Flinders, and E.D. McArthur (compilers), Proc. Symp. Plant-Herbivore interactions. USDA Intermountain Forest & Range Exp. Sta. Gen. Tech. Rep. INT-222.

Pearson, G.A. 1942. Herbaceous vegetation as a factor in the natural regeneration of ponderosa pine in the southwest. Ecol. Monogr. 12:313-338.

Pfister, R.D., B.L. Kovalchik, S.F. Arno, and R.C. Presby. 1977. Forest habitat types of Montana. USDA Intermountain Forest & Range Exp. Sta. Gen. Tech. Rep. INT-34.

Progulske, D.R., and R.H. Sowell. 1974. Yellow ore, yellow hair, yellow pine. A photographic study of a century of forest ecology. South Dakota State Univ. Agri. Exp. Sta. Bull. 616.

Schmutz, E.M., B.N. Freeman, and R.E. Reed. 1968. Livestock poisoning plants of Arizona. Univ. of Arizona Press, Tucson, AZ.

Schubert, G.H. 1974. Silviculture of southwestern ponderosa pine: the status of our knowledge. USDA Rocky Mountain Forest & Range Exp. Sta. Res. Paper RM-123.

Short, R.E., R.B. Staigmiller, R.A. Bellows, K.E. Panter, and L.F. James. 1987. Effects of various factors on abortions caused by pine needles. Fort Keogh Livestock and Range Research Station, Miles City, MT, 1987 Field Day. USDA-ARS and Agri. Exp. Sta. Montana State Univ.

Shue, D.P. 1983. An examination of prostaglandins as causative agents in pine needle (Pinus ponderosa) abortion. M.S. Thesis, South Dakota State Univ., Brookings, SD.

Society of American Foresters. 1980. Forest cover types of the United States and Canada. F. H. Eyre (ed). Washington, DC.

Steele, R. 1984. An approach to classifying seral vegetation within habitat types. Northwest Sci. 58:29-39.

Steele, R., S.V. Cooper, D.M. Ondov, D.W. Roberts, and R.D. Pfister. 1983. Forest habitat types of eastern Idaho-western Wyoming. USDA Intermountain Forest and Range Exp. Sta. Gen. Tech. Rep. INT-144.

Steele, R., R.D. Pfister, R.A. Ryker, and J.A. Kittams. 1981. Forest habitat types of central Idaho. USDA Intermountain Forest and Range Exp. Sta. Gen. Tech. Rep. INT-114.

Stevenson, A.H., L.F. James, and J.W. Call. 1972. Pine needle (Pinus ponderosa) induced abortion in range cattle. Cornell Vet. 62:519-524.

Sudworth, G.B. 1927. Checklist of the forest trees of the United States, their names and ranges. USDA Forest Serv. Misc. Circ. 92.

Sundahl, W.E. 1971. Seedfall from young-growth ponderosa pine. J. Forest. 69:790-792.

Thompson, W.W., and F.R. Gartner. 1971. Native forage response to clearing low quality ponderosa pine. J. Range Manage. 24:272-277.

Toner, F.D. 1971. Implications of nutrient deficiencies and gonadal hormone levels on the etiology of pine needle abortion in rats and cattle. M.S. Thesis, Univ. of Idaho, Moscow, ID.

Tucker, J.S. 1961. Pine needle abortion in cattle. Proc. Amer. Coll. Vet. Toxicol. St. Louis, MO.

Uresk, D.W., and D.G. Lowrey. 1984. Cattle diets in the central Black Hills of South Dakota, pp. 50-52. In: D.L. Noble and R.P. Winokur (eds), Proc. Symp. Wooded draws: characteristics and values for the northern Great Plains. Great Plains Agri. Council Publ. No. 111.

Uresk, D.W., and W.W. Paintner. 1985. Cattle diets in a ponderosa pine forest in the northern Black Hills. J. Range Manage. 38:440-442.

Volland, L.A. 1976. Plant communities of the central Oregon pumice zone. USDA Forest Serv., Pacific Northwest Region. R6 Area Guide 4-2.

Weaver, H. 1951. Observed effects of prescribed burning on perennial grasses in the ponderosa pine forest. J. Forest. 49:267-271.

Williams, C.K., and T.R. Lillybridge. 1983. Forested plant associations of the Okanogan National Forest. USDA Forest Serv., Pacific Northwest Region. RG-Ecol. 132b.

UNPUBLISHED REFERENCES

Cooper, S., K. Neiman, and R. Steele. 1985. Forest habitat types of northern Idaho. USDA Intermountain Forest and Range Exp. Sta., and Northern Rocky Mtn. Region. [Editorial draft].

James, L.F., R.E. Short, K.E. Panter, J. Call, R.J. Molyneux, L. D. Stuart, and R. A. Bellows. 1986. Pine needle abortion in cattle: a review and report of recent research. USDA, ARS, Poisonous Plant Research Lab., Logan, UT. [Editorial draft].

Steele, R., and K. Geier-Hayes. 1982. The grand fir/blue huckleberry habitat type, succession and management. USDA Intermountain Forest and Range Exp. Sta. [Editorial draft].

Steele, R., and K. Geier-Hayes. 1983. The Douglas-fir/ nine-bark habitat type in central Idaho, succession and management. USDA Intermountain Forest & Range Exp. Sta. [Editorial draft].

Steele, R., and K. Geier-Hayes. 1984. The Douglas-fir/pine grass habitat type in central Idaho, succession and management. USDA Intermountain Forest & Range Exp. Sta. [Editorial draft].

Steele, R., and K. Geier-Hayes. 1985. The Douglas-fir/ mountain maple habitat type in central Idaho, succession and management. USDA Intermountain Forest & Range Exp. Sta. [Editorial draft].

8

Ponderosa Pine: Economic Impact

John R. Lacey, Lynn F. James,
and Robert E. Short

INTRODUCTION

The ponderosa pine (Pinus ponderosa) vegetative type extends south from Montana into New Mexico and Arizona and east from California and Oregon into the Plains States. This pine ecosystem occurs in pure or mixed stands on 13.6 million ha. It is the source of 2,383,000 animal unit months (AUMs) of forage annually (USDA Forest Service 1977). However, during the past 60 years there have been many reports of abortions and attendant problems occurring after cows graze pine needles and buds (Albaugh 1974).

Numerous investigators in western Canada and the United States have verified that abortions in cattle are induced by ponderosa pine needles (MacDonald 1952; Stevenson et al. 1972; Kamstra and Cogswell 1975). Abortions usually occurred during the latter stage of gestation (Stevenson et al. 1972). Some calves were born dead, while others died soon after birth (Stevenson et al. 1972; Albaugh 1974). Retained placentas were commonly reported.

Miner et al. (1987) speculated that the annual economic loss from ponderosa pine needle induced abortion approached $20 million. In addition to the aborted calves, other losses include higher cow death losses, loss of income from the premature culling of productive cows, impaired weaning weight of calves, cost of controlling ponderosa pine, sub-optimal utilization of forage on sites infested with pine, cost of leasing additional range or purchasing feed so cows can be maintained in areas where ponderosa pine is not present, fencing and other management costs to keep cows out of the ponderosa pine during critical periods, increased veterinary care for affected cows and calves, and the cost of purchasing or raising additional replacement heifers.

95

All of these factors reduce the profitability of cow-calf operations in ponderosa pine regions.

This study estimated the economic losses resulting from ponderosa pine needle abortion in the 17 western states. The results have implications for the management of range livestock and in establishing research priorities.

METHODS

Data were obtained from published literature, surveys of cattle producers, and personal interviews with producers, veterinarians, researchers, and extension service personnel. The analysis included: 1) determining the number of beef cows in each of the 17 western states; 2) calculating the percentage of the total forest-range acreage comprised by the ponderosa pine vegetation type in each state; 3) estimating the number of beef cows grazing in these pines; 4) estimating the average incidence of ponderosa pine needle abortion; and 5) estimating the direct economic loss from impaired livestock production and the costs of management practices to minimize the abortion problem.

We utilized the Forest Service's acreage and productivity data (USDA Forest Service 1980). The term "forest-range" involves land in native and natural grasslands and commercial and noncommercial forest lands that produce vegetation grazeable by livestock. By Forest Service definition, ponderosa pine forest consists of forest in which one of the pines--ponderosa pine, Jeffrey pine (Pinus jeffreyi), sugar pine (Pinus lambertiana), limber pine (Pinus flexilis), Arizona ponderosa pine (Pinus ponderosa var. arizonica), Apache pine (Pinus engelmanni), or Chihuahua pine (Pinus leiophylla var. chihuahuana) comprises 50% or more of the tree cover. The ponderosa pine unit is open and park-like with excellent ground cover of grasses, sedges, and forbs, or with an understory of shrubs of low to medium height (USDA Forest Service 1972). Areas with less than 50% ponderosa pine were not included in the study. Abortions occur in these areas, but we lacked statistics on the areas involved or utilization by cattle.

RESULTS AND DISCUSSION

Beef Cow Numbers and Characteristics of Western Rangeland

TABLE 8.1
Number of beef cows, total forest-range vegetation, total ponderosa pine vegetative type, percentage of total forest-range vegetation comprised of ponderosa pine, and number of cows grazing in the ponderosa pine type by state.

State	Beef Cows and Heifers that Calved[1] (1,000)	Forest-Range (hectares)[2] (1,000)	Ponderosa Pine (hectares)[2] (1,000)	Extent of Ponderosa Pine in Relation to Forest-Range (%)	Estimated Number of Cows Grazing in Ponderosa Pine[3]
Arizona	293	23,824	1,654	7	6,358
California	950	21,582	3,158	15	44,175
Colorado	773	14,267	809	6	14,378
Idaho	526	9,794	823	8	13,045
Kansas	1,455	6,593	0	0	0
Montana	1,373	21,933	1,059	5	21,282
Nebraska	1,738	9,831	68	1	5,388
Nevada	301	25,408	28	T	0
New Mexico	542	24,232	1,714	7	11,761
North Dakota	958	4,980	0	0	0
Oklahoma	1,837	3,767	0	0	0
Oregon	598	10,046	2,278	23	42,637
South Dakota	1,468	9,478	529	6	27,305
Texas	5,178	37,097	0	0	0
Utah	298	16,044	205	1	924
Washington	407	3,197	947	29	36,589
Wyoming	669	19,281	391	2	4,148
Total	19,364	261,354	13,663		227,990
Average	1,139	15,374	804	5	13,411

[1] Cooperative Extension Service in Agriculture and Home Economics, Montana State University and U.S. Dept. of Ag. Cooperating (1986).

[2] USDA Forest Service (1977).

[3] Estimate considers the relative amount of the ponderosa pine type in relation to other forest/range vegetation, adjusts for the average differences in carrying capacity (0.17 AUM/ha for ponderosa pine vs 0.55 AUM/ha for all 25 vegetation types; 0.17/0.55 = 0.31) and assumes that this percentage of beef cows would be grazing in the ponderosa pine. For example, the ponderosa pine type makes up (1,654/23,824) or 7 percent of the forest/range vegetation in Arizona. By multiplying 293,000 cows by 7 percent and by 31 percent, it is estimated that 6,358 cows graze in the ponderosa pine type in Arizona.

Nearly 20 million beef cows grazed in the 17 western states during 1986 (Table 8.1). The number varied from 293,000 in Arizona to 5,178,000 in Texas. Most of the forage for the cattle was produced on 261,354,000 ha of forest/range land (Table 8.1). Five percent (13,663,000 ha) of this land was dominated by the ponderosa pine vegetative type, which produced an average of 1,822 kg of herbage and browse per ha (Table 8.2). This was slightly less than the average production on 25 types of vegetation in the western states. Average forage harvest of the ponderosa pine vegetative type was 0.17 AUMs/ha, about one-third of the average (0.55 AUMs/ha) for all types of vegetation (Table 8.2). The production of 1,822 kg of herbage and browse per ha and a harvest of 0.17 AUMs/ha suggests that only a portion of the available forage was harvested by livestock. Some of this forage loss may reflect management practices employed to reduce ponderosa pine needle consumption during pregnancy.

It was estimated that 227,990 cows graze in the ponderosa pine type during some part of the year (Table 8.1). This estimate was based on the total number of beef cows in the area and average forage harvest (Table 8.1).

The extent of the ponderosa pine type may have been underestimated. Payne (1973) mapped 6,430,641 ha of ponderosa pine in Montana, six times the estimate (1,059,000 ha) in Table 8.1. The credibility of the larger estimate is supported by a survey that indicated pine needles were a factor in abortions occurring in 19 of Montana's 56 counties (Jacobsen 1959). We used the lower figure in order to maintain a consistent data set. We do not know how this ultimately affected the estimate of economic loss to pine needle abortion.

Estimated forage production in Montana also suggested that the average carrying capacity of the ponderosa pine type may have been underestimated. Payne (1973) estimated the carrying capacity of ponderosa pine range as 2.2 ha/AUM, more than twice as high as the Forest Service's estimated stocking rate of 5.7 ha/AUM. When the larger area (6,430,641 vs 1,059,000 ha) and higher productivity (0.45 vs 0.17 AUM/ha) were used, the estimate for number of cows grazing ponderosa pine for Montana increased from 21,282 to 326,499 (Table 8.1). Even though we did not have data to indicate that cattle use of the ponderosa pine type was similarly underestimated in other states, this 15-fold discrepancy emphasizes the need for better data sets.

TABLE 8.2
Average annual herbage and browse production and average
forage harvested (animal unit months per ha per year) by
ecosystem.

Ecosystem	Annual Herbage and Browse Production (kg/ha)[1]	Average Forage Harvested (AUMs/ha)[2]
Grassland:		
Mountain Grasslands	1,860	0.72
Mountain Meadows	3,163	2.81
Plains Grasslands	1,138	0.77
Prairie	3,716	2.49
Desert Grasslands	344	0.47
Annual Grasslands	2,312	2.59
Wet Grasslands	5,756	
Alpine	632	0.52
Shrublands:		
Sagebrush	1,150	0.30
Desert Shrub	279	0.07
Southwestern Shrubsteppe	547	0.15
Shinnery	2,094	0.57
Texas Savanna	2,399	0.84
Chaparral-Mountain Shrub	2,160	0.17
Pinyon-Juniper	431	0.12
Desert	0	0
Western Forests:		
Douglas-Fir	2,533	0.10
Ponderosa Pine	1,822	0.17
Western White Pine	4,282	0
Fir-Spruce	1,380	0
Hemlock-Sitka Spruce	4,692	0
Larch	2,841	0.02
Lodgepole Pine	1,973	0.02
Redwood	5,355	0
Hardwoods	2,106	0.81
Average	2,198	0.55

[1]USDA Forest Service (1980)
[2]USDA Forest Service (1977)

Incidence of Cattle Abortion From Ponderosa Pine Needle Consumption

Abortions due to consumption of pine needles are
restricted to areas where cows ingest pine needles during

the last trimester of pregnancy. Although abortions are most likely if needles are ingested during the last trimester, some cows may abort earlier in gestation. Not all 227,990 cows grazing in the ponderosa pine type are susceptible to abortion because grazing in these areas is limited by snow and cold temperatures (Currie 1975, Clary 1975). Only 27,679 cows have access to Forest Service ponderosa pine ranges during December, January, and February (Table 8.3). Apparently, most of the abortions induced by pine needles occur on privately-owned lands. Privately-owned ranges provide 38% of the total AUMs harvested from the ponderosa pine type (USDA Forest Service 1980). Privately-owned ponderosa pine ranges are more commonly used for winter grazing because they are usually at lower elevations. Cattle are often concentrated in these areas because they offer protection from winter storms. Ranch headquarters are also located in these areas as are the hay lands, so cows are fed and also calved out in these places. Most cattle grazing private lands during the winter would be in their last trimester of gestation, the period when cows are most susceptible to the abortifacient effects of the pine needles. We therefore assumed that most cattle grazing ponderosa pine on private land would be during the winter, and thus 38% of cattle grazing on ponderosa pine ranges (86,636 cows) would have access to pine needles when they were most susceptible to abortions.

Abortion losses from ingestion of ponderosa pine needles vary within an area and annually on an individual ranch. For example, 96% of the ranchers responding to a survey administered by the Black Hills Area Resource Conservation and Development (RC&D) Committee in the early 1970s had pine trees on their ranches (personal communication, Harold Stoltenberg, coordinator, Black Hills RC&D Office, Rapid City, SD). Eighty percent of the respondents reported abortions they thought were caused by pine needles. An average of 205 calves (or slightly more than one calf per ranch) had been lost during each of the three years preceding the survey. In a more recent survey of one county in the study area, the average abortion rate was 2.6 percent, and ranged from 0 to 20 percent (Jack Robinson, County Agent, Newcastle, WY).

Reasons for the variable losses are not known. Jacobsen (1959) and researchers in Colorado (Anonymous 1960) did not directly associate loses with malnutrition. Uresk and Paintner (1985) reported that 4.5 to 9.4 percent of the diets of cattle grazing ponderosa pine range from June through September consisted of pine needles. Most ranchers

TABLE 8.3
Estimated number of cows that may be grazing on ponderosa pine ranges administered by the U.S. Forest Service during December through February.[1]

USFS Region	Number of Cows
Southwestern(3)	3,672
Intermountain(4)	0
Rocky Mountain(2)	0
Pacific Northwest(6)	14,000
Pacific Southwest(5)	10,007
Northern(1)	0
Total	27,679

[1]Estimates obtained through personal communication with Regional Forest Service personnel, 1986.

and researchers believe cattle readily graze ponderosa pine needles during winter months. More importantly, animals graze needles heavily when they are bored or stressed from weather changes and lack of feed. The situation is aggravated when pine needles from windfall or logging operations are accessible (Panter and James 1987). In feeding trials, 75 percent to 100 percent of the cows fed pine needles have aborted (MacDonald 1952; Short et al. 1987). The two surveys of ranchers indicated that average abortion losses in the ponderosa pine region of the Black Hills are smaller on ranches (Jack Robinson, County Agent, Newcastle WY, Harold Stoltenberg, coordinator, Black Hills RC&D Office, Rapid City, SD). However, these respondents estimated that abortion losses would average 31 percent without special management practices. This is consistent with the abortion rate of 20 to 100 percent reported in some herds (personal communication, Harold Stoltenberg, coordinator, Black Hills RC&D Office, Rapid City, SD; Anonymous 1975; Short et al. 1987).

Members of the National Cattlemen's Association were surveyed in the 1950s about their problems and practices (Ensminger et al. 1955). The report involved 1,588 questionnaires representing about 500,000 cattle. About 2.8 percent of the respondents in the western and northwestern regions reported pine needle abortions in their herds (Table 8.4). Abortions affected 0.04 percent and 0.05 percent of cows in the western and Pacific northwestern regions, respectively.

TABLE 8.4
Incidence of cattle abortion from pine needle ingestion by region.[1]

Region	Percentage of Ranchers Reporting Problem (%)	Percentage of Cattle Aborting (%)
West	2.8	.04
Pacific Northwest	2.7	.05
Great Plains	0.0	.00

[1]Data from Ensminger (1955).

Data of Ensminger et al. (1955) suggests that 2,845 cows abort annually (the average abortion rate multiplied by the current number of beef cows in the West and Pacific Northwest) (Table 8.5). This is substantially lower than the number noted by Panter and James (1987); they reported abortions in most pine areas varied from only a few to 100 percent. The estimate derived from the National Cattlemen's Association survey does not include any abortions from North and South Dakota, Nebraska, and Kansas. Ensminger et al. (1955) reported no abortions in these states, even though pine needle abortions are especially serious in western South Dakota and also occur in western Nebraska.

Data from ranchers in two northeastern Wyoming counties (personal communication, Jack Robinson, county agent, Newcastle, WY) were used to refine the abortion estimate. The 2.6 percent average abortion rate was multiplied by the estimated number of cows grazing in the ponderosa pine vegetative type (227,990). The resulting estimate (5,928 cows) indicated that the original estimate of 2,845 was low. The mean of the two estimates (4,387 cows) was used in this study.

Quantifying Economic Losses of Cattle Abortion from Pine Needle Ingestion

Estimated losses in the western cattle industry due to abortion caused by ingestion of ponderosa pine needles was 2more than $4,500,000 (Table 8.6). This estimate includes the direct costs resulting from 4,387 abortions and the

TABLE 8.5
Number of beef cows in West and Northwest, percent and
number of cows aborting from pine needle ingestion.

State	Beef Cows (1,000)	Percentage of Cows Aborting[1] (%)	Number of Cows Aborting
Northwest:			
Washington	407	.05	204
Oregon	598	.05	299
Idaho	526	.05	263
West:			
California	950	.04	380
Nevada	301	.04	120
Arizona	293	.04	117
Utah	298	.04	119
New Mexico	542	.04	217
Colorado	773	.04	309
Wyoming	669	.04	268
Montana	1,373	.04	549
Total	6,730		2,845

[1]Adapted from Ensminger et al. (1955).

indirect costs associated with managerial practices to
reduce losses.

Assuming 93% of aborted calves died, and their value
was $315/head (204 kg at $1.54/kg--assuming 450 lb calves
sell at $.72/lb), their total loss would be $1,285,200. The
$44,000 cow death loss assumed that 2 percent of the cows
aborting died; cows were valued at $500/head. We assumed
that 7 percent of the aborted calves would survive, a
percentage consistent with a feeding trial reported by
MacDonald (1952). The premature calves are small and do
not gain as well as unaffected calves. We assumed weaning
weights of these calves were 22.7 kg below average (valued
at $1.54/kg). The estimated economic loss due to aborted
calves will vary with the price of calves at the time the
estimate is made.

Losses associated with impaired calving by cows that
had aborted but were retained in the brood herd were
estimated at $270,837. The estimate assumed that all
affected cows would be bred during their third cycle of the

TABLE 8.6
Annual economic loss of cattle abortion from the ingestion of ponderosa-pine needles on western ranges.

Cost	Number	Economic Loss
Direct:		
Calf death loss[1]	4,080	$1,285,200
Death of cow[2]	74	44,000
Small, poor-performing calf[3]	307	10,745
Indirect:		
Impaired calf production[4]		
1st year following abortion	4,299	180,558
2nd year following abortion	4,299	90,279
Pine-induced managerial costs[5]		
(number of cows affected)	86,636	2,900,573
Total Annual Loss		$4,511,355

[1]Ninety-three percent of aborted calves died; calves valued at $315/head (204 kg x $1.54/kg).
[2]Two percent death loss of cows aborting from pine needle consumption (cows valued at $500/head).
[3]Seven percent of total calves that aborted survived; calves born small and weak and average 22.7 kg lighter at weaning ($1.54/kg).
[4]Ninety-eight percent of cows that aborted were retained in brood herd; in first year, cows are bred during third cycle, weight loss of 13.7 kg per missed cycle per calf. In second year, cows bred during second cycle, weight loss of 13.7 kg per missed cycle per calf.
[5]Surveys of producers in N.E. Wyoming, $33.48 extra cost incurred per cow in survey to reduce management problems related to pine needle abortion (includes extra feed, leasing additional pasture, controlling pines, fencing). Assumes 38 percent of the cows grazing in the ponderosa pine type could be affected.

first breeding season following the abortion, the second cycle of the second breeding season, and the first cycle during subsequent seasons. Weaning weights of calves were decreased 13.7 kg for each missed cycle.

The greatest costs may be practices required to minimize or avoid the problem. Ranchers grazing cattle in the Black Hills region of Wyoming spent $33.48 per cow to lease pine-free range, purchase feed, remove pines, and construct fences (personal communication, Jack Robinson, county agent, Newcastle, WY). This annual cost was multiplied by the number of cows (86,6361 grazing on ponderosa pine range during winter) to derive estimated losses of $2,900,573.

CONCLUSION

The paucity of data makes it difficult to quantify the economic losses associated with ponderosa pine needle consumption. The problem varies by location, and its severity throughout the western ranges is not known.

Estimated average annual losses on western rangeland due to ponderosa pine needles exceeded $4,500,000. Additional data are required to verify the affected acreage and the number of cows. Average annual losses may exceed the estimates in this analysis.

LITERATURE CITED

Albaugh, R. 1974. Pine needles--beef cattle. Roundup of Livestock Facts, September 10, pp. 9-10.

Anonymous. 1960. Scientists advise on pine needle abortion. Colorado Farm and Home Research, March-April 10:5.

Anonymous. 1975. Simmental Booster. September, p. 10.

Clary, W.P. 1975. Range management and its ecological basis in the ponderosa pine type of Arizona: The status of our knowledge. USDA For. Serv. Res. Pap. RM-158. Rocky Mt. For. and Range Exp. Sta.

Cooperative Extension Service in Agriculture and Home Economics, Montana State University and USDA Cooperating. 1986. Western Livestock Reporter. March, 1986.

Currie, P.O. 1975. Grazing management of ponderosa pine-bunchgrass ranges of the central Rocky Mountains: The status of our knowledge. USDA For. Serv. Res. Pap. RM-159. Rocky Mt. For. and Range Exp. Sta.

Ensminger, M.E., M.W. Galgan, and W.L. Slocum. 1955. Problems and practices of American cattlemen. Wash. Agr. Exp. Sta. Bull. 562.

Jacobsen, N.A. 1959. Pine needle abortion in range cows. Seminar. November 18. Bozeman, MT.

Kamstra, L.D., and C. Cogswell. 1975. Pine needle abortions among cows grazing foothill ranges. South Dakota Agr. Exp. Sta. A.S. Series 75-44:48-50. Brookings, SD.

MacDonald, M.A. 1952. Pine needle abortion in range beef cattle. J. Range Manage. 5:150-155.

Miner, J.L., R.A. Bellows, R.B. Staigmiller, M.K. Peterson, R.E. Short, and L.F. James. 1987. Montana pine needles cause abortion in beef cattle. Montana Ag Research 4:6-9. Montana Agr. Exp. Sta., Montana State Univ.

Panter, K.E., and L.F. James. 1987. A review of pine needle and broom snakeweed abortion in cattle, pp. 125-130. In: F.D. Provenza, J.T. Flinders, and E.D. McArthur (compilers), Proc. Symposium on plant-herbivore interactions. USDA For. Serv. Gen. Tech. Rep. INT-222, Ogden, Utah.

Payne, G.F. 1973. Vegetative rangeland types in Montana. Montana Agr. Exp. Sta., Montana State Univ., Bozeman. MT. Bull. 671.

Short, R.E., R.B. Staigmiller, K.E. Panter, and L.F. James. 1987. Effect of stage of pregnancy on pine needle induced abortion in cattle. (Abstract). Annual Meeting Canada West Society for Reproductive Biology. Saskatoon, Sask.

Stevenson, A.H., L.F. James, and J.W. Call. 1972. Pine needle (Pinus ponderosa) induced abortion in range cattle. Cornell Vet. 65:519-524.

U.S. Department Agriculture Forest Service. 1972. The nation's range resources--A forest-range environmental study. For. Serv. Rep. No. 19.

U.S. Department Agriculture Forest Service. 1977. The nation's renewable resources: an assessment, 1975. For. Serv. Rep. No. 21.

U.S. Department Agriculture Forest Service. 1980. An assessment of the forest and range land situation in the United States. FS-345.

Uresk, D.W., and W.W. Paintner. 1985. Cattle diets in a ponderosa pine forest in the northern Black Hills. J. Range Manage. 38:440-442.

9

Ecological Considerations
of the Larkspurs

E.H. Cronin, John D. Olsen,
and William A. Laycock

INTRODUCTION

Larkspur, stagger weed, poison weed, and cow poison are a few of the names that have been applied to the members of the holarctic genus Delphinium in North America. The poisonous properties of the larkspurs have been recognized almost since the beginning of agriculture. The ancient herbalists of Greece and Rome acknowledged their toxicity and recognized their medicinal properties (Kingsbury 1964).

Early colonists in North America must have been aware of the toxic properties of the native larkspurs because of the similarities in their appearance to European species. However, the European species and the species native to eastern North America are low in toxicity, and death losses to these species are rare.

Migration of the livestock industry across the Mississippi River began about 1830 and accelerated following the Civil War. As these early pioneers moved into the shortgrass prairie and expanded westward, catastrophic losses attributed to the larkspurs and other poisonous species were reported (Chestnut 1898). Heavy losses to the larkspurs may have been due to increased numbers of species encountered and apparently more toxic species, some seeming to be manyfold more toxic than species from the eastern parts of the continent.

Taxa of the genus Delphinium are readily recognized by their distinctive flowers. Flowers are bilaterally symmetrical with five sepals which may be purple, blue, white, yellow, or red. The upper sepals form the distinctive spur. There are usually four, but sometimes only two, petals. The upper pair of petals project back into the spur. Petals are smaller than the sepals.

The simplicity of their identification to the generic level may have led to the erroneous assumption that the taxa are similar in other respects. Perhaps it is this assumption that accounts for the deficiency of knowledge concerning the larkspurs and the reason the problem continues to plague the livestock industry on western rangeland. Larkspur is a problem of major economic concern and one that will persist until substantial interdisciplinary research has been completed.

Larkspurs are a diverse group of species. Their migrations in North America through geological and climatic changes have enhanced their diversity. Ewan (1945) lists three chief centers of differentiation for the genus on the North American continent: (a) Mexican highlands, more precisely the bounding ranges of the Mesa Central; (b) California coastal ranges; and (c) northern Rocky Mountains, more precisely the Bitter Root Mountains and adjacent drainage basins. In addition, he also states that the Elatopsoid series, which includes the common tall larkspurs, clearly had its origin in Siberia or inner Asia. Its ancestral species migrated across the Bering Strait, and its derivatives moved southward along both the interior Rocky Mountains and the Cascade-Sierran axis in Pleistocene times. Migrations of the taxa have been predominant north to south or south to north along the western cordilleras. He suggests that elements of the Mexican highlands and the northern Rockies are found in Colorado. Species originating in Mexico occupy the more xeric habitats. Taxa of the northern Rockies are found in the montane zone, and Delphinium barbeyi, of the Elatopsoid series, grows in the subalpine zone.

Ewan (1945) divided the North American species of larkspur into thirteen series. He recognized 79 species, only four of them found east of the Mississippi River. West of the Mississippi, the concentration increases, with California containing the largest number of species, including 22 endemic species. Over 231 scientific names have been applied to these 79 species, creating some confusion concerning the identification of the species discussed in the older literature.

Identification is also complicated by integration and hybridization among the taxa in North America. Ewan (1945) lists eleven presumed hybrids between native species. He also recorded four possible hybrids between garden varieties and species native to North America. Many of our garden varieties are the result of willful interspecific crosses which should increase expectations of natural hybridization

between native species and between native species and garden
varieties. He also lists three species that have escaped
from our gardens to become naturalized components of the
North American flora.

LIFE HISTORY

A surprisingly meager amount of information exists in
the literature on the life histories of larkspur taxa. This
deficiency is most obvious for species associated with the
three centers of differentiation in North America.
Comprehensive studies of the life histories of these taxa
will form the basis for sound management of the resources
and ecologically acceptable control practices.

Information is needed on: (1) seed dispersal, mode of
transport, and distance of potential travel; (2) the
viability, dormancy, longevity, and conditions required for
germination of seed; (3) growth and development, above and
below the soil surface, of the seedling stage as well as
their mortality; (4) the duration, mortality, and the
environmental conditions required by the juvenile stage of
growth; (5) factors that initiate reproductive maturity; (6)
the reproductive potential and the longevity of the mature
plants; and (7) possible asexual reproduction. Holman's
(1973) dissertation on the ecological life histories of D.
occidentale and D. barbeyi is an example of the type of
study needed as a basis for the solutions of the larkspur
problems.

Seed production of D. barbeyi and D. occidentale varies
with the number of stems but can be characterized as
prolific. This is especially true when it is considered
that the same plant produces an abundance of seed almost
annually. Most of the seed are shaken from the partially
open follicles to fall near the parent plant (Holman 1973).
Bending of the flexible dry stems to near the breaking point
can cause the seed to be catapulted several meters (Cronin
1971). Cattle can deplete seed production through their
preference for the maturing follicles.

Long distance dispersal of larkspur seed depends on
wind driving them along a smooth surface, such as on crusted
snow (Fitz 1972). This mode of travel would be possible for
species growing in the montane or subalpine zones but not
for species growing in the more arid habitats since seed
dispersal occurs in late July or August.

A long dormant period has been recorded for seed of D.
barbeyi, D. occidentale, D. nelsonii (Williams and Cronin

1968, Holman 1973) and D. geyeri (Hyder and Sabatka 1972). All four species appear to require a long cold period (near freezing) with adequate moisture to break dormancy. Germination usually occurs under the snow or shortly after snowmelt.

Seedlings of D. barbeyi enter the juvenile stage during the summer of germination by producing their first true leaves. Observations suggest that the seedlings of D. nelsonii also follow this pattern. Hyder and Sabatka (1972) do not address this change directly but report that plants with one true leaf accounted for 7% of the population of D. geyeri on their study plots while "new seedlings" accounted for less than 1% of the populations. Seedlings of D. occidentale do not produce a true leaf until the second year. Most of the energy available to the seedlings of D. barbeyi and D. occidentale appears to be devoted to root development.

The juvenile is also a stage where the energy budget is devoted to root development. D. barbeyi roots begin to develop the typical tap root form and extend down into the soil to depths of 10 cm in the second year of growth and down to 20 cm in the fourth year (Holman 1973). D. occidentale retains a shallow fibrous root system into its third year of growth before developing the tap root form which may penetrate the soil to a depth of 10 cm (Holman 1973). Aerial growth of both species tends to match root growth. Leaf production for both species remains low during the juvenile stage with D. occidentale producing about three leaves and D. barbeyi eight leaves in the third year of growth. The duration of the juvenile stage depends on the canopy cover. Juveniles growing under a closed canopy lag substantially behind juveniles growing in the open in root and aerial growth and development. A minimum of about four years is required for D. barbeyi to reach reproductive maturity and an estimated eight to ten years for D. occidentale (Holman 1973).

Mortality rates for both the seedling and juvenile stages of growth are very high. Only about 2% of the established seedlings reach the fourth year of growth. The high mortality rate is not surprising, considering the relatively small size of the root and its limited capacity to store reserves. It is probable that the distribution of larkspur species is determined by the tolerance limits of the seedling and juvenile stages rather than limits of the mature stage. Certainly mature plants can be transplanted far outside of their normal distribution and will flourish for at least a few years.

Holman (1973) suggests that D. barbeyi and D. occidentale may live 50 years. There is evidence that some D. barbeyi plants on the Wasatch Plateau may exceed 75 years of age (Cronin and Nielsen 1979). A five-year study of marked plants of D. andersonii and D. nelsonii failed to record the death of a single mature plant (E.H. Cronin, unpublished data). Hyder and Sabatka (1972) suggest populations of D. geyeri are all composed of many size classes and may indicate a higher mortality rate for mature plants of this species.

All species discussed produce an annual seed crop. The reproductive potential of the larkspurs can be awesome if measured over their lifetime. Plants of D. occidentale averaged 15,120 seed and plants of D. barbeyi averaged 35,700 seed annually (Holman 1973). The evidence indicates that the seed do not accumulate in the soil, although annual seed production exceeds the availability of suitable living space for new plants. This is not an unusual strategy for long-lived perennials.

Only the root of larkspur remains physiologically active throughout the year. It is the organ that stores the energy and produces buds for new aerial growth each spring. D. barbeyi, D. geyeri, and D. occidentale have a tap root, although it is doubtful they can be said to be typical tap roots. They do not appear to be unified into a single continuous organ. Cronin (1971) described the root as consisting of scattered strands of living tissue embedded in compacted dead tissue and reported the absence of herbicide injury to other stems on the same caudex when only a single stem was treated with a herbicide. Hyder and Sabatka (1972) noted that one to five small plants often appeared in the place previously occupied by one large plant the year following applications of herbicides. Holman (1973) found connective tissues between a root clump and a small root, producing a single stem, that were 15 cm apart, suggesting the possibility of vegetative reproduction.

Vegetative reproduction, if it occurs, could be an important aspect in the life cycle. There is no evidence to suggest it is of significant importance; however, it is informative to note that commercial growers make use of vegetative reproduction to expand their stocks of prized cultivars (Wilde 1931). By dividing the roots longitudinally, they produce a number of separate plants. Heel cuttings which include a few buds, part of the caudex, and some of the adjacent root can be used to produce a new plant. They also use two methods of layering. After removing the aerial growth, soil is mounded over the caudex to produce a number of small shoots, complete with root tissue, that can be excised from the old plant. Another method is used where an incision is made on the underside of a stem and covered with soil. A callus tissue forms on the cut surface, which may produce root tissue on the selected stem. Vegetative reproduction of commercial cultivars suggests the possibility that some or all of our native species may have the same capacity.

TOXIC PRINCIPLE

Poisonous principles in larkspurs are complex diterpenoid alkaloids. Many of these compounds have been named and characterized and their structural formulas determined (Pelletier et al. 1984). New techniques and instruments permit more rapid and accurate description of these compounds than that of previous studies. As isolation and evaluation of specific alkaloids becomes more routine, it will be possible to correlate plant alkaloid content, plant toxicity, and evolutionary ecological relationships. Cataloging of alkaloids by species could make a substantial contribution to our understanding of taxonomic relationships among the larkspur species. Such data would be useful in studying the ecology of larkspur and planning effective management strategies to reduce poisoning of cattle by larkspur.

All larkspur species probably contain alkaloids; however, all larkspurs are not equally toxic. Early investigators fed animals measured amounts of larkspur to determine the toxicity of the species. Recently, the trend has shifted to estimating the alkaloid contents of a plant (Williams and Cronin 1963). This relatively simple procedure may provide an estimate of the relative toxicity of the various plant parts and to follow changes through the growing season.

Olsen (1977, 1983) estimated the toxicity of larkspur by injecting plant extracts into rats and mice. Significant differences in toxicity were noted for D. barbeyi, D. glaucescens, and D. occidentale, each growing in a different ecological habitat, with D. barbeyi being the most toxic and D. occidentale the least. Further research is needed to determine if such differences exist among larkspur species growing in similar habitats. Toxicity may not correlate directly with total alkaloid content (Olsen 1979, 1983) because the mix of alkaloids may change during growth. Changing concentrations of individual alkaloids in D.

occidentale measured over a growing season suggested they are active metabolic compounds rather than waste products of the metabolic process (Kreps 1969). Aiyar et al. (1979) demonstrated that toxicity of alkaloids extracted from D. brownii varied and suggested toxicity of the species appeared to be largely due to a single alkaloid.

SIGNS OF POISONING

Clinical signs of larkspur poisoning apparently result from general weakness thought to be brought about primarily by action of the alkaloids on neuromuscular function. In order of chronological appearance, they include uneasiness, stiffness of gait, and a characteristic straddled stance, with the hindlimbs far apart. Very early signs of weakness are accompanied by muscle twitches, cervical muscle tremors in sheep, and femoral muscle tremors in cattle.

Muscle tremors, once evident, generally increase in severity until the animal collapses, usually forelimbs first. After varying periods of quiet, the animal may regain its feet. The period of time that the animal can remain standing between episodes of tremors becomes shorter as the degree of poisoning intensifies and, conversely, is longer as recovery occurs from a nonlethal dose.

If vomiting occurs (not observed in sheep but not unusual in cattle), death occurs immediately, usually by asphyxiation. Otherwise, death occurs from respiratory paralysis. Constipation and bloating are universal (Kingsbury 1964, Olsen 1978).

Under range conditions, it is seldom possible to approach animals to observe these early clinical signs. In the mountains north of Dubois, Idaho, one cow on a range containing D. glaucescens was observed in a weakened condition, presumably from eating larkspur. The animal could stand, but would fall down when it tried to run upon the approach of the observer. The animal rested and, in a few hours, showed no signs of the toxicity (W.A. Laycock, personal communication). The results of poisoning in cattle on the Wasatch Plateau were not observed until animals were dead. Death appeared to occur swiftly. The evidence indicated the animals fell suddenly and hard. In a few instances, where death occurred on soft moist soil, the nose and the sternum consistently left distinct impressions on the soil. There was no evidence to suggest that the larkspur-poisoned animals struggled after they had fallen. Severe bloating was rapid and was observed before body heat had been lost or joints stiffened (Cronin et al. 1976).

CONDITIONS OF POISONING

The toxicity of larkspurs varies with the amount ingested, seasonal changes in the alkaloid content, the duration of the ingestion period, the species ingested, the part of the plant ingested, and the animal species ingesting the larkspur (Kingsbury 1964).

Research into the reasons livestock utilize larkspur and the conditions under which they utilize it are almost totally lacking. The literature contains a surplus of opinions; some are logical, and some may prove to be accurate, but all lack documentation. It is a complex problem because of the variation among individual animals. In areas where all cattle have an opportunity to ingest lethal levels of larkspur, only 12% or less are actually lost.

CONTROL OF LARKSPUR

Five methods of controlling the populations of larkspur have been attempted: (1) grubbing, (2) removal of aerial growth, (3) applications of nitrogen fertilizers, (4) applications of selective herbicides, and (5) applications of soil sterilants. All five methods require substantial expenditures of labor and funds. All require post-treatment vigilance to insure that the larkspur has, indeed, been destroyed. A severely injured larkspur plant can be difficult to see because evidence that it has survived may consist of a few small leaves on short petioles. However, these plants often require surprisingly few growing seasons to regain their former robust vigor.

Digging roots from the soil is an effective method of destroying larkspur, providing the root is removed to a depth of 20 to 30 cm. The root must be burned or removed from the site. If the root is left lying on the soil, the plant can reestablish itself. In rocky soil it is difficult to remove enough of the root to be effective. Most writers have concluded that this is not a practical method of control; however, no one has attempted to make an economic study of its benefits.

Clipping the aerial growth at the soil surface should receive more attention. Cronin (1971) reported results of clipping the same D. barbeyi plants at 7-, 14-, or 28-day intervals during 1968. Clipping at 7- or 14-day intervals substantially reduced production of aerial growth as compared to plants clipped at 28-day intervals. In 1969, none of the plants appeared to have sustained any visual reduction in vigor except that none of the clipped plants produced flowers. Laycock (1975) clipped D. occidentale plants at the soil line for two consecutive years at the vegetative, bud, early flower, or the late flower to early seed stage of growth. Plants clipped in the vegetative growth produced only leaves, were significantly smaller, and contained significantly lower concentrations of total alkaloids. He also reported that he was unable to find a number of marked plants during the third year and concluded they died as a result of the clipping treatment.

Clipping would be a tedious, time consuming process. Mowing with conventional machinery would speed the process but would be impractical on rough or rocky surfaces. Recent development and improvements in the gasoline powered line trimmers used by homeowners to trim the edges of their laws and to cut down weeds may be a practical tool for clipping larkspur. They are relatively safe for the operator, they are readily available, and they are relatively inexpensive. A heavy duty model using a 2 mm diameter line should prove adequate to cut the thickest larkspur stems. Their portability would allow use on steep terrain and among trees, shrubs, and rocks. This method of control should be researched more extensively.

Broadcast applications of ammonium nitrate at rates of up to 1,344 kg N/ha produced only minor injury to D. barbeyi (Cronin et al. 1977). Others have reported that placing a handful of ammonium sulfate among the stems and on the caudex of D. barbeyi (Binns et al. 1971) and D. occidentale (Little 1977) killed over 95% of the treated plants. The costs of treating individual plants would depend on the density of the larkspur. The effectiveness of this treatment depends on the production of toxic concentrations of nitrogen around the root. Perhaps it would be less expensive to place a handful of sodium chloride on the crown of the plants.

The literature contains numerous papers on the control of many larkspur taxa with applications of selective herbicides. A partial list of species would include: D. barbeyi (Hervey and Klinger 1961, Binns et al. 1971, Cronin and Nielsen 1972, Cronin 1974, 1976; Cronin and Nielsen 1979); D. geyeri (Alley and Lee 1971a, 1971; Hyder 1972, Hyder and Sabatka 1972); D. nelsonii (Cronin and Nielsen 1979); and D. occidentale (Baker 1958, Torrel and Haas 1963). Rates of herbicide applications and the stage of growth treated are critical to the success of herbicide treatments.

Applications of some herbicides can result in an apparent increase in total alkaloids in the treated larkspur (Williams and Cronin 1963). Their application may also increase the palatability of treated vegetation, including larkspur (Cronin and Nielsen 1979). Treatments with selective herbicides have been shown to be an effective method of reducing cattle losses (Cronin et a. 1976). Applications of effective herbicide treatments can also produce very good economic returns on the costs of applying the treatments (Nielsen and Cronin 1977).

A number of the herbicides that have been shown to control larkspurs are no longer available. Hopefully, evaluations of new herbicides will continue and will result in alternative treatments that will prove effective and economical.

Torrel and Haas (1963) applied soil sterilants for the control of D. occidentale but concluded the treatments were unsuitable. The compounds they applied either failed to produce a total kill of larkspur or eliminated the associated vegetation. Removal of the associated vegetation exposes the soil to erosion. Larkspur often grows on steep sites where runoff water is abundant for at least part of the year. Larkspur often grows on land that is more valuable as a watershed than as a grazing resource.

Reductions in larkspur production are indicative of the success of a control effort. However, the reductions in livestock losses are the critical measurement of success. Measuring changes in livestock losses requires records of losses over a number of years prior to the application of the treatment and following the treatment. Losses fluctuate widely from year to year and would introduce serious errors in the evaluation of the economic analyses if based on only a few years of records. However, any method used to control losses, whether its application involves treatment of the animals or of the vegetation, should be subjected to the same long term evaluation of results.

LITERATURE CITED

Aiyar, V.H., M.H. Benn, T. Hanna, J. Jacyno, S.H. Roth, and J.L. Wilkens. 1979. The principal toxin of Delphinium brownii Rydb., and its mode of action. Experientia 35:1367-1368.

Alley, H.P., and G.A. Lee. 1971a. Larkspur (Delphinium geyeri Greene) control resulting from applications of

Picloram + 2,4-D and Picloram + 2,4,5-T combinations. Res. Prog. Rep., West. Soc. Weed Sci., p. 17.

Alley, H.P., and G.A. Lee. 1971b. Larkspur (Delphinium geyeri Greene) control resulting from three successive years treatment with two formulations of 2,4-D. Res. Prog. Rep., West. Soc. Weed Sci., p. 18.

Baker, L.O. 1958. Control of tall larkspur, Delphinium occidentale, with herbicides. Res. Prog. Rep., West. Weed Control Conf., p. 36-37.

Binns, W., L.F. James, and A.E. Johnson. 1971. Controlling larkspur with herbicides plus nitrogen fertilizer. J. Range Sci. 24:110-113.

Chestnut, V.K. 1898. Thirty poisonous plants of the United States. USDA Farmers Bull. 86.

Cronin, E.H. 1971. Tall larkspur: some reasons for its continuing preeminence as a poisonous weed. J. Range Manage. 24:258-263.

Cronin, E.H. 1974. Evaluations of some herbicide treatments for controlling tall larkspur. J. Range Manage. 27:219-222.

Cronin, E.H. 1976. The impact of controlling tall larkspur on the associated vegetation. J. Range Manage. 29:202-206.

Cronin, E.H., J.E. Bowns, and A.E. Johnson. 1977. Herbicides, nitrogen, and control of tall larkspur under aspen trees. J. Range Manage. 30:420-422.

Cronin, E.H., and D.B. Nielsen. 1972. Controlling tall larkspur on snowdrift areas in the subalpine zone. J. Range Manage. 25:213-216.

Cronin, E.H., and D.B. Nielsen. 1979. The ecology and control of rangeland larkspurs. Utah Agr. Exp. Sta. Bull. 499.

Cronin, E.H., D.B. Nielsen, and Ned Madsen. 1976. Cattle losses, tall larkspur, and their control. J. Range Manage. 29:364-367.

Ewan, J.A. 1945. A synopsis of the North American species of Delphinium. Univ. of Colo. Studies (Ser. D.) 2:55-244.

Fitz, F.K. 1972. Some factors affecting the distribution and abundance of two species of tall larkspur in the Intermountain Region, with thoughts on biological control. Ph.D. Diss., Utah State Univ., Logan, Utah.

Hervey, D.F., and B. Klinger. 1961. Herbicidal control of tall larkspur (Delphinium barbeyi Huth Huth). Res. Prog. Rep., West. Weed Control Conf., p. 12.

Holman, J.R. 1973. A comparison of the ecological life histories of Delphinium occidentale S. Wats. and

Delphinium Huth. Ph.D. Diss., Utah State Univ., Logan, Utah.

Hyder, D.N. 1972. Paraquat kills Geyer larkspur. J. Range Manage. 24:460-464.

Hyder, D.N., and L.D. Sabatka. 1972. Geyer larkspur phenology and response to 2,4-D. Weed Science 20:31-33.

Kingsbury, J.M. 1964. Poisonous plants of the United States and Canada. Prentice-Hall Inc., Englewood Cliffs, NJ.

Kreps, L.B. 1969. The alkaloids of _Delphinium occidentale_ S. Wats. Ph.D. Diss., Utah State Univ., Logan, Utah.

Laycock, W.A. 1975. Alkaloid content of duncecap larkspur after two years of clipping. J. Range Manage. 28:257-259.

Little, Bill. 1977. Controlling tall larkspur with ammonium sulfate fertilizer. (Abstract). 30th Annual Meeting of the Soc. for Range Manage., p. 26.

Nielsen, D.B., and E.H. Cronin. 1977. Economics of tall larkspur control. J. Range Manage. 30:434-438.

Olsen, J.D. 1977. Rat bioassay for estimating toxicity of plant material from larkspur (_Delphinium_ sp.). Amer. J. Vet. Res. 38:277-279.

Olsen, J.D. 1978. Tall larkspur poisoning in cattle and sheep. J. Amer. Vet. Med. Assoc. 173:762-765.

Olsen, J.D. 1979. Toxicity of larkspur (_D. occidentale_) from two sites. Proc. of the 60th Annual Meeting of the Conference of Research Workers in Animal Disease. Abst. 155. Chicago, IL.

Olsen, J.D. 1983. Relationship of relative total alkaloid concentration and toxicity of duncecap larkspur during growth. J. Range Mange. 36:550-52.

Pelletier, S.W., N.V. Mody, B.S. Joshi, and L.C. Schramm. 1984. ^{13}C and proton NMR shift assignments and physical constants of C_{19}-diterpenoid alkaloids, pp. 205-462. In: S.W. Pelletier (ed), Alkaloids: chemical and biological perspectives, Vol. 2. John Wiley & Sons, New York.

Torrel, P.J., and R.H. Haas. 1963. Herbicidal control of tall larkspur. Weeds 11:10-13.

Wilde, E.I. 1931. Studies of the genus Delphinium. Cornell Univ. Agr. Exp. Sta. Bull. 519.

Williams, M.C., and E.H. Cronin. 1963. Effects of silvex and 2,4,5-T on alkaloid content of tall larkspur. Weeds 11:317-319.

Williams, M.C., and E.H. Cronin. 1968. Dormancy, longevity, and germination of seeds of three larkspurs and western false hellebore. Weed Sci. 16:381-384.

10

Larkspur: Economic Considerations

Darwin B. Nielsen and Michael H. Ralphs

INTRODUCTION

In 1913, Marsh et al. (1934) made the statement that larkspur (Delphinium spp.) caused more cattle deaths on western ranges than any other poisonous plant, with the exception of locoweed. Kingsbury (1964) made a similar statement during the mid-part of this century. Today, tall larkspurs remain the most severe poisonous plant problem on mountain cattle ranges in the western states.

Marsh et al. (1934) and Kingsbury (1964) classified the larkspur species into two categories: low and tall (Table 10.1). Low larkspurs grow from 0.2 to 0.5 m tall. They generally occur on foothill and low mountain ranges, start growth early in the spring, and rapidly complete their annual growth cycle. They remain uniformly toxic throughout the growth cycle. The low larkspurs experience wide population fluctuations due to varying range conditions and weather patterns. Cattle losses are also sporadic and cyclic.

Plains larkspur (Delphinium geyeri) falls between the low and tall larkspur classifications. Plants range from 0.3 to 2 m tall (Ewan 1945). It occurs on the high windswept short grass plains of Wyoming and into Nebraska and in sagebrush and juniper woodlands of the Colorado Plateau. It is the most severe larkspur problem in Wyoming, with annual cattle losses ranging from 2 to 15% of the cattle grazing larkspur-infested ranges (Alley and Lee 1970).

The remainder of our discussion will deal with the tall larkspurs. Species of tall larkspur are generally robust, multi-stemmed plants 1 to 2 m tall. They occupy mesic habitats at elevations greater than 2,400 m (Ewan 1945).

Delphinium glauscences is somewhat shorter (0.2 to 2 m), with one to a few stems per plant, but is closely related to D. occidentale (Ewan 1945). It occupies a similar ecological niche and causes severe poisoning problems in mountain valleys in southern Montana and northeastern Idaho.

MAGNITUDE OF LOSSES

Aldous (1917) reported an average of 5,500 cattle died annually from plant poisoning on national forest lands between 1913 and 1916. Ninety percent of losses were due to tall larkspur, averaging from 3 to 5% of all cattle grazing

TABLE 10.1
Classification and distribution of larkspur species

Low Larkspur
  D. bicolor
    (Wyoming, Montana, Idaho, Alberta, Saskatchewan)
  D. nelsonii
    (Idaho, Utah, Wyoming, Colorado, South Dakota)
  D. andersoni
    (Oregon, California, Nevada, Idaho, Utah)
  D. virescens
    (Great Plains east of Rocky Mountains)

Plains Larkspur
  D. geyeri
    (Wyoming, Utah, Colorado, Nebraska)

Tall Larkspur
  D. glaucum
    (Washington, Oregon, California, Nevada, Idaho, Montana)
  D. trolliifolium
    (Washington, Oregon, California)
  D. glaucescens
    (Idaho, Montana)
  D. occidentale
    (Washington, Idaho, Wyoming, Nevada, Utah, Colorado)
  D. barbeyi
    (Utah, Wyoming, Colorado, New Mexico)
  D. robustum
    (Colorado, New Mexico)

Source: Marsh et al. (1934); Kingsbury (1964).

on tall larkspur-infested ranges. An early distribution map (Figure 10.1) showed tall larkspur occurring on every national forest in the western states with the exception of the northern Cascades and the extreme northern Rocky Mountains in Montana and Idaho. The average annual cattle loss on each national forest between 1913 and 1916 is typed in.

A survey of U.S. Forest Service districts in the Intermountain Forest Service Region 4 was conducted in 1986, and the district ranger was asked to verify the number of reported cattle losses to poisonous plants in 1985 and 1986 and estimate the percentage due to tall larkspur. Of 76 districts in the region, 45 percent reported cattle losses to tall larkspur, representing losses on every national forest in the region. Seventy-two percent (532 cows) of the total losses to poisonous plants in 1986 was attributed to tall larkspur (Table 10.2). U.S. Forest Service officials recognized problems of permittees reporting losses and estimated that reported losses are probably only about half of the actual losses. Using these estimates, total cattle losses from tall larkspur in Region 4 probably exceed 1,000 head.

The problems associated with grazing cattle on a tall larkspur range have been studied for several years on the Manti Canyon Cattle Allotment, Manti-LaSal National Forest. This grazing allotment will be used as a case study to discuss the ways larkspur poisoning can impact a rancher. Baseline data were gathered on losses for 15 years, 1956 to 1970 (Cronin et al. 1976). The death loss caused by tall larkspur consumption varied from 103 to 13 cattle over these 15 years, with an annual average of 36 cattle. This allotment carries about 837 cows and produces about 4,225 AUMs of forage. The percentage of the herd lost while on the allotment varied from 12.3% to 1.5% with an average of 4.3% over the 15-year period.

Let us examine the economic impact these losses have on ranchers. Assume solid-mouthed mid-aged cows with an average value of $500/head. All normal costs of production would be the same with or without larkspur problems (excluding extra cost of managing around larkspur). For simplicity, assume they purchase replacement cows in the fall and carry them through the winter. The rancher incurs essentially all of the normal production costs whether he loses cattle to larkspur or not. Thus, the value of the cattle lost comes directly off the top or out of the family living income. The year in which 103 cows died, it cost the families $51,500 because of larkspur. They probably could

122

not absorb this loss in one year, so it would be spread over several years by borrowing money and paying interest or running fewer cows and having lower income. On the average year, family living income would be reduced $18,000 (36 cows x $500/cow).

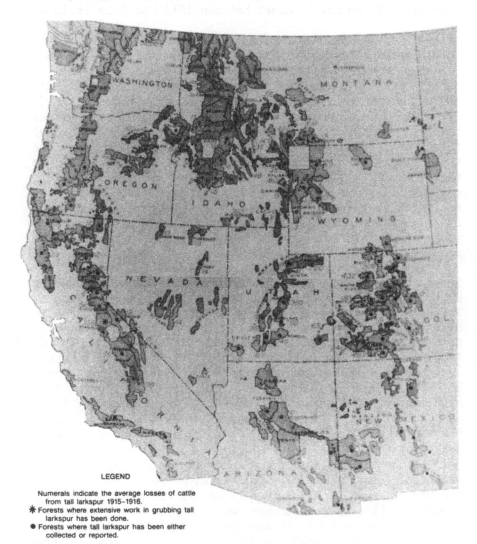

LEGEND

Numerals indicate the average losses of cattle from tall larkspur 1915-1916.
✳ Forests where extensive work in grubbing tall larkspur has been done.
● Forests where tall larkspur has been either collected or reported.

FIGURE 10.1
National forests, showing distribution of tall larkspur and average annual cattle losses from grazing it (taken from Aldous 1917).

TABLE 10.2
Number of cattle lost to poisonous plants and to tall larkspur on the national forests in the Intermountain Region 4.

| National | Poisonous Plant | | Larkspur | |
Forest	1985	1986	1985	1986
Ashley	2	20	0	6
Boise	40	0	23	0
Bridger Teton	107	152	60	85
Caribou	0	12	0	6
Challis	15	36	15	36
Dixie	3	33	0	28
Fishlake	87	89	71	73
Humbolt	67	66	42	39
Manti-LaSal	62	111	62	111
Payette	38	48	29	26
Salmon	16	25	8	12
Sawtooth	119	92	81	61
Targhee	47	49	42	43
Toiyabe	5	3	3	1
Uinta	6	5	3	3
Wasatch-Cache	1	2	1	2
TOTAL	615	743	440	532

Nursing calves (136 to 205 kg) have been observed eating enough larkspur to cause death. An average of 11 calves died per year on the Manti allotment (Cronin and Nielsen 1979). Some losses could be attributed to lower weaning weights of calves. The seriousness of this would depend on how old the calf was when its mother died. If the calf was too young, larkspur could be blamed for the death of both cow and calf.

Normally, one would not expect the consumption of larkspur to have an effect on reproductive efficiency. However, if a bull died early in the season, it could significantly impact next year's calf crop.

Ranchers who graze cattle on larkspur ranges often encounter extra costs of management as they attempt to prevent losses. Because of the larkspur problem, the Manti Canyon Allotment has more miles of fence and a higher quality fence than normally would be required. Fence construction and maintenance is very expensive, especially on steep mountain ranges where the snow drifts accumulate. On some allotments, ranchers fence off areas where larkspur is the thickest or where their losses have been the heaviest in the past. The Sheep Creek Allotment on the Fishlake National Forest has a "poison pasture" fenced to keep cattle out. This pasture consists of about 194 ha. A 4,630 m log fence was constructed around the pasture. The replacement cost of the fence is $5.40/meter, or $25,000. Annual maintenance requires eight man-days at $50 per day, or $400. Even with this expense, a few cattle get in and are killed by larkspur.

Grazing systems often must be modified because of larkspur. Manti ranchers try to hold their cattle off the higher elevation ranges where the tall larkspur grows until after the plants become mature and less toxic. This puts a lot of grazing pressure on the mid-elevation pastures. Thus, the mid-elevation pasture is the "bottleneck" and limits the number of cattle that can be grazed on the allotment. Controlling larkspur would allow ranchers to move cattle onto the higher elevation pastures earlier in the grazing season and relieve grazing pressure on the lower area. It would seem possible that after the mid-elevation pastures have a chance to recover that additional livestock could be grazed on the allotment.

The quality of forage available to cattle is another factor to be considered in looking at the full cost of grazing larkspur-infested ranges. Marsh et al. (1934) recommended keeping cattle off larkspur-infested ranges until after larkspur flowered and its toxicity declined. Other forage would also mature and decline in nutrient quality. Cook and Harris (1968) reported that crude protein content of cattle diets on early summer mountain ranges was 10.2% compared with 8.4% during the latter part of the summer. This level of crude protein in calf diets would support daily gains of 0.227 kg (NRC 1980) early in the summer but would not support gains, and even cause weight losses, later in the summer. The permittees in the Manti Canyon allotment put their cattle on the upper pastures about August 1. Most of these ranges could be grazed as early as July 20 if they did not fear the consequences of grazing tall larkspur. During August, the larkspur plants vary in growth from the vegetative stage to full bloom. To achieve the greatest risk reduction from larkspur poisoning, they should hold off until September 1 when all of the larkspur plants have produced seed pods. If calves and yearlings could be managed such that they used the mid-elevation range in its early growth stages and then moved to the high elevation range in July to take advantage of high quality immature vegetation, they may be able to gain an additional 9 kg during the grazing season. Assuming an 85% calf crop from 837 cows in the Manti Canyon Allotment, 711 calves gaining an additional 9 kg each could possibly produce an extra 6,450 kg for sale at $1.43/kg, or $9,243 per year.

LOSS PREVENTION

Herbicide Control

The upper end of the Manti Canyon Allotment, about 3,238 ha, had an estimated 139 ha of dense stands of tall larkspur (D. barbeyi). The high loss in 1958 of 103 head triggered a move on the part of the ranchers to see if a solution to their problem could be found.

After several years of research, it was determined that the tall larkspur could be controlled with the application of 4.5 kg a.e./ha of 2,4,5-T [(2,4,5-trichlorophenoxy) acetic acid] (Cronin and Nielsen 1972). In order to get adequate control of the larkspur, the herbicide had to be applied at the 4.5 kg/ha rate for two consecutive years (Cronin 1974).

It was hypothesized that control of the dense patches of larkspur would significantly reduce cattle death losses. The larkspur-infested area had been divided into several pastures for grazing management. One of these pastures was grazed sparingly in most years because of the threat of severe larkspur losses. This pasture was selected as the area to test the hypothesis. Prior to control, cattle losses in this pasture averaged 13 head per year. These losses were reduced 94% when grazing was resumed after herbicide control. Assuming that this same level of death loss reduction could be accomplished across the entire upper part of the allotment, losses could be reduced from the expected average of 36 cows per year to three cows per year, or a savings of 33 cows and ten calves annually.

The costs of control estimated for the research area were $57/ha for the first year's treatment and $42/ha for the second year's treatment. Thus, it would cost $7,923 in year one and $5,838 in year two to spray the 139 ha of dense patches of tall larkspur. The expected benefits from this control would be the value of cows and calves saved. In this case, 33 cows and ten calves are valued at $9,450. One must also make an estimate of the number of years that the control will last in order to determine the length of the benefit stream. In this case, it was assumed that the control would be effective for at least ten years. It has been over 15 years since some of the first patches of larkspur were sprayed, and they still do not need retreatment.

The internal rate of return on this spray project is between 60 and 72%, depending on the assumption made about how the spraying takes place and whether the value of calves saved is included in the benefits. Ranchers have very few investment opportunities that have expected returns this high. There was another significant benefit for this spray project that was not considered. As larkspur losses decreased, ranchers were able to move their cattle onto the upper pastures earlier and relieve some of the grazing pressures on their mid-elevation pastures, which could result in substantial range improvement.

The Environmental Protection Agency banned use of 2,4,5-T in 1983. Research is ongoing to find a herbicide that will control tall larkspur. At this point in time, Roundup [glyphosate, N-(phosphonomethyl) glycine] appears to give good control at 2.2 kg a.e./ha with a single year's application, but it is nonselective. The nonselectivity may not be a serious problem on dense patches of larkspur or for spot treatment of scattered plants. When the cost for Roundup ($86.50/ha) was substituted for the chemical cost in the 2,4,5-T treatment, the analysis indicated a 50% plus yield [IRR (internal rate of return)].

Sequential Grazing of Sheep before Cattle

Sheep grazing of larkspur patches prior to cattle turn-in has shown varied results in reducing the threat of cattle poisoning. Sheep consumed sufficient low larkspur to make the range safe for cattle (Marsh et al. 1934). Mixed results were obtained on tall larkspur ranges (Aldous 1917). Heavy grazing of young larkspur by sheep early in the grazing season on the Mono (now Toiyabe) National Forest in Nevada eliminated the threat to cattle. Sheep profitably utilized forage that was deadly for cattle. However, on the Ruby Mountains in Nevada, sheep did not graze larkspur but trampled a sufficient amount that made it unacceptable to cattle. On the Fishlake National Forest in Utah, however, sheep grazed all other forage in preference to larkspur, leaving a greater threat to cattle. Alexander and Taylor (1986) reported that sheep preferred D. glaucescens and grazed it in amounts that prevented cattle deaths.

If sheep can be closely herded through dense concentrations of larkspur and selectively graze it or trample it down, sequential grazing of sheep before cattle on larkspur-infested rangelands may be an effective means of reducing the larkspur threat to cattle. Common use grazing of sheep and cattle can also reduce grazing pressure (Cook et al. 1967; Cook 1954), more evenly utilize the available forage (Ruyle and Bowns 1985), or even increase carrying capacity (Schlundt 1980).

Conditioned Feed Aversion

Conditioned feed aversion may allow cattle to graze larkspur ranges without the threat of poisoning. Preliminary results indicate that cattle can be conditioned to avoid eating tall larkspur. Cattle were fed fresh larkspur and infused with a strong emetic (lithium chloride) which made them ill. They associated the induced illness with the taste of larkspur and subsequently refrained from eating larkspur in the pen and in field grazing studies (M.A. Lane, M.H. Ralphs, and J.D. Olsen, unpublished data). If the technology can be developed, replacement heifers may be conditioned for life by feeding pelleted larkspur plant material with an emetic at an impressionable period in their early development. Over a period of years, the whole herd would eventually be conditioned to avoid eating larkspur.

Grazing Management

Grazing strategies also may be developed to reduce cattle losses. Research is proposed at the USDA/ARS Poisonous Plant Research Laboratory to determine the toxicity of tall larkspur at various stages of development, on various soils or sites where it grows, and the influence of weather changes on toxicity. Cattle grazing behavior with respect to tall larkspur also will be quantified. Toxicity of a particular larkspur population could be predicted to determine when it would be most dangerous and when cattle would most likely consume it. Models could then be developed to generate grazing strategies to minimize risk of poisoning yet obtain optimum utilization from the productive mountain rangelands.

128

Feed Supplements

Various feed supplements have been used by ranchers who graze cattle on larkspur ranges. These supplements vary from "top secret" formulations of various minerals and other substances, including ground larkspur, to iodized salt. A local feed company in Utah sells a supplement called "Larkspur Alleviator" at $232 per ton. Iodized salt costs $68 per ton. The "jury" is not in on the effectiveness of these supplements. It is a popular preventive that everyone hopes will work because it would be rather simple to carry out.

LITERATURE CITED

Aldous, A.E. 1917. Eradicating tall larkspur on cattle ranges in the national forests. USDA Farmers Bull. 826.

Alexander, J.D., III, and J.E. Taylor. 1986. Sheep utilization as a control method on tall larkspur infested cattle range. (Abstract). 39th Annual Meeting, Society for Range Management.

Alley, H.P., and G. Lee. 1970. What can be done about controlling larkspur on western rangelands. Down to Earth 26:31-32.

Cook, C.W. 1954. Common use of summer range by sheep and cattle. J. Range Manage. 7:10.

Cook, C.W., and L.E. Harris. 1968. Nutritive value of seasonal ranges. Utah Agr. Exp. Sta. Bull. 472.

Cook, C.W., L.E. Harris, and M.C. Young. 1967. Botanical and nutritive content of diets of cattle and sheep under single and common use on mountain range. J. Anim. Sci. 26:1169-1174.

Cronin, E.H., and D.B. Nielsen. 1979. The ecology and control of rangeland larkspurs. Utah Agr. Exp. Sta. Bull. 499.

Ewan, J. 1945. A synopsis of the North American species of Delphinium. University of Colorado Studies, Series D 2:55-244.

Kingsbury, J.M. 1964. Poisonous plants of the United States and Canada. Prentice-Hall, Englewood Cliffs, NJ.

Marsh, C.D., A.B. Clawson, and H. Marsh. 1934. Larkspur or poison weed. USDA Farmers Bull. 98 (revised 1934).

National Research Council. 1980. Nutrient requirements of beef cattle, 4th ed. National Academy of Sciences, Washington, D.C.

Ruyle, G.B., and J.E. Bowns. 1985. Forage use by cattle and sheep grazing separately and together on summer range in southwestern Utah. J. Range Manage. 38:299-302.

Schlundt, A.F. 1980. Common use grazing studies on southern Utah summer range. Ph.D. Diss., Utah State Univ., Logan.

USDA Forest Service. 1986. Grazing statistics for 1985. Washington, D.C.

11

Ecology and Toxicology of Bitterweed (Hymenoxys odorata)

Darrell N. Ueckert and Millard C. Calhoun

INTRODUCTION

Bitterweed (Hymenoxys odorata) (Compositae) was recognized as being poisonous to sheep about 60 years ago. The problem is most serious on limestone-derived soils in western Texas. Bitterweed has been a significant factor in the decline of numbers of sheep and sheep ranchers in this region. Considerable effort has been expended to better understand the bitterweed problem and to devise ways to continue producing sheep on bitterweed-infested rangelands. The objective of this treatise was to synthesize the results from numerous studies dealing with the ecology, toxicology, and management of this economically important weed.

THE PLANT

Bitterweed is a much-branched annual plant with a slender tap root. Mature plants vary in height from about 7 to 60 cm, depending on environmental conditions. The species is locally referred to as bitterweed, western bitterweed, and bitter rubberweed. The stem is usually purplish towards the base. The leaves are alternate, one to three times alternately divided into very narrow, glandular, woolly divisions not distinct from the petiole. Composite flower heads are located at tips of stem branches. Heads have both ray and disk flowers which are golden yellow with 60 or more disk florets and six to ten ray florets. The rays are tipped with three lobes. Each flower head produces from 50 to 75 seeds under normal growth conditions. The plant has a bitter taste, and crushed or bruised leaves have an aromatic odor (Kingsbury 1964; Sperry et al. 1964; James et al. 1980).

DISTRIBUTION AND ECOLOGY OF BITTERWEED

Populations of bitterweed containing the ancestral number of chromosomes (N=15) occur in a gypsiferous highland area in northcentral Mexico, and the species has apparently spread from this area (Sanderson and Strother 1973). The species is common in the semiarid range areas from southwest Kansas through central and western Texas into Mexico, Arizona, New Mexico, and southeastern California (Kingsbury 1964). The species occurs in almost every county of Texas west of the 99th meridian, but heaviest infestations occur in about 12 counties in the western portion of the Edwards Plateau in southwestcentral Texas and the adjacent Trans-Pecos region of west Texas (Sperry et al. 1964). It occurs in the southern parts of Coconino and Navajo and in Greenlee, Cochise, Pima, and Yuma Counties of Arizona at elevations below about 1829 m. It is particularly abundant on moist, alluvial soils along the Gila River drainage (Schmutz et al. 1968). Bitterweed grows in clay, clay loam, alkali, adobe, sandy, and alluvial soils, but is most common in drainage areas such as flood plains, lake beds, and along roadsides (Parker 1972).

Year-to-year variation in the abundance of bitterweed is largely a function of the amount of precipitation received during autumn and winter. The species is generally abundant in years with above normal, autumn-winter precipitation, whereas it may not occur (or be rare or only occasional) in dry years. Bitterweed seeds may germinate any time of the year, but autumn, winter, and early spring are the normal periods of germination and seedling establishment. Optimal temperatures for germination are 20 to 25 C, but germination occurs over the temperature range of about 10 to 40 C. Bitterweed seed germination decreases significantly as moisture tensions increase. The seed may remain viable in the soil for several years (Whisenant and Ueckert 1982).

Prior to disturbances of rangelands associated with the activities of man and, in particular, overgrazing with livestock, bitterweed was largely confined to shallow lakebeds in the Edwards Plateau of Texas, where standing water occasionally destroyed the grass cover, or in low places where grass was destroyed by repeated depositions of silt. Such places were conspicuous with dense stands of green bitterweed in winter and yellow flowers in spring. Overgrazing and drought during 1913 to 1928 were major factors contributing to the increased abundance of bitterweed on rangeland (Hardy et al. 1931). Bitterweed densities are invariably greater where the cover of desirable grasses has been reduced by overgrazing and drought.

Bitterweed is a prolific seed producer with 50 to 75 seeds per head (Rowe et al. 1973) and 100 to 3,000 heads per plant (Cory 1951). However, there may be only one or two heads per plant in extremely dense bitterweed stands, in dry periods, or where the weeds occur within dense grass stands. Bitterweed seeds are disseminated by overland flow during heavy rain storms (Jones et al. 1932), on the hooves of animals and undercarriages of vehicles during wet periods, and in the fleeces of sheep and goats. Bitterweed establishment is almost always greater on sites that have been seriously disturbed, such as along roadways, pipeline rights-of-way, drilling sites, livestock bedgrounds, sacrifice areas, lakebeds, etc., compared to well-managed rangeland. Bitterweed frequently is the dominant vegetation for decades in these early seral stages. Bitterweed densities are greatest in the seedling and early vegetative growth stages and decrease as the plants mature (Pfeiffer 1982). Bitterweed densities of 1.8 to 18.5 million plants/ha have been reported from the Edwards Plateau of Texas (Cory 1951, Pfeiffer 1982), but it should be recognized that bitterweed grows in defined areas and is generally not found evenly distributed across pastures.

TOXICOLOGY OF BITTERWEED

The area with heaviest infestations of bitterweed in western Texas is also the area with the greatest concentration of sheep and, consequently, the area where most sheep poisoning problems occur. Bitterweed is also toxic to cattle and goats grazing these same areas, but these species do not consume the weed in sufficient quantities to cause poisoning problems.

Clinical Signs and Pathology

Hardy et al. (1931) first demonstrated experimentally that bitterweed was poisonous to sheep. Sheep with acute or chronic poisoning stop eating, salivate, vomit, become depressed and weak, and lie down most of the time. A green salivary discharge and stain about the muzzle is common. Affected animals often lag behind the flock and may stand with backs arched.

There is gaseous and fluid distension of the ruminoreticulum, abomasum, and cecum in sheep poisoned by bitterweed. The ruminoreticulum and abomasum are severely congested and occasionally eroded on the mucosal surface with an accumulation of edematous fluids along the ruminoreticular folds and ruminal sulci. The most common histopathological finding in subacute bitterweed poisoning is severe glomerulonephrosis characterized by proteinaceous casts, swollen and degenerated glomerular tufts, and degeneration and necrosis in the inner renal cortex and outer medulla. Mild to moderate toxic hepatosis, characterized by vacuolar degeneration of hepatocytes primarily around the central vein, is also a consistent finding (Witzel et al. 1977).

Increases in hematocrit, hemoglobin, total serum protein, blood urea nitrogen, serum glutamic-oxaloacetic transaminase, and blood levels of pyruvic and lactic acids have been reported in acutely poisoned sheep (Dollahite et al. 1973; Witzel et al. 1974), as have decreases in blood glucose and pH (Witzel et al. 1974). In subacute studies, bitterweed depressed feed intake and serum concentrations of total protein, albumin, and thiols; however, serum urea nitrogen, creatinine, bilirubin, lactic dehydrogenase, glutamic-oxaloacetic transaminase, gamma glutamyl transferase, hematocrit, and rectal temperatures were increased (Calhoun et al. 1978; Baldwin and Calhoun 1979; Calhoun et al. 1981b). An acquired coagulation factor X activity deficiency has been reported in sheep given a subacute bitterweed treatment (Steel et al. 1976).

Toxic Dosage of Bitterweed

The amount of bitterweed necessary to produce signs of poisoning varies greatly among individual sheep (Witzel et al. 1977; Calhoun and Baldwin 1980; Calhoun et al. 1981a). As little as 500 g was adequate to kill some sheep. At the other extreme, a sheep consumed 14,514 g of immature green bitterweed plants over a 50-day period without exhibiting signs of poisoning (Hardy et al. 1931). Acquired tolerance following repeated exposure may be a factor in the individual variation in tolerance that has been observed (Post and Bailey 1980; Calhoun et al. 1981a), although variability has also been observed in sheep previously not exposed to bitterweed (Calhoun et al. 1981a).

The acute LD50 of fresh, seedling bitterweed in sheep averaged 1.3% of body weight, but was as low as 0.5% if the plants were drought stressed (Boughton and Hardy 1937). Acute values for air-dried bitterweed of 3.6 to 8.5 g/kg (0.36 to 0.85% of body weight) have been reported (Dollahite et al. 1973). The effects of the toxin are cumulative when less than the acute oral LD50 is consumed over a period of time (Dollahite et al. 1973). Blood constituents return to normal in about seven to ten days when bitterweed intake ceases. Unless sheep are severely poisoned, removal from bitterweed results in fairly rapid recovery.

The Toxic Principle

Isolation of sesquiterpene lactones and lactone glycosides from Hymenoxys species (Herz et al. 1970) lead to the suggestion that a sesquiterpene lactone might be the toxic principle in bitterweed. A poisonous lactone was subsequently isolated from bitterweed which, when administered orally, intravenously, or intramuscularly to experimental animals, produced signs of poisoning similar to those of natural bitterweed intoxication (Kim et al. 1974a). The nuclear magnetic resonance (NMR) and infrared (IR) spectra revealed the presence of an α-methylene-γ-lactone grouping.

Complete characterization of this poisonous lactone was reported by Kim et al. (1975) and Ivie et al. (1975). The empirical formula of the chemical isolated by both research groups was identical ($C_{15}H_{55}O_5$). The chemical isolated by Kim et al. (1975) was called hymenoxon, and the melting point (mp) was reported as 135 to 142 C. The compound isolated by Ivie et al. (1975) was called hymenovin (mp 115 to 137 C) and appeared to be a mixture of two hemiacetals epimeric at the C-3 and C-4 positions. Subsequently, Pettersen and Kim (1976) elucidated definitively the structure of hymenoxon by x-ray crystallography (Figure 11.1). Hymenolane, another sesquiterpene lactone isolated from bitterweed, has been found to be relatively nontoxic (Kim 1980; Terry et al. 1981). No other toxic compounds have been isolated from bitterweed, and hymenoxon appears to be the principle, if not the only, toxic compound present in the plants.

Hymenoxon concentrations were highest in seedling bitterweed and decreased as the plant matured in a December through June study on the Edwards Plateau. Hymenoxon levels ranged from 1.1 to 4.5% (dry matter basis) in bitterweed collected in December (early vegetative stage) and from 0 to 0.8% in June collections (mature, mostly desiccated plants)

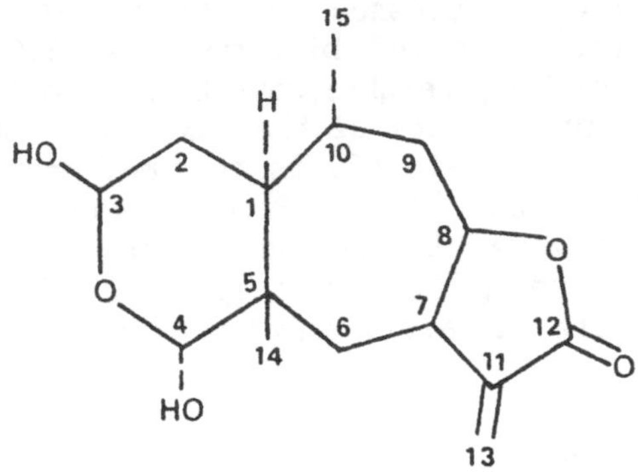

FIGURE 11.1
Chemical structure of hymenoxon.

(Pfeiffer 1982). There were significant differences in hymenoxon content of bitterweed in the same growth stage at different locations within an area 220 by 80 km in size. These differences appeared unrelated to environmental conditions. Plants growing south and southwest of San Angelo contained higher hymenoxon levels throughout the growing season than those growing north of San Angelo.

Hymenoxon is concentrated in bitterweed leaves and flowers. Stems contain about one-fifth the hymenoxon levels found in leaves and flowers; roots contain only a trace of hymenoxon. Other Compositae which are toxic to sheep and contain hymenoxon or hymenovin include pingue bitterweed (Hymenoxys richardsonii), desert baileya (Baileya multiradiata), and orange sneezeweed (Helenium hoopesii) (Ivie et al. 1976; Hill et al. 1977).

BITTERWEED MANAGEMENT

Growth of bitterweed in late fall and early winter coincides with the period of greatest physiological demand for nutrients (late gestation and early lactation) of sheep on the Edwards Plateau and Trans-Pecos range areas of Texas. Most sheep are very reluctant to eat bitterweed, even in the seedling and early vegetative stages. The unavailability of suitable alternative forage and hunger are the main reasons sheep consume the plant. Greatest losses have been reported in years where late-autumn rains follow an extended drought during which range vegetation is depleted or very low in quality. With late-autumn rains, there is often not sufficient time for forage recovery to provide competition to germinating bitterweed and to produce an alternative feed source for sheep.

Several practices are available for managing bitterweed problems on rangeland. Any one of these practices will not be an acceptable alternative to all ranchers (or even to neighboring ranchers) because of inherent differences in their land and financial resources or management objectives. Effective bitterweed management will most often be achieved when two or more carefully selected practices are applied in a planned sequence so that the characteristic weaknesses of one practice are compensated for by the characteristic strengths of the subsequent treatment(s) and to take advantage of treatment synergisms.

Antidotes

A report that the amino acid cysteine reacted with cytotoxic alphamethylene lactones to form less toxic compounds (Kupchan et al. 1970) lead to efforts to find an antidote for bitterweed and focused on compounds known to increase thiol levels in the body. Cysteine was found to be an effective antidote for bitterweed poisoning in sheep when either the bitterweed plant or the isolated lactone were administered for relatively short time periods (Kim et al. 1974b; Ivie et al. 1975; Calhoun et al. 1979; Rowe et al. 1980). However, cysteine is not suitable for use in pasture supplements because it is expensive, unstable, and not readily available. Also, cysteine is toxic to sheep when administered over prolonged periods at levels required to reduce toxicity of bitterweed (about 25 molecules of cysteine per molecule of hymenoxon) (Calhoun et al. 1979).

Diets containing high levels of natural proteins, such as cottonseed meal and soybean meal, also increase thiol levels in the body (M.C. Calhoun, unpublished data) and provide some protection against bitterweed poisoning (Bridges et al. 1980; Post and Bailey 1980; Calhoun et al. 1986). The addition of a sulfur source, such as sodium sulfate, has also provided some protection when sheep were dosed with bitterweed (Post and Bailey 1980). The greatest protection has been obtained with the addition of 0.5% of the commercially available antioxidant Santoquin (6-ethoxy-1,2-dihydro-2,2,4-trimethylquinoline) to sheep diets (Kim et al. 1982, 1983; Calhoun et al. 1986). Santoquin did not appear to affect body thiol levels in sheep. Its activity may be due to alterations in the activities of several drug-metabolizing enzymes, such as glutathione-S-transferases and epoxide hydrolase. Santoquin decreases the palatability of feeds when added at 0.5% and is currently cleared for use in animal feeds only as an antioxidant at a level not to exceed 0.015%.

Livestock and Grazing Management

Flash grazing by sheep (turning many sheep into a pasture with bitterweed for a few days and removing them before poisoning occurs) may reduce bitterweed vigor and allow desirable plants to predominate. Several ranchers have avoided toxicity by shifting sheep from bitterweed-infested to bitterweed-free pastures in seven-day cycles (Landers et al. 1981).

Bitterweed-poisoned sheep usually recover if confined and fed alfalfa hay for ten days. Drylot feeding with alfalfa or peanut hay may be a temporary alternative if bitterweed-free pastures are not available (Landers et al. 1981).

Bitterweed poisoning can be reduced or eliminated in some areas by managing grazing through rotation and proper stocking. Pasture deferment, for example, increases vigor of grasses and forbs that compete with bitterweed (Merrill and Schuster 1978). Combining cattle, sheep, and goats at moderate stocking rates greatly reduces sheep losses, compared to stocking with sheep alone or to heavy stocking rates with all three livestock species. A four-pasture, deferred rotation system, grazed moderately with cattle, sheep, and goats, was established at Texas A&M University's Agricultural Research Station at Sonora in 1949. Despite heavy stands of bitterweed, no sheep have died from bitterweed poisoning in this grazing system for about 30 years. Annual losses on pastures heavily grazed with sheep alone have averaged 8%. There were no losses on lightly grazed pastures stocked with cattle, sheep, and goats. Similar grazing systems and stocking rates at the Texas Range Station at Barnhart have reduced, but not eliminated, death losses from bitterweed (Merrill and Schuster 1978). Because bitterweed grows from seed during cool, moist periods, it does not compete successfully with perennial plants under proper grazing management. Long-term control depends on maintaining the vigor of desirable range grasses and forbs through deferment and use of proper stocking rates with combinations of livestock.

Many ranchers have observed that feeding supplemental protein aggravates bitterweed problems, although results from feeding experiments have not substantiated these observations. Concentrated natural protein sources such as soybean and cottonseed meal, as well as nonprotein nitrogen (urea), may overwhelm livers and kidneys of sheep already damaged by bitterweed (Parkins et al. 1973).

Herbicides

Bitterweed control with herbicide sprays depends on air and soil temperatures, soil moisture, plant phenology, and herbicide type (Ueckert et al. 1980). Control of bitterweed in the vegetative stage has been successful with sprays of ester or amine formulations of 2,4-D [(2,4-dichlorophenoxy) acetic acid] applied at 1.12 kg acid equivalent/ha when temperatures and soil moisture favor bitterweed growth. Excellent short-term control resulted with 2,4-D applied at 22 C air temperature, 16 C surface soil temperature, and when surface soil-water content averaged 20%. Control of bitterweed with 2,4-D was severely reduced when air temperatures were below 14 C, surface soil temperatures were below 12 C, and when bitterweed plants were flowering. Factors which reduce the effectiveness of 2,4-D for control of bitterweed, including low temperatures, advanced phenological stage of the weed, and low residual activity, can be overcome by the addition of dicamba (3,6-dichloro-2-methoxybenzoic acid) or picloram (4-amino-3,4,6-trichloro-2-pyridinecarboxylic acid) to 2,4-D or by applying picloram alone at 0.28 kg/ha (Ueckert et al. 1980). Desirable forbs that are important in sheep diets are severely reduced for a year or longer after herbicides are applied for bitterweed control. Forb standing crops were reduced 49 and 92% five months after aerial sprays of 2,4-D at 1.1 kg/ha and picloram at 0.56 kg/ha were applied, respectively, in a western Texas experiment (Ueckert et al. 1977b).

The effects of 2,4-D on hymenoxon concentration of bitterweed and its toxicity to sheep were studied after it was observed that spraying with 2,4-D at 1.1 kg/ha appeared to increase its palatability and decrease its toxicity to sheep (L.B. Merrill, personal communication). Field and feeding studies failed to substantiate this field observation, and at present there is no evidence that bitterweed sprayed with 2,4-D is less toxic than unsprayed bitterweed (Calhoun et al. 1982). Spraying with 2,4-D increased hymenoxon concentration during the first 15 days post-treatment in one study (Hill 1976), but hymenoxon concentration decreased following 2,4-D applications in other studies (Ueckert et al. 1977a; Calhoun et al. 1982).

Herbicides can effectively control localized infestations of bitterweed, prevent the species from spreading, and be used to maintain bitterweed-free pastures where animals can be grazed during critical bitterweed periods. Manually clearing or hand spraying small patches and isolated plants may help reduce future infestations, although controlling plants for many years may be essential to deplete the viable seed reserve in the soil. Disturbance of soil with machinery or excessive hoof action of grazing animals should be avoided (Landers et al. 1981).

LITERATURE CITED

Baldwin, B.C., Jr., and M.C. Calhoun. 1979. Dietary protein and subacute bitterweed poisoning in sheep. Texas Agr. Exp. Sta. Prog. Rep. 3571.

Bridges, G.W., E.M. Bailey, Jr., and B.J. Camp. 1980. Prevention of bitterweed intoxication of sheep. Vet. Human Toxicol. 22:87-90.

Boughton, I.B., and W.T. Hardy. 1937. Toxicity of bitterweed (Actinea odorata) for sheep. Texas Agr. Exp. Sta. Bull. 552.

Calhoun, M.C., and B.C. Baldwin, Jr. 1980. Sheep tolerance to bitterweed poisoning--variation between animals. Texas Agr. Exp. Sta. Prog. Rep. 3695.

Calhoun, M.C., B.C. Baldwin, Jr., S.W. Kuhlmann, and H.L. Kim. 1986. Bitterweed antidote research. Texas Agr. Exp. Sta. Prog. Rep. 4382.

Calhoun, M.C., B.C. Baldwin, Jr., and C.W. Livingston, Jr. 1978. Reduction in blood thiols in subacute bitterweed toxicity. Texas Agr. Exp. Sta. Prog. Rep. 3512.

Calhoun, M.C., B.C. Baldwin, Jr., and C.W. Livingston, Jr. 1979. Effect of abomasal cysteine addition on response to subacute bitterweed (Hymenoxys odorata) poisoning. Texas Agr. Exp. Sta. Prog. Rep. 3570.

Calhoun, M.C., B.C. Baldwin, Jr., and F.A. Pfeiffer. 1981a. Bitterweed adaptation in sheep. Texas Agr. Exp. Sta. Prog. Rep. 3892.

Calhoun, M.C., D.N. Ueckert, C.W. Livingston, Jr., and B.C. Baldwin, Jr. 1981b. Effects of bitterweed (Hymenoxys odorata) on voluntary feed intake and serum constituents of sheep. Amer. J. Vet. Res. 42:1713-1717.

Calhoun, M.C., D.N. Ueckert, C.W. Livingston, Jr., and B.J. Camp. 1982. Effect of 2,4-D on hymenoxon concentration and toxicity of bitterweed (Hymenoxys odorata) force-fed to sheep. J. Range Manage. 35:489-492.

Cory, V.L. 1951. Increase of poison bitterweed (Hymenoxys odorata) on Texas range lands. Field and Lab. 19:39-44.

Dollahite, J.W., L.D. Rowe, H.L. Kim, and B.J. Camp. 1973. Control of bitterweed (Hymenoxys odorata) poisoning in sheep. Texas Agr. Exp. Sta. Prog. Rep. 3149.

Hardy, W.T., V.L. Cory, H. Schmidt, and W.H. Dameron. 1931. Bitterweed poisoning in sheep. Texas Agr. Exp. Sta. Bull. 433.

Herz, W., K. Aota, M. Holub, and Z. Samck. 1970. Sesquiterpene lactones and lactone glycosides from Hymenoxys species. J. Org. Chem. 35:2611-2624.

Hill, D.W. 1976. Analysis of hymenoxon in plant tissue and disposition of hymenoxon in the rabbit. Ph.D. Diss., Texas A&M Univ., College Station.

Hill, D.W., H.L. Kim, C.L. Martin, and B.J. Camp. 1977. Identification of hymenoxon in Baileya multiradiata and Helenium hoopesii. J. Agr. Food Chem. 25:1304-1307.

Ivie, G.W., D.A. Witzel, W. Herz, R. Kannan, J.O. Norman, D.D. Rushing, J.H. Johnson, L.D. Rowe, and J.A. Veech. 1975. Hymenovin. Major toxic constituent of bitterweed (Hymenoxys odorata DC.). J. Agr. Food Chem. 23:841-845.

Ivie, G.W., D.A. Witzel, W. Herz, R.P. Sharma, and A.E. Johnson. 1976. Isolation of hymenovin from Hymenoxys richardsonii (Pingue) and Dugaldia hoopesii (orange sneezeweed). J. Agr. Food Chem. 24:681-682.

James, L.F., R.F. Keeler, A.E. Johnson, M.C. Williams, E.H. Cronin, and J.D. Olsen. 1980. Plants poisonous to livestock in the western states. USDA Agr. Info. Bull. 415.

Jones, S.E., W.H. Hill, and T.A. Bond. 1932. Control of the bitterweed plant poisonous to sheep in the Edwards Plateau Region. Texas Agr. Exp. Sta. Bull. 464.

Kim, H.L. 1980. Toxicity of sesquiterpene lactones. Res. Commun. Chem. Pathol. Pharmacol. 28:189-192.

Kim, H.L., A.C. Anderson, B.W. Herrig, L.P. Jones, and M.C. Calhoun. 1982. Protective effects of antioxidants on bitterweed (Hymenoxys odorata DC.) toxicity in sheep. Amer. J. Vet. Res. 43:1945-1950.

Kim, H.L., B.W. Herrig, A.C. Anderson, L.P. Jones, and M.C. Calhoun. 1983. Elimination of adverse effects of ethoxyquin (EQ) by methionine hydroxy analog (MHA).

Protective effects of EQ and MHA for bitterweed poisoning in sheep. Toxicol. Letters 16:23-29.

Kim, H.L., L.D. Rowe, and B.J. Camp. 1975. Hymenoxon, a poisonous sesquiterpene lactone from Hymenoxys odorata DC. (bitterweed). Res. Commun. Chem. Pathol. Pharmacol. 11:647-650.

Kim, H.L., L.D. Rowe, M. Szabuniewicz, J.W. Dollahite, and B.J. Camp. 1974a. The isolation of a poisonous lactone from bitterweed Hymenoxys odorata DC. (Compositae). Southwest. Vet. 27:84-86.

Kim, H.L., M. Szabuniewicz, L.D. Rowe, B.J. Camp, J.W. Dollahite, and C.H. Bridges. 1974b. L-Cysteine, an antagonist to the toxic effects of an α-methylene-γ-lactone isolated from Hymenoxys odorata DC. (bitterweed). Res. Commun. Chem. Pathol. Pharmacol. 8:381-384.

Kingsbury, J.M. 1964. Poisonous plants of the United States and Canada. Prentice-Hall, Inc., Englewood Cliffs, NJ. pp. 414-415.

Kupchan, S.M., O.C. Fessler, A. Eakin, and T.J. Giacobbe. 1970. Reactions of the alpha methylene lactone tumor inhibitors with model biological nucleophiles. Science 168:376-378.

Landers, R.Q., Jr., D.N. Ueckert, and L.B. Merrill. 1981. Managing bitterweed to reduce sheep losses. Texas Agr. Ext. Serv. Fact Sheet L-1845.

Merrill, L.B., and J.L. Schuster. 1978. Grazing management practices affect livestock losses from poisonous plants. J. Range Manage. 31:351-354.

Parker, K.F. 1972. An illustrated guide to Arizona weeds. Univ. Arizona Press, Tucson.

Parkins, J.J., R.G. Hemingway, and N.A. Brown. 1973. The increasing susceptibility of sheep to dietary urea toxicity associated with progressive liver dysfunction. Res. Vet. Sci. 14:130.

Pettersen, R.C., and H.L. Kim. 1976. X-ray structures of hymenoxon and hymenolane: pseudoquaianolides isolated from Hymenoxys odorata DC. (bitterweed). J. Chem. Soc. Perkin Transactions II: 1399-1403.

Pfeiffer, F.A. 1982. Effects of environmental, site and phenological factors on hymenoxon content of bitterweed (Hymenoxys odorata). M.S. Thesis, Angelo State Univ., San Angelo, Texas.

Post, L.O., and E.M. Bailey, Jr. 1980. The effects of dietary supplements on chronic bitterweed (Hymenoxys odorata) poisoning in sheep. Texas Agr. Exp. Sta. Prog. Rep. 3698.

Rowe, L.D., J.W. Dollahite, H.L. Kim, and B.J. Camp. 1973. Hymenoxys odorata (bitterweed) poisoning in sheep. Southwest. Vet. 26:287-293.

Rowe, L.D., H.L. Kim, and B.J. Camp. 1980. The antagonistic effect of L-cysteine in experimental hymenoxon intoxication in sheep. Amer. J. Vet. Res. 41:484-486.

Sanderson, S.C., and J.L. Strother. 1973. The origin of aneuploidy in Hymenoxys odorata. Nature New Biol. 242:220-221.

Schmutz, E.M., B.N. Freeman, and R.E. Reed. 1968. Livestock-poisoning plants of Arizona. Univ. Arizona Press, Tucson.

Sperry, O.E., J.W. Dollahite, G.O. Hoffman, and B.J. Camp. 1964. Texas plants poisonous to livestock. Texas Agr. Exp. Sta. Bull. B-1028.

Steel, E.G., D.A. Witzel, and A. Blanks. 1976. Acquired coagulation factor X activity deficiency connected with Hymenoxys odorata DC. (Compositae) bitterweed poisoning in sheep. Amer. J. Vet. Res. 37:1383-1386.

Terry, M.K., H.L. Kim, D.E. Corrier, and E.M. Bailey, Jr. 1981. The acute oral toxicity of hymenoxon in sheep. Res. Commun. Chem. Pathol. Pharmacol. 31:181-184.

Ueckert, D.N., B.J. Camp, and L.B. Merrill. 1977a. Hymenoxon concentration in bitterweed after herbicide application. Texas Agr. Exp. Sta. Prog. Rep. 3457.

Ueckert, D.N., C.J. Scifres, S.G. Whisenant, and J.L. Mutz. 1980. Control of bitterweed with herbicides. J. Range Manage. 33:465-469.

Ueckert, D.N., S.G. Whisenant, and R.C. Papasan. 1977b. Bitterweed herbicides: effect on cool-season forbs. Texas Agr. Exp. Sta. Prog. Rep. 3458.

Whisenant, S.G., and D.N. Ueckert. 1982. Factors influencing bitterweed seed germination. J. Range Manage. 35:243-245.

Witzel, D.A., L.P. Jones, and G.W. Ivie. 1977. Pathology of subacute bitterweed (Hymenoxys odorata) poisoning in sheep. Vet. Pathol. 14:73-78.

Witzel, D.A., L.D. Rowe, and D.E. Clark. 1974. Physiopathologic studies on acute Hymenoxys odorata (bitterweed) poisoning in sheep. Amer. J. Vet. Res. 35:931-934.

12

Impact of Bitterweed on the Economics of Sheep Production in the Texas Edwards Plateau

J.R. Conner, J.L. Schuster,
and E.M. Bailey, Jr.

INTRODUCTION

Bitterweed (Hymenoxys odorata) occurs from southwestern Kansas to northern Mexico and southeastern California (Clawson 1931). It is common in Texas west of the 99th meridian (Sperry et al. 1968). The most severe infestations occur in the western portion of the Edwards Plateau of Texas (Landers et al. 1981). The area where bitterweed concentrations are of most concern in Texas is bordered on the west and south by the Pecos and Rio Grande Rivers and on the north and east by the Colorado River. This region also delineates the heaviest concentration of sheep production in Texas (Dollahite et al. 1973).

Bitterweed produces poisoning in sheep and occasionally in goats and cattle (Bailey 1978). It is recognized as a major poisonous plant on the Edwards Plateau of Texas (Sperry 1949) and is the most detrimental poisonous plant problem for sheep production in Texas (Dollahite et al. 1973). Death losses in sheep vary but have been reported to average between 1 and 6% annually across the Edwards Plateau (Sultemeier 1961). It is reported that bitterweed, increased predation, and rising labor costs are primarily responsible for the more than 50% reduction in the number of sheep produced in West Texas over the past 30 years (Ueckert et al. 1980).

BITTERWEED AND SHEEP PRODUCTION IN THE WESTERN EDWARDS PLATEAU OF TEXAS

Bitterweed is an invader plant on most range sites of the Edwards Plateau. It is found in greatest abundance in lake beds, draws, along roadsides, around pens and watering places, and in other areas where the grass stands have been thinned because of drought, overgrazing, or soil disturbance (Sperry 1949; Dollahite et al. 1973; Ueckert et al. 1980). While isolated plants can be found in almost any rangeland pasture, many pastures, and some entire ranches, do not have sufficient infestations to result in problems with sheep poisoning. Sheep do not generally select bitterweed if other green forbs or grasses are available. In a long-term study at the Sonora Research Station, sheep losses due to bitterweed occurred in 12 of 20 years on continuously grazed pastures heavily stocked with sheep only. They were encountered in eight years under moderate stocking (Merrill and Schuster 1978).

Bitterweed grows from seed during cool, moist periods, thus populations are greatest in mild, wet winters and early springs. However, climatic conditions can also restrict growth of other plants, and so poisoning in sheep may be more severe under these conditions than in years more favorable to bitterweed growth, due to the limited availability of alternative green forage. Sheep poisoning from bitterweed commonly occurs from early December to late March, but problem periods will vary from year to year in intensity and duration. Producers in bitterweed-infested areas must exercise intensive scrutiny of herds throughout the winter months in order to detect early signs of bitterweed poisoning.

CONTROL OF BITTERWEED POISONING

Bitterweed can be controlled with 2,4-D [(2,4-dichlorophenoxy) acetic acid], picloram (4-amino-3,5,6-trichloropicolinic acid), and other herbicides (Ueckert et al. 1980). However, since bitterweed is an annual, control must be repeated annually. For the same reason, manual clearing by pulling or hoeing is also usually effective only for a single season.

Herbicides and manual control costs are high, thus most ranchers find these practices economically feasible in clearing only infestations that are isolated in a few relatively small areas of larger pastures or in pastures infested with a few isolated plants (Landers et al. 1981).

Bitterweed poisoning problems in sheep production are most commonly handled through alterations of grazing and livestock management practices. Where possible, removing sheep from pastures with heavy bitterweed infestation during problem periods is an efficient and effective practice on ranches. Unfortunately, many ranches with bitterweed poisoning problems have significant bitterweed infestations in most of their pastures. In such cases, the problem is harder to handle, and total avoidance of death losses is impractical. The usual management practices when sheep must be grazed in bitterweed-infested pastures during the winter months include frequent (once every day or every other day) observation of the entire flock to detect those individuals showing signs of bitterweed poisoning. Individual sheep found exhibiting symptoms are caught and transported to pens where they are fed a high quality ration. After a 10- to 14-day feeding period, most of the sheep will have recovered and can be returned to pasture. Unfortunately, not all sheep are detected early enough to prevent them from succumbing to the poison. Also, lambing percentages are generally reduced after a severe bitterweed poisoning episode.

Other management practices that are sometimes successful include grazing combinations of cattle, sheep, and goats in the same pasture or pasture system. Merrill and Schuster (1978) reported no sheep losses due to bitterweed poisoning over a 20-year period in a moderately stocked four-pasture, deferred rotation grazing system with cattle, sheep, and goats. Regardless of the stocking rate, death losses were greatest on pastures with sheep only and least when pastures were stocked with a combination of cattle, sheep, and goats. Also, flash grazing (very high stocking rates for very short periods) in bitterweed-infested pastures will sometimes allow sheep to utilize a bitterweed-infested area without being killed by bitterweed poisoning. Proper grazing management which encourages competition from more desirable grasses and forbs will also help to reduce problems with bitterweed poisoning (Merrill and Schuster 1978; Landers et al. 1981).

IMPACT OF BITTERWEED ON NET RETURNS FROM WOOL-LAMB OPERATIONS

During 1986, information on costs, returns, and operating practices associated with wool-lamb production was obtained from a selected sample of ranches in the western Edwards Plateau and from Texas Agricultural Experiment Stations at Sonora and Barnhart, Texas. The data were used to develop two cost/return budgets of wool lamb operations: one representing an operation where bitterweed poisoning was not a problem, and one representing a similar operation with bitterweed-infested rangeland (Table 12.1).

In both operations, cull ewes are replaced by retaining ewe lambs. Replacement rates for both operations are 17%, with death losses 3 and 9%, respectively, for the without and with bitterweed operations. The 6% greater death loss for the operation with bitterweed represents the annual average percent deaths due to bitterweed poisoning at the Texas Range Station at Barnhart for the past 8 years (Dusek, personal communication). Lambing percentages average 100 and 85%, respectively, for the operations without and with bitterweed problems (Dusek and Taylor, personal communication).

Costs used in Table 12.1 are representative of 1986 prices for feed, services, etc., for the region. Revenues represent the average of prices received over the five-year period 1981-1985. Wool prices also include average U.S. government production incentive payments.

Costs and returns per animal unit in Table 12.1 are detailed by cost category and revenue source for the typical wool-lamb operations without and with bitterweed poisoning problems. Revenue for the operation with bitterweed is reduced by 14% compared to the operation without bitterweed. The revenue reductions occur in all sources. Less wool and cull ewes are sold due to the higher death losses, and fewer lambs are sold due to the reduced lambing percentage in the bitterweed-infested operation.

Operating costs are 7% higher and ownership costs 6% higher for the bitterweed-infested operation. Principal cost increases are for additional feed and labor associated with observing, penning, and drylot feeding the ewes showing symptoms of bitterweed poisoning. There are also some slight cost decreases, e.g., marketing, shearing, etc., associated with the bitterweed-infested operation due to the increased death loss and reduced lambing percentage.

CONCLUSIONS

As can be seen from Table 12.1, the net returns to land, management, and profit for the bitterweed-infested operation are reduced by 97% compared to the operation without bitterweed. Without the bitterweed problem, the wool-lamb operation is an attractive ranch enterprise, highly competitive with cattle and Angora goats. The impact of the bitterweed poisoning problem is to reduce the wool-lamb operation to a noncompetitive status with other enterprises.

TABLE 12.1
Annual costs and returns per animal unit sheep in the absence and presence
of bitterweed-infested rangeland for a typical wool-lamb operation.

	Without Bitterweed			With Bitterweed		
Investment Requirements	Number (head)		Value/AU ($)	Number (head)		Value/AU ($)
Ewes	5.00		140.00	5.00		140.00
Rams	0.10		15.00	0.10		15.00
Yearling Ewes	0.85		23.80	0.85		23.80
Horse	0.01		10.00	0.01		10.00
Dog	0.01		5.00	0.01		5.00
Total			193.80			193.80
Production	Number	Unit	Returns/AU ($)	Number	Unit	Returns/AU ($)
Wool	18.14	kg	120.00	17.22	kg	113.88
Tags	2.27	kg	2.00	2.04	kg	1.80
Lambs	122.36	kg	167.51	100.25	kg	137.24
Cull Ewes	31.75	kg	10.85	18.14	kg	6.20
Total			300.36			259.12
Operating Inputs	Number	Unit	Costs/AU ($)	Number	Unit	Costs/AU ($)
Supplemental Feed	163.30	kg	36.00	178.72	kg	39.40
Creep Feed	27.22	kg	5.40	23.13	kg	4.59
Salt & Minerals			10.50			10.50
Shearing			11.16			10.73
Veterinary Medicine			5.50			5.05
Marketing			8.40			6.51
Miscellaneous Expenses				12.00		12.00
Equipment, Fuel & Lube			3.53			4.06
Equipment Repair			.74			.85
Labor	11.2	hr	78.40	12.9	hr	90.30
Total			171.63			183.99
Capital Investment	Amount ($)		Costs/AU ($)	Amount ($)		Costs/AU ($)
Livestock	193.80		19.38	193.80		19.38
Equipment	277.61		27.76	277.61		27.76
Total			47.14			47.14
Ownership (depreciation, taxes, insurance)			Costs/AU ($)			Costs/AU ($)
Livestock			4.34			4.34
Equipment			20.21			21.72
Total			24.55			26.06
Residual Returns to Land, Management & Profit			Net/AU ($)			Net/AU ($)
			57.04			1.93

To the extent that the "with bitterweed" scenario depicted here is representative of ranches in the western Edwards Plateau, it is easy to see why, under such conditions, a rancher might decide to get out of the sheep production business. However, since there are still relatively large numbers of sheep producers in the western Edwards Plateau, one must conclude that the bitterweed poisoning scenario depicted in this example represents relatively few of the remaining sheep production operations in the region.

Unfortunately, the extent of the problem and the overall impact that bitterweed has had on the economics of the sheep industry and the economy of the Edwards Plateau Region of Texas are likely to remain largely unknown. There are several factors which contribute to the difficulty of assessing these overall impacts. First, there is the extreme variation from pasture to pasture and year to year in the abundance of bitterweed and the relative availability of alternative green forage at critical times. Second, there is the large number of ways in which ranchers may choose to combat the problem and the resulting variations in costs and revenue losses. Another confounding factor is the variation in the incidence of sheep death losses in the region due to other factors such as predation and consumption of other problem plants. Despite the lack of precision regarding the magnitude of its economic impact, it is safe to assume that bitterweed is a serious problem in the western Edwards Plateau, and there is a need to continue to search for economical means of reducing its detrimental impacts.

LITERATURE CITED

Bailey, E.M. 1978. Major poisonous plants of the Southwest. (Mimeo). Fourth Biennial Veterinary Workshop, Logan, Utah, June 19-24, 1978.

Clawson, A.B. 1931. A preliminary report on the poisonous effects of bitter rubber weed (Actinea odorata) on sheep. J. Agr. Res. 43:693-701.

Dollahite, J.W., L.D. Rowe, H.L. Kim, and B.J. Camp. 1973. Control of bitterweed (Hymenoxys odorata) poisoning in sheep. Texas Agr. Exp. Sta. PR-3149.

Landers, R.Q., Jr., D.N. Ueckert, and L.B. Merrill. 1981. Managing bitterweed to reduce sheep losses. Texas Agr. Ext. Serv. L-1845.

Merrill, L.B., and J.L. Schuster. 1978. Grazing management practices affect livestock losses from poisonous plants. J. Range Manage. 31:351-354.

Sperry, O.E. 1949. The control of bitterweed (<u>Actinea odorata</u>) on Texas ranges. J. Range Manage. 2:122-127.

Sperry, O.E., J.W. Dollahite, G.O. Hoffman, and B.J. Camp. 1968. Texas plants poisonous to livestock. Texas Agr. Exp. Sta./Ext. Serv. B-1028.

Sultemeier, G.W. 1961. Responses of bitterweed (<u>Hymenoxys odorata</u>) to 2,4-D in relation to soil moisture. MS Thesis. Texas A&M Univ., College Station, TX.

Ueckert, D.N., C.J. Scifres, S.G. Whisenant, and J.L. Mutz. 1980. Control of bitterweed with herbicides. J. Range Manage. 33:465-469.

13

Astragalus and Related Genera— Ecological Considerations

Phil R. Ogden, Stanley L. Welsh, M. Coburn Williams, and Michael H. Ralphs

INTRODUCTION

This paper is a review of literature and a summary of the experiences and observations of the authors on the ecology of species in the Astragalus and Oxytropis genera and three toxicoses (swainsonine, nitro toxins, and selenium) associated with these genera.

The genus Astragalus includes approximately 2,000 species and is indigenous to all continents except Australia and Antartica. The closely related Oxytropis genus is found in the temperate and frigid regions of the northern hemisphere and is represented by about 300 species. A third genus, Swainsona, which is limited to Australia and New Zealand, is not discussed but is of interest as the indolizidine alkaloid, swainsonine, was first isolated in a species of this genus in Australia and later confirmed as the toxin in locoweed species (Molyneux and James 1982).

DESCRIPTION OF GENERA

The generic descriptions are from Welsh et al. (1987) and are modified by additional information taken from Bentham and Hooker (1872), Bunge (1874), Komarov (1946, 1948), Barneby (1952, 1964) and Welsh (1974).

Astragalus

Plants are annual or perennial, caulescent or acaulescent, suffruticose or rarely fruticose in some, arising from a taproot, a caudex commonly developed, rarely rhizomatous; pubescence basifixed or malpighian (dolabriform); leaves alternate, odd-pinnate (rarely even-pinnate), trifoliolate, or simple; stipules adnate to the petiole base, sometimes connate-sheathing around the stem; petiole rachis persistent as a spine in some or sometimes marscescent (and forming a thatch at the plant base); flowers papilionaceous, borne in axillary racemes, each subtended by a single bract; bracteoles one or two or lacking, when present attached at the base of the calyx or on the pedicel; calyx cylindrical or campanulate, sometimes inflated or bladdery-inflated in fruit, ruptured by the pod, or not ruptured and sometimes enclosing the pod, bearing five apical teeth; five petals, pink, lavender, pink purple, ochroleucous, or white, or variously suffused, the keel shorter than the wings, rounded to attenuate apically; stamens diadelphous, the vexillar stamen positioned along the ventral suture of the ovary; ovary enclosed in the staminal sheath, the style glabrous; pods variable in size, shape, and dehiscence, unilocular to bilocular (by intrusion of the ventral suture, dorsal suture, or both), the valves membranous, coriaceous, cartilaginous, or ligneous, sometimes inflated; seeds several to numerous, borne along the ventral suture, sessile, subsessile, or stipitate (or with a gynophore); $x = 8, 11, 12, 13$.

Species of Astragalus fit into several of the life-form categories as elaborated by the late Professor C. Raunkiaer of Denmark (Polunin 1960). They are mainly hemicryptophytes or therophytes, with some fitting into the chamaephyte and geophyte categories. Plants that fall into the hemicryptophyte, chamaephyte, and geophyte groups are typically mesophytes; the therophytes mainly are adapted to growth in areas with hot dry summers.

Oxytropis

Perennial, caulescent, or acaulescent herbs from a taproot and caudex, or subshrubs or shrubs, unarmed or armed, sometimes glandular-viscid, in part; pubescence basifixed or malpighian (dolabriform); petioles sometimes persistent as a marscescent that on the caudex or less commonly persistent as pungent spines; stipules adnate to the petiole base or free; leaves alternate or basal, odd-pinnate; stipules adnate to the petiole base, often connate-sheathing; flowers papilionaceous, scapose or borne in axillary, bracteate racemes, each subtended by a single bract; bracteoles none (rarely two); calyx cylindric to campanulate, rarely inflated and enclosing or ruptured by the pod, five-toothed; petals five, pink, pink purple, white, or ochroleucous, the keel shorter than the wings, the keel tip produced into a porrect beak; stamens ten, diadelphous; ovary enclosed in the staminal sheath, the style glabrous; pods sessile or stipitate, straight, erect, ascending, or spreading-declined, one- or two-loculed, or partially two-loculed by intrusion of the ventral (upper, seed-bearing) suture, two-valved, dehiscent apically or throughout; x = 8. Members of Oxytropis are mainly hemicryptophytes, but some are chamaephytes, and a small number are phanerophytes.

DISTRIBUTION

Geographical

Astragalus is distributed mainly in arctic and temperate portions of the Northern Hemisphere but extends south into both the Paleotropics and Neotropics. It extends through the Neotropics to South America. Centers of distribution include portions of the Middle East, Asia, and western North America. The Flora Europaea includes 133 species; Flora of the USSR treats 869 species; and the Atlas of North American species of Astragalus lists 368 species. The genus is represented in much of North America, but most of the taxa occur in the west, centering about the Colorado Plateau and Great Basin.

Oxytropis is represented in temperate and frigid regions of Asia, Asia Minor, Europe, and America. It is lacking in Africa and in the Southern Hemisphere.

Soils and Habitats

Soils occupied by plants of Astragalus vary texturally from silts and clays through sands, gravels, and boulders. Many of them are specifically adapted for growth in crevices in rock, especially along the erosional features associated with joint systems. Some species are specifically adapted to substrates rich in soluble salts. Others require selenium since they thrive only where selenium is present in the substrates (Rosenfeld and Beath 1964). Still other species occupy river gravels and naturally (or human-induced) disturbed sites, but few, if any, are truly weedy. Rich organic soils tend to support fewer species than

mineral soils, at least in western North America. Some
species are psammophytes and are essentially restricted to
sandy substrates.

Habitats occupied by species of Astragalus include
niches in tundra, taiga, boreal forest, summer deciduous
forest, prairie, steppe, and desert. Speciation in the
genus has been possible, in some part, because of the
availability of a great variety of exposures of raw, only
slightly modified, geological strata exposed in places such
as the western United States. Soil development and
permafrost at high latitudes and elevations tend to insulate
roots from highly variable parental materials. Hence,
species of Astragalus in those areas are generalists that
often have rather wide areas of distribution. Specialists,
on the other hand, have the ability to grow on the raw
geological substrates where they flourish without
competition. Such habitats are mostly available at low
elevations where annual precipitation is much surpassed by
evaporation, soil development is at a minimum, and
permafrost is generally lacking. Raw geological substrates
are available at high elevations and at high latitudes,
mainly on ridges, rock outcrops, or on river gravels.
Species in the Arctic can often be found on gravels of
braided stream courses or on alluvial fans where total
vegetative cover is low.

Many species of Oxytropis occur on river gravels,
alluvial fans, ridge crests, and other places where mineral
soils are present. Competition on such sites is generally
low, and the plants are adapted to reoccupation of the sites
following disturbance. Other species occur as components of
vegetative types with well-developed, organic soils.
Lambert locoweed (Oxytropis lambertii), white locoweed (O.
sericea), and yellow locoweed (O. campestris) often occur
within grasslands, forests, and tundra communities.
However, even those species take advantage of disturbed
sites and frequently grow on stream gravels. Some species
occupy peculiar substrates, where parental material crops to
the surface, e.g., mountain oxytrope (O. oreophila) grows
from low to high elevations, often on fine-textured, harsh,
arid sites.

Several arctic species of Oxytropis compete well in
tundra, often occupying microhabitats such as margins of
frost boils and polygons. Others, including Scamman oxytrope (O. scammaniana) and Maydell oxytrope (O. maydelliana)
occur mainly within the tundra communities proper. Lake
shores, stream banks, and meadows are common sites for the
caulescent phases of stemmed oxytrope (O. deflexa).

Dominance in Plant Communities

Astragalus species are seldom dominants or codominants in a vegetative type. Typically, they represent minor components except where they constitute the principal vegetation on some peculiar substrate. In some years, for example, freckled milkvetch or spotted locoweed (Astragalus lentiginosus) is the dominant species on openings in pinyon-juniper woodland in the Colorado Plateau, and the narrowly endemic dinosaur milkvetch (A. saurinus) is often the dominant species on the Morrison, Carmel, and Moenkopi formations in eastern Utah. Seleniferous species, such as stinking milkvetch (A. praelongus) of the Colorado Plateau, often form extensive stands on portions of the otherwise essentially barren exposures of the Mancos Shale Formation.

Plants of Oxytropis species are seldom dominant within other vegetative types except in disturbed sites, such as braided gravels along streams and on road shoulders, where they are often the principal plant species. Typically they are minor components of other vegetative types.

PLANT POPULATION CYCLES AND ENVIRONMENTAL INFLUENCES

Ten to fifteen percent of the Astragalus species are annual or biennial with the balance primarily herbaceous perennials. Annual species of Astragalus tend to germinate in autumn, over-winter as rosettes of tiny eophylls, grow rapidly following warmth of early springtime, produce fruit, and die prior to the heat of summer. Some annuals persist in summer, however, on moist alluvium along stream courses and on lake shores in the arid southwestern portions of the United States. Perennials also tend to germinate in autumn, following storms of late summer and early autumn. They persist through the winter, and some of them will flower as annuals in the first springtime of their existence. Many do not flower in the first year but continue active growth as long as water is available, become dormant in the hot, dry portions of the year, and often grow again in the autumn of the year. In the spring of the second year, they are sufficiently mature to produce flowers and fruit. If the spring of the second year is dry and moisture is inadequate, many of the potentially perennial plants die, having been functually biennial. When conditions of moisture are adequate, however, those that have survived to flower in the second year become dormant following fruiting and persist to the spring of the third year.

Seldom is precipitation adequate for continued perennation of more than a few years. The populations, which often are continuous among other vegetation for miles, gradually shrink in numbers as the dry seasons rule out subsequent germination of seeds. Skeletonized dead plants then exist as bleached remains for a few years. The seeds persist in the ground until conditions for germination are again adequate. The cycle of moist to dry to moist conditions might require few to many years.

The situation as described above applies mainly to the potential perennials in arid or semiarid sites. There is evidence to indicate long life to some of the perennials, especially in areas of more equable rainfall. An example of this latter type is ground plum (A. crassicarpus), which occurs in the prairies and plains of the middle western states. A mesophyte of meadowlands in the west, mesic milkvetch (A. diversifolius) is likewise a perennial of long duration. The taproot is crowned by a well-developed caudex that persists for many years. The plant is adapted to survive among rushes, grasses, and forbs typical of meadow conditions. The nearly allied lesser rushy milkvetch (A. convallarius) also functions as a long-lived perennial from a deeply seated root crown with much elongated caudex branches that elevate the branches to ground level. This taxon is typical of sagebrush and grassland communities over wide expanses.

Many species of Astragalus tend to set an initial series of flowers that then begin to mature fruit. Additional racemes produced above the fruiting ones tend to abort as energy is directed to the fruiting inflorescences. In some species, however, additional racemes also result in fruit, and the plants will continue to flower and produce fruit far into the season. Wetherill milkvetch (A. wetherillii) and Rydberg milkvetch (A. perianus) are examples of plants with extended flowering and fruiting periods.

Except where swept away by action of flooding rivers and freshets from canyon mouths, the plants of Oxytropis species tend to be long-lived. They do well in seral environments where pressures from competition with other plants are not great. They do not invade native communities easily, but persist following land management practices such as burning or chaining.

SWAINSONINE

Description and Signs of Poisoning

The locoweed toxin in Astragalus and Oxytropis is the indolizidine alkaloid swainsonine (Molyneux and James 1982). It exists in very low levels (0.007% of dry weight) in the plant and causes a chronic poisoning characterized by neurological damage, emaciation, abortions, and birth defects (James et al. 1981). High altitude brisket disease occurs when plants containing swainsonine are grazed on high elevation ranges (James et al. 1983). Swainsonine inhibits the enzyme α-mannosidase in glycoprotein metabolism in most body systems. It prevents cleavage of mannose molecules from complex oligosaccharides, resulting in hybrid glycoproteins which are not metabolized by the cell. These hybrid glycoproteins build up in vacuoles in the cytoplasm and physically retard the cell function (Broquist 1985).

Cytoplasmic vacuolization due to locoweed poisoning has been observed in many organs and systems throughout the body (Van Kampen and James 1970). Van Kampen et al. (1978) discussed the correlation between clinical signs and pathological changes. Neurological damages (incoordination, impaired motor function, and prehension) are due to neuronal and axonal lesions in the central nervous system. Emaciation results from reduced assimilation and metabolism of food due to lesions in the liver, parathyroid, pancreas, and thyroid and to degeneration of neurons in the autonomic ganglia in the gut, resulting in reduced intestinal motility. Impaired vision and the "dull" eye results from cytoplasmic vacuolation in ganglion cells of the retina. Reproductive problems (abortion, fetal malformation, and small size at birth) result from lesions in the fetus, placenta, and ovaries. Cytoplasmic vacuolations also affect normal hormone levels which suppress estrus in the female, reduce libido in the male, and decrease spermatogenesis (Panter and James 1985).

Neuronal changes were observed in ewes fed a ration consisting of 20 to 25% locoweed after eight days; damage occurred after 16 days (Van Kampen and James 1970). However, Van Kampen et al. (1978) later stated that lesions in the parenchymal organs can be repaired without permanent damage if the animals are taken off locoweed before the onset of clinical signs. Clinical signs appeared in Shetland ponies fed 0.34 kg locoweed per day after 21 days (James and Van Kampen 1971). Marsh (1909) stated that cattle and horses became intoxicated after grazing locoweed for 30 to 60 days.

Lesions in most of the body organs disappeared after locoweed was removed from the diet (James and Van Kampen 1971). However, vauoles persisted in neurons in the central nervous system after one month and in Purkinje cells in the cerebellum (responsible for motor control) for as long as a year after intoxication. There was also a definite loss of Purkinje cells (more pronounced in horses than in cattle) which would result in a permanent partial loss of motor control.

Conditions of Poisoning

Locoweed populations are very cyclic and so are the incidence of losses. Sheep losses to Green River milkvetch (A. pubentissimus) in eastern Utah and southwestern Wyoming were especially severe in 1957-1958 and 1965-1966, as were losses to spotted locoweed (A. lentiginosus) in western Utah and parts of Nevada in 1964-1965 (James et al. 1968). Cattle losses to wooly locoweed (A. mollissimus var. earlei) were also severe in the Big Bend Country of Texas in 1976 (Lackey 1976). These semidesert locoweed species germinated in the fall, remained green over winter, and resumed rapid growth early in the spring. During this period, other forage was dormant and dry and so livestock were enticed to graze locoweed. Marsh (1909) also stated that cattle and horses grazed white locoweed (Oxytropis sericea) in the early spring when other feed was scarce and continued grazing it even when other feed became abundant.

Ralphs (1987) found the immature seed pod to be the only part of white locoweed that was palatable and actively selected by cattle. However, under abnormal environmental conditions (hard freeze in early July which caused all grasses to senesce while white locoweed remained green), cows began consuming the locoweed leaves and continued to do so throughout the grazing season. Three out of five calves on the study developed symptoms of locoism/brisket disease intoxication (Ralphs, unpublished data).

Grazing management may force livestock to consume locoweed. Ralphs et al. (1984) reported that cattle were forced to uniformly graze all forage, including locoweed, in the heavy use pasture of a rest rotation grazing system. They continued eating locoweed even when moved to a new pasture.

The dry senescent stalks of locoweed also present a threat to livestock. Cattlemen in the Henry Mountain area of Utah are more concerned with poisoning from dry spotted locoweed stalks in the fall and winter than they are of cattle consuming the green locoweed in spring (M.H. Ralphs, personal observation). There have been several reports throughout the West of cattle starting to graze dry locoweed stalks and then moving to graze the new green leaves as they start growth (L.F., personal communication).

NITRO-TOXINS

Description and Signs of Poisoning

Nitro-containing species of Astragalus are indigenous throughout the temperate and subalpine regions of the Northern Hemisphere and from Ecuador southward to Argentina in South America (Williams 1981). More than half of the species of Astragalus in North America synthesize nitro compounds (Williams and Barneby 1977).

The toxic compounds in nitro-containing species of Astragalus are principally miserotoxin, karakin, cibarian, and coronarian (Stermitz and Yost 1978). Miserotoxin, the β-D-glucoside of 3-nitro-1-propanol, is catabolized to 3-nitro-1-propanol (3-NPOH) by microflora in the rumen (Williams et al. 1970). The other nitro compounds catabolize to 3-nitropropionic acid (3-NPA). The 3-NPOH is more rapidly absorbed than 3-NPA from the digestive tract into the circulatory system. When administered orally, at least five to six times more 3-NPA than 3-NPOH (calculated as mg NO_2/kg of body weight) is required to produce the same degree of toxicity in birds and mammals (Williams and James 1978). Virtually all nitro intoxication of livestock on the range by nitro-containing species of Astragalus is caused by species which contain miserotoxin.

Under experimental conditions, cattle and sheep gavaged with Wasatch milkvetch (A. miser var. oblongifolius) have been acutely poisoned with doses of 25 and 37.5 mg NO_2/kg of body weight, respectively (Williams and James 1978). The approximate lethal dose of Wasatch milkvetch that contains 10 mg NO_2/g of plant is 625 g (dry wt.) for a 250 kg cow and 175 grams (dry wt.) for a 46 kg sheep.

Toxic signs include depression, drooling, rapid heartbeat, muscular incoordination, grinding of teeth, frothing at the mouth, frequent urination, and complete prostration. Death occurs in acute cases from respiratory failure in four to five hours.

The disease in livestock associated with chronic intoxication by nitro-containing species of Astragalus has been called roaring disease, knocking disease, and cracker heels (James et al. 1980a). One of the first signs of intoxication is labored and rapid breathing which, as the poisoning becomes more severe, produces a wheezing or roaring sound. Other signs of chronic intoxication include depression, general body weakness, incoordination in the hindquarters, and knuckling of the fetlocks. Knocking disease and cracker heels refer to sounds made by the rear hooves as they strike or rub together due to loss of motor control in the hindquarters. The animal drools saliva, the coat becomes rough, and constipation, diarrhea, and frequent urination are observed. In chronic cases, animals in advanced stages may linger for weeks or months before death. If forced to move rapidly, the animal may collapse and die. Cattle, sheep, and goats are most susceptible to nitro poisoning. Most losses, however, occur among cattle.

Methemoglobin is formed in the blood when nitro-containing plants are eaten. Methemoglobin levels may reach 30% or more within four to five hours after a lethal dose of Astragalus is ingested (James et al. 1980a). Animals rarely survive if methemoglobin levels exceed 20%.

Gross pathological changes in cattle poisoned in the field or under experimental conditions include swelling and congestion of the liver, pulmonary emphysema and pneumonia, excessive amounts of pericardial fluid, ulceration of the cardiac region of the abomasum, and excessive amounts of cerebro-spinal fluid (James et al. 1980a). Histopathological changes include dilated lateral ventricles and numerous focal hemorrhages of the brain, focal malacic lesions in the thalamus, mild wallerian degeneration in the spinal cord, and mild to marked alveolar emphysema, constricted bronchioles, and interlobular edema or fibrosis in the lungs.

Although many species of North American Astragalus contain miserotoxin, losses in the field are caused primarily by Wasatch milkvetch (Williams and Norris 1969), Emory milkvetch (A. emoryanus var. emoryanus) (Williams et al. 1979), Columbia milkvetch (A. miser var. serotinus) Yellowstone milkvetch (A. miser var. hylophilus) and Kelsey milkvetch (A. atropubescens) (Cronin et al. 1981).

Forty head of cattle died overnight after grazing a meadow infested with Wasatch milkvetch that contained 10 to 12 mg NO_2/g of plant as miserotoxin. Widespread chronic poisoning and some death losses were reported among cattle and sheep that grazed range infested with Emory milkvetch near Roswell, New Mexico. The plant assayed 5 to 9 mg NO_2/g of plant as miserotoxin (Williams et al. 1979).

Conditions of Poisoning

Nitro compounds are found primarily in the green leaves of nitro-containing species of <u>Astragalus</u>. The miserotoxin concentration in Wasatch milkvetch averages 9 to 13 mg NO_2/g of plant and reaches its highest concentration during the immature pod stage of growth (Williams and James 1978). The plant is highly toxic from emergence through seed production. Nitro compounds are associated with green color in the leaves, and once leaves begin to desiccate and bleach, the concentration of miserotoxin drops rapidly to 1 mg of NO_2 or less. Poisoning in the field occurs most often when plants are in the prebloom to bloom stage of growth. Plants are apparently not distasteful to livestock, which will graze them even if adequate desirable forage is available.

Canada milkvetch (<u>A</u>. <u>canadensis</u>) contains 40 to 60 mg NO_2/g of plant as 3-NPA and is the only 3-NAP-containing species indigenous to North America reported to poison livestock (Williams and James 1975). A closely related introduced species, sicklepod milkvetch (<u>A</u>. <u>falcatus</u>), contains up to 84 mg NO_2/g of plant and has acutely poisoned cattle under experimental conditions (Williams et al. 1976; James et al. 1980a).

Livestock, particularly cattle, sheep, and goats, should not be allowed to graze range infested with species of <u>Astragalus</u> that contain miserotoxin. Animals removed from areas as soon as signs of nitro intoxication appear may recover, but if chronic poisoning persists, vital organs are irreversibly damaged and the animals will never recover.

SELENIUM

Description and Signs of Poisoning

Selenium is one of the few elements that can be absorbed in toxic concentrations by plants from soils. The element is widely distributed in trace amounts in the earth's crust, but the highest concentrations of selenium are found in the rocks of certain geologic formations and in soils derived from them. Sedimentary rocks such as shales are highest in selenium. Lesser concentrations of selenium are found in limestones, sandstones, and phosphorite rocks. Nearly 300 ppm selenium has been reported in black shales of the Permian age from Wyoming (Rosenfeld and Beath 1964).

Approximately 24 species of North American <u>Astragalus</u> are known to accumulate selenium in concentrations toxic to livestock (Rosenfeld and Beath 1964, Barneby 1964). These species thrive only on soils that contain selenium, and since their presence indicates a seleniferous soil, they have been called "indicator plants." While these plants tend to be more vigorous when selenium is available, a definite requirement for the element has not been proven. These plants can, however, accumulate several thousand ppm selenium in their foliage when rooted on soils that contain only a few ppm selenium. The selenium in accumulator plants is volatile and emits an odor resembling rotten eggs. Much of the selenium in accumulator species of <u>Astragalus</u> is found as selenium-methylselenocysteine and selenocystathionine (Mayland 1985).

Acute selenium poisoning occurs when animals eat a sufficient quantity of highly seleniferous <u>Astragalus</u> to cause severe toxic signs. Death may follow in a few hours. The minimum lethal dose of selenium (administered orally as selenite) is 3 mg/kg of body weight for cattle and 3.3 mg/kg for horses and mules (Rosenfeld and Beath 1964, Miller and Williams 1940). A single subcutaneous dose of 2 mg selenium/kg of body weight (as selenite) was lethal to swine in four hours (Orstadius 1960). Plant material that contained 400 to 800 ppm selenium was fatal to sheep when fed in amounts ranging from 8 to 16 g/kg of body weight.

Affected animals stand with head lowered and ears drooping. The temperature is elevated to 39.4 to 40.6 C, and the pulse is rapid and weak (Rosenfeld and Beath 1964). Urine excretion is increased. The mucous membranes are pale and discolored. The pupils are dilated. Death may occur in a few hours to a few days and is preceded by collapse and prostration. Death is caused by respiratory failure.

Selenium enters the circulatory system and accumulates primarily in the blood, liver, kidney, spleen, and brain. Virtually every organ of the body is affected and congestion and hemorrhage are evident in the liver, lungs, omasum, pancreas, gall bladder, spleen, and lymph nodes. No effective treatment is available for acute selenium poisoning, primarily because death occurs quickly before the disease can be diagnosed.

Chronic selenium poisoning occurs when animals eat sublethal amounts of seleniferous plants for extended periods. Toxic signs include loss of the long hair of the body, increased hoof growth, lameness, and emaciation. There can be a marked effect on the reproductive function of the animal. Toxic signs may not appear for several weeks or months, depending upon the amount of plant consumed. The animal may eventually become paralyzed, and death soon follows.

Steers fed comb milkvetch (A. pectinatus) in daily doses of 2.2 mg selenium/kg of body weight developed chronic poisoning after 6 to 16 doses (Rosenfeld and Beath 1964). A steer fed A. bisulcatus developed chronic poisoning after seven doses, each supplying 3.4 mg of selenium/kg of body weight.

Conditions of Poisoning

The selenium content of accumulator species of Astragalus is influenced by climatic conditions, selenium content of the soil, and stage of growth (Rosenfeld and Beath 1964). Selenium concentration is high during seasons of high rainfall since plants grow more vigorously and absorb greater concentrations of selenium with adequate moisture. Rainfall in late summer or early fall often promotes regrowth, and green succulent shoots are particularly high in selenium. The selenium content is very high in soils derived from rocks of certain geologic formations, and Astragalus and other accumulator species become very toxic when rooted in them. Seleniferous soils are found from the Great Plains westward to the Pacific coast, but the major seleniferous areas are found in North and South Dakota, Montana, Wyoming, Utah, and Colorado (James et al. 1980b).

Seleniferous species of Astragalus are highest in selenium in the prebloom stage of growth (Rosenfeld and Beath 1964). Selenium levels begin to decrease during full bloom and may drop as much as 90% after the seed stage. Two-grooved milkvetch (A. bisulcatus) has been found to contain 8,840 ppm selenium in the prebloom stage of growth, 6,590 ppm at full bloom, and 595 ppm after seeding. Seeds often contain 3,000 to 4,000 ppm selenium.

Highly seleniferous plants are unpalatable, and livestock usually avoid plants that contain enough selenium to give an offensive odor. Livestock losses from seleniferous plants can be reduced by preventing livestock from grazing ranges infested with accumulator species. Affected animals that are removed from seleniferous areas as soon as toxic signs appear may recover. The rate of recovery is dependent upon the degree of intoxication (James et al. 1980b).

MANAGEMENT RECOMMENDATIONS

Ingestion of species of Astragalus and Oxytropis by livestock produces highly variable poisoning responses. There are many species within these genera that vary greatly in amount of toxin in the plant material, and there are at least three major poisonous principles involved. In addition, seasonal and annual plant populations of many species vary greatly.

The overall variability appears overwhelming, but the poisoning problem on a specific ranch may be caused by a single or a few species growing on very specific range sites and with a specific toxin most likely to be a problem at a specific season. Management recommendations, then, are to first identify the specific problem for the management unit involved. One should first identify which species of Astragalus or Oxytropis are present on the management area. This may require technical assistance from a taxonomist familiar with these genera. The species should then be tested to determine what toxin, if any, is present. If the species is toxic, then range sites on which the species grow should be designated so that the area distribution of the species in question is known. The plant population within the range sites can then be monitored on a seasonal or annual basis to determine if plant density increases, which may be a warning of potential poisoning problems. Also, monitor use of poisonous species by livestock on a regular basis. Most poisoning can be averted if animals are removed from the areas where they have access to the poisonous species when the plants are palatable to them and the toxin concentration is high.

Finally, develop a plan and be prepared to rotate animals to less dangerous areas if hazardous conditions are identified.

LITERATURE CITED

Barneby, R.C. 1952. A revision of the North American species of Oxytropis DC. Proc. California Acad. Sci. IV. 27:177-312.

Barneby, R.C. 1964. Atlas of North American Astragalus. Memoirs of the New York Botanical Garden, Bronx, NY, Vol. 13.

Bentham, G., and J.D. Hooker. 1872. Genera Plantarum. 1(1):i-xii, 1-1040.

Broquist, H.P. 1985. The indolizidine alkaloids, slaframine and swainsonine: contaminants in animal forages. Ann. Rev. Nutr. 5:391-409.

Bunge, A. 1874. Species Generis Oxytropis, DC. Mem. Acad. Imper, Sci. St.-Petersb. VII. 22:1-166.

Cronin, E.H., M.C. Williams, and J.D. Olsen. 1981. Toxicity and control of Kelsey milkvetch. J. Range Manage. 34:181-183.

James, L.F., K.L. Bennett, K.G. Parker, R.F. Keeler, W. Binns, and B. Lindsay. 1968. Loco plant poisoning in sheep. J. Range Manage. 21:360-365.

James, L.F., W.J. Hartley, M.C. Williams, and K.R. Van Kampen. 1980a. Field and experimental studies in cattle and sheep poisoned by nitro-bearing Astragalus or their toxins. Amer. J. Vet. Res. 41:377-382.

James, L.F., W.J. Hartley, and K.R. Van Kampen. 1981. Syndromes of Astragalus poisoning in livestock. J. Amer. Vet. Med. Assoc. 178:146-152.

James, L.F., W.J. Hartley, K.R. Van Kampen, and D. Nielsen. 1983. Relationship between ingestion of the locoweed Oxytropis sericea and congestive right-sided heart failure in cattle. Amer. J. Vet. Res. 44:254- 259.

James, L.F., R.F. Keeler, A.E. Johnson, M.C. Williams, E.H. Cronin, and J.D. Olsen. 1980b. Plants poisonous to livestock in the western states. USDA Agr. Infor. Bull. 415.

James, L.F., and K.R. Van Kampen. 1971. Acute and residual lesions of locoweed poisoning in cattle and horses. J. Amer. Vet. Med. Assoc. 158:614-618.

Komarov, V.L., (ed). 1946. Leguminosae: Astragalus. Flora USSR 12:i-xxviii, 1-681.

Komarov, V.L., (ed). 1948. Leguminosae: Oxytropis, Hedysarum. Flora of the USSR. 13:i-xxiii, 1-455.

Lackey, J. 1976. Locoweed in Big Bend. Cattleman 63(4):62-64.

Marsh, C.D. 1909. The loco-weed disease of the plains. USDA Bur. Animal Ind. Bull. 112. p. 130.

Mayland, H.F. 1985. Selenium in soils and plants, pp. 5-10. In: Selenium responsive diseases in food animals. Proc. of symposium, Western States Veterinary Conf., Las Vegas, NV.

Miller, W.T., and K.T. Williams. 1940. Minimum lethal dose of selenium as sodium selenite for horses, mules, cattle and swine. J. Agr. Res. 60:163-173.

Molyneux, R.J., and L.F. James. 1982. Loco intoxication: indolizidine alkaloids of spotted locoweed (Astragalus lentiginosus). Science 216:190-191.

Molyneux, R.J., L.F. James, and K.E. Panter. 1985. Chemistry of toxic constituents of locoweed (Astragalus and Oxytropis) species, pp. 266-278. In: A.A. Seawright, M.P. Hegarty, L.F. James, and R.F. Keeler (eds), Plant Toxicology. Proc. Aust-U.S.A. Poisonous Plants Symp., Brisbane. Queensland Poisonous Plant Comm., Ani. Res. Inst., Yeerongpilly.

Orstadius, K. 1960. Toxicity of a single subcutaneous dose of sodium selenite in pigs. Nature 188:1117.

Panter, K.E., and L.F. James. 1985. Effects of locoweed on reproduction in livestock, pp. 349-353. In: A.A. Seawright, M.P. Hegarty, L.F. James, and R.F. Keeler (eds), Plant Toxicology. Proc. Aust.-U.S.A. Poisonous Plants Symp., Brisbane. Queensland Poisonous Plant Comm., Ani. Res. Inst., Yeerongpilly.

Polunin, N. 1960. Introduction to plant geography. McGraw-Hill Book Co., Inc.

Ralphs, M.H. 1987. Cattle grazing white locoweed: influence of grazing pressure and palatability associated with phenological growth state. J. Range Manage. (In press.)

Ralphs, M.H., L.F. James, D.B. Nielsen, and K.E. Panter. 1984. Management practices reduce cattle loss to locoweed on high mountain range. Rangelands 6:175-177.

Rosenfeld, I., and O.A. Beath. 1964. Selenium. Academic Press, New York.

Stermitz, F.R., and G.S. Yost. 1978. Analysis and characterization of nitro compounds from Astragalus species, pp. 371-378. In: R.F. Keeler, K.R. Van Kampen and L.F. James (eds), Effects of poisonous plants on livestock. Academic Press, New York.

Van Kampen, K.R., and L.F. James. 1970. Pathology of locoweed (Astragalus lentiginosus) poisoning in sheep. Path. Vet. 7:503-508.

Van Kampen, K.R., R.W. Rhees, and L.F. James. 1978. Locoweed poisoning in the United States, pp. 465-471. In: R.F. Keeler, K.R. Van Kampen, and L.F. James (eds), Effects of poisonous plants on livestock. Academic Press, New York.

Welsh, S.L. 1974. Anderson's flora of Alaska and adjacent parts of Canada. Brigham Young University Press, Provo, Utah.

Welsh, S.L., N.D. Atwood, S. Goodrich, and L.C. Higgins. 1987. A Utah flora. Mem. Great Basin Nat. 10:1-894.

Williams, M.C. 1981. Nitro compounds in foreign species of Astragalus. Weed Sci. 29:261-269.

Williams, M.C., and R.C. Barneby. 1977. The occurrence of nitro-toxins in North American Astragalus (Fabaceae). Brittonia 29:310-326.

Williams, M.C., and L.F. James. 1975. Toxicity of nitro-containing Astragalus to sheep and chicks. J. Range Manage. 28:260-263.

Williams, M.C., and L.F. James. 1978. Livestock poisoning from nitro-bearing Astragalus, pp. 379-389. In: R.F. Keeler, K.R. Van Kampen, and L.F. James (eds), Effects of poisonous plants on livestock. Academic Press, New York.

Williams, M.C., L.F. James, and A.T. Bleak. 1976. Toxicity of introduced nitro-containing Astragalus to sheep, cattle, and chicks. J. Range Manage. 29:30- 33.

Williams, M.C., L.F. James, and B.O. Bond. 1979. Emory milkvetch (Astragalus emoryanus var. emoryanus) poisoning in chicks, sheep, and cattle. Amer. J. Vet. Res. 40:403-406.

Williams, M.C., and F.A. Norris. 1969. Distribution of miserotoxin in varieties of Astragalus miser Dougl. ex Hook. Weed Sci. 17:236-238.

Williams, M.C., F.A. Norris, and K.R. Van Kampen. 1970. Metabolism of miserotoxin to 3-nitro-1-propanol in bovine and ovine ruminal fluids. Amer. J. Vet. Res. 31:259-262.

14

Locoweeds: Assessment of the Problem on Western U.S. Rangelands

Lynn F. James and Darwin B. Nielsen

INTRODUCTION

As many as 300 species of Astragalus may grow in North America, making the genus the largest of any of the legume family in this part of the world (Kingsbury 1964).

The toxic species of Astragalus cause three different syndromes of poisoning according to the toxin they contain (James et al. 1981). These three groups are: (1) nitro-containing Astragalus, acute and chronic poisoning; (2) selenium-accumulating Astragalus, acute and chronic poisoning; and (3) swainsonine-containing Astragalus and Oxytropis (locoweeds), chronic poisoning.

NITRO-CONTAINING ASTRAGALUS

These species of Astragalus contain miserotoxin, the β-D-glycoside of 3-nitro-1-propanol or the less toxic 3-nitropropionic acid (Williams and Barnaby 1977). Species of this group of Astragalus can be found on ranges from western Canada throughout the western United States and into northern Mexico (James et al. 1981). Examples of these species include A. emoryanus (red-stemmed pea vine) in New Mexico, Texas, Arizona, and northern Mexico; A. tetrapterus in Utah, Arizona, and Nevada; A. hylophilus in Utah, Colorado, Idaho, Montana, and Wyoming; A. canadensis in most of the western states; and three varieties of A. miser--serotinus in western Canada, hylophilus in southwestern Montana and northwestern Wyoming, and oblongifolius along the Wasatch Plateau in Utah. Williams and Barnaby (1977) have listed the nitro-containing plants in detail.

Cattle, sheep, and horses are susceptible to poisoning by Astragalus plants containing these toxins (Williams and James 1976). The signs of poisoning in cattle include general body weakness, knuckling at the fetlocks, interference between the hind limbs (cracker heels) as the animal moves, and respiratory distress (roaring disease) (James et al. 1980). Microscopic lesions of the spinal cord are responsible for the problems of mobility. Respiratory distress results from pulmonary alveolar emphysema and bronchiolar constriction. As the intoxication proceeds, the animal may collapse on exertion. If grazing of these plants continues, the animal wastes and dies. Animals may make a short-term, apparent recovery after removal from grazing, but then waste and die.

Field observations indicate that intoxication occurs in most areas where these plants grow. The total extent of the problem is not known, as many livestock producers often do not associate intoxication with these plants. Some animals die; others waste and become of low market value or are unmarketable. There may be considerable weight losses in mature livestock, and a reduction in offspring weight due to a decrease in milk flow.

In one case of intoxication by a member in this group of plants (Astragalus emoryanus), 500 of 1,500 cows were visibly ill (Williams et al. 1979). Several other groups of cattle and sheep were likewise ill in the same general area. Although there were no death losses that could be attributed to these plants, there was a loss of weight and subsequent decrease in market values, increased labor costs, alteration of grazing programs, increased feed costs, and increased veterinary fees.

Economic losses attributable to these plants vary from year to year because the growth of the plant varies with moisture and other environmental conditions. At present, there are too many unknowns to make an intelligent estimate of the economic impact on livestock production. However, the large area over which the plants grow and knowledge of the effects of this group of plants on livestock indicate that the overall economic effect may reach several million dollars.

SELENIUM-ACCUMULATING PLANTS

Selenium, an element found in some soils, may be taken up by many plants in sufficient quantities to render them toxic. Among the selenium-accumulating plants, the

Astragalus species shows the greatest diversity of form and the greatest geographic distribution (Rosenfeld and Beath 1964). About 24 species of Astragalus are known to accumulate selenium. Some of the more notable are A. bisulcatus, A. pattersoni, and A. praelongus. These plants may accumulate selenium from 50 to 100 to several thousand parts per million (Rosenfeld and Beath 1964). These plants are referred to as indicator plants because they usually grow in the more seleniferous soils and have been used to identify areas where soils are high in selenium.

Acute selenium poisoning from Astragalus does not occur frequently, but losses of several hundred sheep in a single incident have been recorded (James et al. 1982). However, the grazing of seleniferous forage of lower selenium content (5 to 10 ppm) by livestock can result in appreciable economic loss to a livestock enterprise because of decreased reproduction (Olsen 1978). At the present time, the effect of selenium on reproduction in sheep and cattle is not understood. There are no statistics on the acute losses from seleniferous Astragalus poisoning. Therefore, there is no basis on which to base an estimate on the economic impact of seleniferous Astragalus on livestock production.

CHRONIC LOCOWEED INTOXICATION

According to Marsh (1929), the locoweeds are the most destructive of the poisonous plants. Loco is a term derived from the Spanish word for crazy. The term "locoed" is properly applied only in relation to the poisoning that occurs when livestock ingest plants of this group. The following is a partial list of Astragalus and Oxytropis that are classed as locoweeds: Astragalus lentiginosus, A. mollissimus, A. wootonii, A. thurberi, A. nothoxys, Oxytropis sericea, and O. lambertii. The Astragalus locoweeds grow only under appropriate moisture and other environmental conditions. Thus their growth on arid ranges is irregular and unpredictable.

The toxic principle is an indolizidine alkaloid called swainsonine. The toxin inhibits the enzyme α-mannosidase which functions in the processing of glycoproteins (Molyneux and James 1982).

Locoweed poisoning is chronic, and livestock must graze the plant for long periods of time before outward signs of poisoning are obvious. The locoweeds cause many different effects in livestock poisoned by them. These include (1) neurological effects, (2) emaciation, (3) habituation, (4) reproductive effects, and (5) cardiopulmonary effects (James et al. 1981). Signs of poisoning include depression, problems with proprioception, rough hair coat, dull staring eyes, emaciation, and nervousness when stressed. If stressed, an animal poisoned on locoweeds may show signs of poisoning after being removed from the plant. Draft horses and riding horses are of no value except as breeding animals once they have been poisoned on locoweed (Marsh 1909, Mathews 1932, James et al. 1967).

Animals that graze on locoweed may appear to thrive for a short period of time, then lose condition and become emaciated. They may then become recumbent and die.

Much has been said regarding the habituating effects of locoweed (James et al. 1981). Tradition and the accounts of many livestock producers would indicate that habituation may be a factor in locoweed poisoning, but current research has cast some doubt on this aspect of locoweed intoxication (Ralphs 1987). If it is true that habituation is an aspect of locoweed, then animals that have grazed locoweed are at greater risk to future exposure to these plants. There may also be increased risk to associated animals. It is thought by some that an animal that is grazing locoweed may teach others to eat it.

The reproductive effects include abortion, skeletal birth defects, birth of small and weak offspring, decreased sexual activity in the male and female, decreased spermatogenesis and oogenesis, and loss of libido. There can be a delay in placentation. In addition, some pregnant cows may develop hydrops amnii, which is the accumulation of fluid in the uterus. Abortion in cattle and sheep may approach 100% under conditions where locoweed is grazed heavily, but the rate of abortion is highly variable.

Locoweed enhances congestive right-sided heart failure (high mountain disease) when grazed at high elevations (James et al. 1983). Field and experimental observations suggest that locoweed may increase the incidence of congestive heart failure from about 5% to nearly 100%.

In evaluating the economic impact of locoweed on livestock production, the many variables discussed in the intoxication of livestock, plus factors associated with management and the plant's growth, must be considered. Costs associated with potential outbreaks of locoweed poisoning can be high. These costs include increased labor, which may entail the total removal of livestock from a range, herding, fencing, feed, and weed control programs.

The economic losses due to the presence of locoweed on rangelands are very difficult to measure. Death loss caused by locoweed consumption is probably not as important as the other economic losses caused by the plant. In some areas, locoweed problems occur every year, while in other locations they sporadically cause problems.

Three case studies will be used to illustrate the various economic impacts that locoweed can have on a ranch. These ranches are located in Wyoming, New Mexico, and Utah (D.B. Nielsen, unpublished data).

Wyoming Case Study

This Wyoming ranch has 500 cow-calf pairs and feeds hay during part of the winter. The rancher reported that because of locoweed he was faced with selling out, leasing his ranch, or reducing the locoweed problem.

The following problems were reported as being associated with locoweed poisoning in the cattle: 250 calves per year were affected by locoweed poisoning. These calves weighed an average of 22.5 kg less than other calves in the herd. The locoed calves would regain the weight, but it would take an additional two months on a higher quality feed to reach the weight of normal calves. The rancher estimated that this loss amounted to $15,000 per year ($60 per calf). Additionally, 20 to 50 calves would be held over the winter because of extreme intoxication from which a few died. Cows poisoned on locoweeds were difficult to manage, and their reproductive efficiency was reduced. Many poisoned cows aborted their calves. Others seemed to lose their mothering instinct. Over 5% of the cows (excluding 2-year-olds) were open at the end of the breeding season; many of these were culled. This required a higher number of replacement heifers. The normal breeding season was extended, resulting in 60 to 75 calves born after the normal calving period. Cows that were poisoned as calves would immediately graze locoweed when given the opportunity, even after as much as four years with no access to the plant. About 5% of newborn calves had malformations of various types and severity, of which 50% usually died. Neighboring ranches reported intoxicated bulls; however, this ranch did not. Labor and management were intensified to observe cattle for signs of locoweed poisoning and to move intoxicated cattle (2 to 4%) to locoweed-free meadows thus reducing hay production. Two to three percent of the cows had foot rot problems probably resulting from a depressed immune response (Sharma et al. 1984).

In the mid-1970s, the ranch was sprayed fence-line to fence-line with 2,4-D to control the locoweed. The following year the number of open cows was reduced to less than 0.5%. Less than 3% of the cows calved beyond the normal calving period; there were no malformed calves born, the number of replacement heifers was reduced by 30%, average weaning weight of the calves was higher, and no foot rot problems were observed. No cows had to be put in the hay meadows, which increased yields and lessened the management problems. Spraying resulted in control of not only the locoweed but other nonpalatable brush species, increasing the carrying capacity of the ranch by 33% which allowed an extended grazing season. This increase in grazing resulted in 500 to 600 tons of hay being leftover in the spring of 1978. Marketing of this hay would increase the benefit from spraying.

New Mexico Case Study

This New Mexico ranch consists of 80,940 ha. Cattle are grazed 12 months of the year with limited winter supplemental feeding when necessary. There is some locoweed (Oxytropis sericea) throughout the entire ranch. About 4,047 ha were sprayed for locoweed control. Other positive vegetative responses did not appear to occur on the sprayed range; thus, the only benefits were from reduced cattle poisoning. It was determined to be not economically feasible to spray the entire ranch.

From 1964 to 1972, no locoweed poisoning problems were reported. Between 1972 and 1978 the cow herd was reduced from 2,500 cows to 1,400 cows. A reduction in the calf crop from 89% in 1972 to 81% in 1978 was observed.

Reproductive problems were observed in cows exposed to locoweed. In 1970, 14% of the cows were open at the end of the breeding season. In 1975, only 41% of a group of 106 locoweed-intoxicated cows produced a calf. This same year 75% of these cows missed at lease one estrous cycle resulting in lengthened calving intervals.

Five to eight percent more replacement heifers were kept or purchased to replace intoxicated cows. Heifers from intoxicated cows were not kept as replacements. Replacement heifers poisoned between weaning and first calving are subsequently sold. Thus the number of heifers saved as replacements was influenced more by locoweed problems than the number needed to maintain the cow herd.

Estimated death loss on this ranch was 8% compared to 1.5% estimated death loss on their other ranches without locoweed. Based on cow numbers of 1,450 at a cost of $500 per head, the death loss equaled about $47,000 per year.

There were extra management costs ($12.50 per locoed cow) incurred from supplemental feeding and transportation to move locoed cows, heifers, and calves to other locations. Labor costs are estimated to be 30% higher because of locoweed.

Cows that were culled because of locoweed poisoning usually are lighter in weight and bring a lower price. Normal cull cow weights on this ranch average about 450 kg per head; one group of locoed cows averaged 350 kg per head, while another group averaged 272 kg per head.

There are many similarities between the Wyoming and New Mexico cases. However, the Wyoming case demonstrated an economic success because of the large positive forage response after spraying, whereas in the New Mexico case it was determined from the smaller acreage sprayed that it was not economically feasible to spray the entire range.

Utah Case Studies

In northern Utah, three ranchers graze cattle on high mountain summer range (3090 m) that is infested with locoweed (Oxytropis sericea). Losses in cows and calves were occurring from death, lower productivity, and increased management costs. In the early 1980s, a study was conducted to assess the economic impact on this range due to locoweed (Barnard 1983). At the time of this study, the grazing program consisted of 394 cow-calf pairs grazed on these pastures from July 1 until about September 10.

Weaning weights, culling practices, death losses, calving percentages, and management practices were observed, recorded, and a dollar value assigned. Weaning weights from calves on the high mountain range averaged 9.1 kg lighter than calves of the same age and genetic background on ranges without locoweed. This resulted in a loss of $14 per head. Cows culled because of locoweed poisoning were more numerous and averaged 150 kg less than nonpoisoned cows, resulting in a $115 loss per cow sold because of locoweed poisoning. Death losses due to locoweed poisoning, averaged 15 calves and 1.5 cows per year over a two-year period, resulting in a monetary loss of $4,182 per year. In 1983, it was shown that locoweed predisposes calves to high mountain disease resulting in a higher death loss in calves (James et al.

1983). These ranchers estimated a reduced calf crop of 15% which resulted in 59 fewer calves from this herd, a loss of $15,000 per year. Additionally, the calving period was 162 days long, which increased management costs. It is difficult to estimate management costs; however, one can estimate the time necessary to check cattle on this locoweed-infested range and to drive sick cows or cows with sick calves, many of which died due to the stress of moving or hauling them off the high mountain locoweed pastures to lower locoweed-free range.

The economic analysis showed that for each of the 394 cows grazed on this range, $75 per cow ($29,550 total) was lost due to locoweed. Since this study was completed, one pasture was sprayed to reduce the locoweed, and a new grazing program was implemented. More cattle are grazed on the pasture, but the grazing season is reduced; thus the AUMs remain the same, but the grazing pressure is reduced (Ralphs et al. 1984) resulting in a substantial reduction in losses and concomitant increase in net income.

Other cases have been reported in Utah from locoweed poisoning. In 1958, 6,000 sheep died from locoweed (Astragalus pubentissimus) in eastern Utah; in addition, there was a marked decrease in lamb crop from the surviving ewes. Some of these sheep operations went out of business because of the loss. In 1964, a rancher in this same area estimated his loss at $125,000 due to grazing locoweed, another operator estimated a $45,000 loss in dead sheep and decreased lamb crop. In southern Utah in the early 1970s, a big game outfitter lost his entire string of horses to locoweed (Astragalus lentiginosus) and a neighboring rancher sold his herd of cattle of which two-thirds had aborted.

Most locoweeds are biennials that flourish periodically under appropriate environmental conditions. Thus, periodic regional problems usually occur, causing heavy livestock losses.

OTHER COSTS DUE TO LOCOWEED

Other losses due to locoweed may include emotional or sentimental losses, such as the loss of the family ranch, a favorite horse, or a child's pony. It is not possible to attach a dollar value to these losses.

Management time and effort expended to prevent locoweed losses, as reported in the case studies, can be a substantial cost to a rancher. If these ranchers could devote their time to other ranch management problems, the resulting gains might be very high. Thus, there is an opportunity cost of management because one has a locoweed problem. The cost of hiring additional labor to remove locoweed-poisoned animals, grubbing locoweed from grazing areas, fixing fences, rotating grazing animals, and other tasks caused by locoweeds can be estimated from ranch interviews. However, no attempt has been made to estimate the impact that released management time would have on ranch income if the locoweed problem did not exist.

CONCLUSION

The scenarios as described in these cases are known to occur in other areas of the western United States. The list of examples could go on and on. These examples have pointed out dramatic losses to individual livestock operations and when the entire western U.S. is considered the loss is monumental. It is doubtful that any study encompassing anything less than the entire western U.S. over a period of 50 years could give a reliable estimate of the true economic impact of locoweed on the livestock industry. At present we are left with Marsh's generality of 1929 in which he states that locoweed is one of the most destructive of the poisonous plants.

LITERATURE CITED

Barnard, John E. 1983. Locoweed poisoning in cattle: an overview of the economic problems associated with grazing these ranges. M.S. Thesis, Utah State University, Logan.

James, L.F., W.J. Hartley, and K.R. Van Kampen. 1981. Syndromes of Astragalus poisoning in livestock. J. Amer. Vet. Med. Assoc. 178:146-150.

James, L.F., W.J. Hartley, K.R. Van Kampen, and D.B. Nielsen. 1983. Relationship between ingestion of the locoweed Oxytropis sericea and congestive right-sided heart failure in cattle. Amer. J. Vet. Res. 44:254-259.

James, L.F., W.J. Hartley, M.C. Williams, and K.R. Van Kampen. 1980. Field and experimental studies on cattle and sheep poisoned by nitro-bearing Astragalus or their toxins. Amer. J. Vet. Res. 41:377-382.

James, L.F., J.L. Shupe, W. Binns, and R.F. Keeler. 1967. Abortion and teratogenic effects of locoweed on sheep and cattle. Amer. J. Vet. Res. 28:1379-1388.

James, L.F., R.A. Smart, J.L. Shupe, J. Bowns, and J. Schoenfeld. 1982. Suspected phototoxic selenium poisoning in sheep. J. Amer. Vet. Med. Assoc. 180:1478-1481.

Kingsbury, John M. 1964. Poisonous plants of the United States and Canada. Prentice-Hall, Inc., Englewood Cliffs, NJ.

Marsh, C.D. 1909. The loco-weed disease of the Plains. Bull. 112. USDA Bur. Animal Industry.

Marsh, C.D. 1929. Stock-poisoning plants of the range. USDA Bull. 1245. pp. 22-44.

Mathews, F.P. 1932. Locoism in domestic animals. Texas Agric. Expt. Sta. Bull. 456.

Molyneux, R.J., and L.F. James. 1982. Loco intoxication: indolizidine alkaloids of spotted locoweed (Astragalus lentiginosus). Science 216:190-121.

Olsen, O.E. 1978. Selenium in plants as a cause of livestock poisoning, pp. 121-133. In: R.F. Keeler, K.R. Van Kampen, and L.F. James (eds), Effects of poisonous plants on livestock. Academic Press, New York.

Ralphs, M.H., L.F. James, D.B. Nielsen, and K.E. Panter. 1984. Management practices reduce cattle loss to locoweed on high mountain range. Rangelands 6:175-177.

Rosenfeld, I., and O.A. Beath. 1964. Selenium, geobotany, biochemistry, toxicity and nutrition. Academic Press, Inc., New York.

Sharma, R.P., L.F. James, and R.J. Molyneux. 1984. Effect of repeated locoweed feeding on peripheral lymphate function and plasma protein in sheep. Amer. J. Vet Res. 45:2090-2093.

Williams, M.C., and R.C. Barnaby. 1977. The occurrence of nitro-toxin in North American Astragalus (Fabaceae). Brittonia 29:310-326.

Williams, M.C., and L.F. James. 1976. Poisoning in sheep from Emory milkvetch and nitro compounds. J. Range Manage. 29:165-167.

Williams, M.C., L.F. James, B.O. Bond. 1979. Emory milkvetch (Astragalus emoryanus var. emoryanus) poisoning in chicks, sheep, and cattle. Amer. J. Vet. Res. 40:403-406.

15

Ecology and Toxicology of
Senecio Species with Special Reference
to Senecio jacobaea and Senecio longilobus

S.H. Sharrow, D.N. Ueckert, and A.E. Johnson

Senecio is a large and varied genus. Estimates of the number of Senecio species worldwide range from approximately 1,000 (Hitchcock et al. 1969) to 3,000 (Correll and Johnston 1970). Confusion concerning the number of species is understandable, considering the relatively low degree of distinctness separating some species (Hitchcock et al. 1969) and the tendency of some Senecio species to freely hybridize. Morphologically, plants in this genus share three common characteristics: (1) the presence of a single, major whorl of nonimbricate bracts, sometimes subtended by a second whorl of smaller bracts, beneath the individual flower heads (Kingsbury 1964); (2) a tendency to produce seed from both disc and ray flowers, where present (Kingsbury 1964); and (3) the presence of a pappus of soft white bristles on disc flower achenes. The name "senecio" is believed to be derived from the Latin "senex" (old man), likely referring to the prominent white bristles comprising the pappus of many Senecios (Hitchcock et al. 1969). Senecio seed may be disseminated by wind, animals, water, or man. However, wind-dispersal of achenes is a function of wind velocity, air turbulence, humidity, aerodynamic properties of the achene-pappus unit, and drop height (Sheldon and Burrows 1973). Given the combination of these parameters which need to occur for long-distance travel of achenes, the effectiveness of this mechanism for the rapid spread of Senecio to distant areas is probably grossly exaggerated. Type species for the genus is Senecio vulgaris (Britton and Brown 1970).

As one might expect from such a large genus, Senecio species differ substantially in both their growth form and habitat requirements. For example, Muntz and Keck (1968) list 37 Senecio species occurring in California that varied

in growth habit from annual herbs to perennial shrubs and
vines. One or more species are present in almost all of
California's major vegetation types, from desert scrub to
alpine fell-fields. Senecio species are found on most U.S.
pasture and rangeland areas, where they are commonly
referred to as ragworts, groundsels, or butterweeds (Dayton
1937; Garrison et al. 1976).

TOXICITY

Although some Senecios are known to be toxic to humans
and livestock, toxicity does not appear to be a problem with
the vast majority of them. Of the 112 Senecio species
listed for North America (USDA 1982), only seven are known
to have caused poisoning in livestock (Kingsbury 1964): S.
jacobaea, S. longilobus, S. integerrimus, S. plattensis, S.
riddellii, S. vulgaris, and S. glabellus. Of these species,
S. jacobaea, S. longilobus, and S. riddellii, in that order,
account for the majority of reported cases of Senecio
poisoning. Some of the nontoxic species are important
forage plants for both native (Crawford et al. 1969; Kufeld
1973) and domesticated (Dayton 1937) large herbivores,
especially during times when other forage is not available.
Toxic Senecios generally are of low palatability to
livestock (Kingsbury 1964). This may be an important factor
limiting economic losses incurred in grazing lands where
Senecios are present. Senecio poisoning has occasionally
been a significant problem in human populations where food
supplies have been contaminated with Senecio flowers and
achenes inadvertently harvested with grain crops (Cheeke and
Shull 1985). Herbal teas and folk medicines are another
common cause of Senecio poisoning in humans. This problem
has apparently been exacerbated by use of the common name
"tansy" to refer to both Tanacetum vulgare, a nontoxic plant
often used in herbal teas, and S. jacobaea, which is highly
toxic (Cheeke and Shull 1985).

Although toxic Senecios are poisonous to most animals
and to man (McLean 1970), they derive their greatest
economic importance from the losses they cause to the cattle
industry. Cattle that eat toxic Senecios may be affected
either acutely or chronically. However, acute intoxications
seldom occur because the low palatability of Senecios
usually results in only small quantities of the plant being
consumed per day. Acute cases that do occur result in
massive generalized hemorrhage and acute liver toxicity that
is often fatal in one to two days (Johnson and Molyneux
1984).

Pyrrolizidine alkaloids (PA) are the toxic agents in Senecio species. Total PA content within a Senecio species can vary seasonally and among different locations by as much as five- to ten-fold (Johnson et al. 1985). In general, PA content of S. jacobaea varies most, while that of S. longilobus and S. riddellii is more constant. Studies of PA toxicity in animals should take this variation into account by expressing dosage rates as mg of toxic PA/kg body weight rather than as kg of plant eaten/kg live weight, as is commonly done. There are both toxic and nontoxic forms of PA, all of which have the same basic double, five-membered ring structure. Toxic PA are esters of unsaturated amino alcohols. They may be mono or diesters with a wide variety of branched chain aliphatic acids. Thus, there are numerous toxic PA, often with several closely related forms occurring in the same plant. For example, Senecio jacobaea has six or more PA, S. longilobus four, and S. riddellii two. Nonester PA, of which there are also many, are nontoxic and may occur in a variety of nontoxic Senecios.

PA capable of inducing toxicity occur in plants in two forms, the free base and the N-oxide (Mattocks 1971). Some Senecios contain principally free base, while others contain mostly N-oxide forms of PA (Johnson et al. 1985). Through studies in rats, both forms have been determined to be relatively nontoxic. However, when the free base is absorbed and reaches the liver, it is converted to highly reactive, toxic pyrroles by liver mixed function oxidases (Mattocks 1971). Since only free base is converted to toxic pyrroles, the proportion of free base to N-oxide in the plant, together with the animal's ability to convert N-oxide to free base before absorption, is an important consideration in establishing the toxicity of the various Senecio species.

Very little is known of the specific enzymes which convert free base to PA toxic pyrroles. Even though liver oxidases vary considerably among different animal species (Smith et al. 1984), it is assumed that they are the same mixed-function oxidases as those in rats. Mixed-function oxidases are susceptible to a variety of inducers and depleters, thus the amount of available reactive enzyme in a given cow liver or the amount of toxic pyrrole produced is difficult to estimate.

While the above variables complicate the overall picture of Senecio toxicity, several important points concerning Senecio toxicity in livestock are known:

1. Ruminants apparently are able to detoxify small quantities of PA-containing <u>Senecios</u> daily (Johnson and Molyneux 1984) and therefore have a threshold level that must be exceeded for toxicity to occur.
2. Cattle and horses are much more susceptible to pyrrolizidine alkaloid poisoning than are sheep or goats. Sheep can often consume great quantities of <u>Senecios</u> without apparent injury.
3. Young animals are more susceptible to injury by <u>Senecios</u> than are older animals.

Most <u>Senecio</u> intoxications that cause economic loss are chronic and may take many months to develop. They result from a gradual loss of liver function that eventually develops into a cirrhosis-like condition (Peterson and Culvenor 1983). Apparently, an increasing number of liver cells is damaged with each successive toxic dose of plant consumed until a sufficient number are injured to eventually lead to death of the animal. The size of the daily dose determines the number of liver cells damaged and the degree of damage to each cell. A chronic lethal dose can be eaten by a cow in 10 to 20 days under most conditions.

Liver cells may be affected by <u>Senecios</u> in several ways. One of the most significant is that the affected cell, even though capable of performing many of its biological functions, loses its ability to reproduce and thus assumes the characteristics of a senile cell (Jago 1969). As the cell dies, it is not replaced. When a sufficient number of liver cells either become senile and partially functional or die, the liver can no longer support the animal and death occurs.

<u>Senecios</u> may also directly affect the endothelium of certain liver blood vessels causing eventual vessel blockage and disintegration. This, in conjunction with a biliary blockage that often occurs, may lead to death of functional liver cells, fibrosis, and a complete disruption of the structural integrity of the liver (McLean 1970).

Functionally impaired livers are not able to withstand stresses tolerated by normal livers. The immediate cause of death in <u>Senecio</u> intoxicated cattle is often the result of secondary infection or stress.

Signs of intoxication may vary considerably, depending on the rate and total quantity of plant intake, content of the toxin in the plant, the rapidity with which the intoxication develops, and other stresses to which the animal is exposed. Since Senecio intoxication is primarily a liver malady, signs are similar to those of other hepatic disorders. Signs typically seen in heifers experimentally fed chronic, lethal doses of Senecio plants are as follows (Johnson 1979):

1. Slight disinterest in food; if manger-fed with other animals, the animal may mouth its food without eating much.
2. Loss of interest in surroundings, stands off by itself, and is reluctant to move; may stand much of the day with its head lowered.
3. Signs of discomfort evidenced by occasional kicking at the belly.
4. Gradual weight loss or failure to gain weight; the hair coat becomes dull and rough.
5. Diarrhea or constipation, depending on the type of diet fed.
6. Behavioral changes, most typically belligerency or stubbornness.
7. As intestinal edema develops, the animal may strain severely, sometimes causing prolapsed rectum.
8. In terminal stages, the animal may pace incessantly, bumping into fences, etc.; at this stage it may appear to be blind, is weak from lack of food, and is therefore unsteady in the hindquarters when walking; the rear knees become springy.
9. As further weakness develops, the animal becomes recumbent; lying on its brisket and straining, it usually dies without struggle.

Animals that survive a year or longer may show no signs other than alternate periods of mild depression until a few weeks before they die (Johnson and Molyneux 1985a). Gross signs most commonly seen in post mortem examinations include a greatly enlarged gall bladder, a tough fibrotic liver, and edema of the digestive tract that begins at the true stomach (abomasum) and extends posteriorly. Ascites (accumulation of fluid in the abdominal cavity) may form; many liters of fluid may be present.

ECONOMIC ASSESSMENT

The economic impact of toxic Senecios in range and pasture lands is difficult to accurately assess. Much of the monetary loss probably accrues from subclinical levels of poisoning in which animals do not appear to be in distress but do not gain weight or reproduce as expected. When affected livestock die, it may be as long as 18 months after ingestion of Senecio (Johnson and Molyneux 1984), and complications with other diseases, parasites, or nutritional factors often make accurate diagnosis of cause of death difficult. Extensive losses of livestock to Senecios have occurred in the Southwest (western Texas, New Mexico, and Arizona) due to S. longilobus and in the Pacific Northwest (Oregon, northern California, and Washington) due to S. jacobaea. Cattle losses in the Trans Pecos and Davis Mountains areas of Texas where S. longilobus is abundant were 5 to 30% during 1927 to 1933 (Mathews 1933, Norris 1951, Dollahite 1972). Threadleaf groundsel (S. longilobus) continues to be a serious toxic plant for both cattle and horses in the Southwest (Reagor 1981). Cattle losses in Oregon due to tansy ragwort (S. jacobaea) poisoning in 1971 were estimated at $1.2 million (Snyder 1972). Tansy ragwort infestations may also compete with valuable forage species. Annual economic losses in western Oregon due to forage displacement by tansy ragwort were estimated to exceed $600,000 (Isaacson and Schrumpf 1979).

ECOLOGICAL CONSIDERATIONS

Because tansy ragwort and threadleaf groundsel account for most of the Senecio poisoning of livestock in the United States, we will examine their ecology more closely.

Tansy Ragwort

Tansy ragwort is a cool-season herb. In many ways, it presents a microcosm of the variability seen in the genus Senecio as a whole. Although tansy ragwort is a climax plant in coastal sand dune communities in Britain and western Europe, its major distribution is as an early to mid seral weed on disturbed sites (Harper and Wood 1957). Tansy ragwort is often present as a weed in poorly managed pastures, recently logged forests, road edges, and railroad rights-of-way. Its worldwide distribution appears to follow the activities of Europeans as they colonized new areas, inadvertently transporting ragwort with them. The species is found in most of Europe, as far north as Norway, as far east as Siberia, and as far south as Asia Minor (Harper

1958). It is a common weed on both the east and west coasts
of the United States and Canada, as well as in New Zealand
and Australia.   Ragwort also has been introduced to
Argentina and South Africa.   The plant seems to be rather
insensitive to soil texture and acidity.   lt grows on soils
ranging from gravels to clays with pH ranging from 3.95 to
8.2 in Britain (Harper and Wood 1957).   Harper (1958)
suggested that soil moisture plays a role in ragwort
distribution.   He observed that seedling establishment was
low on waterlogged soils.   This agreed with his field
observations that ragwort is often replaced by S. aquaticus
on marshy sites.   The lower moisture limit required by tansy
ragwort is less clearly defined.   The most vigorous stands
in Oregon generally occur on moist, well-drained sites in
western Oregon where annual precipitation exceeds 800 mm.
However, populations have been observed in subirrigated
sites in eastern Oregon where annual precipitation is less
than 400 mm.   Similar patterns of distribution of ragwort
have been reported for Australia, New Zealand, and Tasmania
(Harper 1958).

Tansy ragwort hybridizes freely with other Senecio
species, such as S. aquaticus, S. squalidus, S. cinceraria,
S. vulgaris, and S. erucifolius in Britain (Harper and Wood
1957; Benoit et al. 1975).   The presence of numerous hybrids
has greatly confused the taxonomy of S. jacobaea.   Some
hybrids are recognized as such, while others have been
described as varieties of S. jacobaea or even as distinct
species.   The widespread distribution of S. jacobaea in many
different climatic regimes suggests that many distinct
ecotypes may exist.

Tansy ragwort displays a high degree of reproductive
plasticity.   lt is often classified as a biennial plant,
but may behave as a biennial, a winter annual, or a
perennial, depending upon environmental conditions.   In a
ragwort population studied by Forbes (1977), 8% of the
plants flowered in the first year as annuals, 39% behaved as
biennials, and 53% were perennial.   Analysis of flora
presented by Hart (1977) indicates that such reproductive
plasticity within species of biennial plants is not unusual.
Hart (1977) contends that annual plants are best adapted to
unstable environments, perennial plants to stable
environments, and biennial plants to intermediate,
semistable situations.   Logical extension of this concept
would suggest that the high reproductive plasticity of tansy
ragwort provides a mechanism by which the plant may inhabit
environments varying considerably in their temporal
stability.   Moreover, one may expect that the proportion of

annual, biennial, and perennial ragwort plants within a population will vary as community temporal stability varies.

Most reproduction of tansy ragwort is by seed. Seed is produced from both ray and disc flowers. Total number of seeds produced per plant ranges from 5,000 to 200,000 (Cameron 1935). Achenes produced by disc flowers have a deciduous pappus of soft bristles which aid in wind and animal dissemination. Wind dissemination can move seed great distances from the mother plant so that new safe sites may be colonized as they become available. In the case of tansy ragwort, however, relatively few achenes are carried more than 40 m from the parent (Poole and Cairns 1940). Presumably, updrafts or very strong, dry winds may carry a few achenes great distances (Sheldon and Burrows 1973). Ragwort achenes may be transported by animals, either attached to their coats or passed undigested in feces (Eadie and Robinson 1953). Ray flowers produce heavy achenes which lack a pappus. Ray flower achenes are shed after disc flower achenes and generally fall to the ground beneath the parent (Greene 1937). Both Greene (1937) and McEvoy (1984a) speculated that seed dimorphism is a form of specialization enabling ragwort to ensure reproduction on site through the presence of ray flower achenes, while new areas are colonized by wind-, water-, and animal-borne disc flower achenes. Ragwort seeds lack innate dormancy. Dormancy is induced after achene maturation by restrictive environmental conditions such as drought or extremes in temperature (Van Der Meijden and Van Der Waals-Kooi 1979). Ragwort seeds require light for germination. Once buried in the soil, they may remain viable for 10 to 16 years (Thompson and Makepeace 1983).

Seeds of tansy ragwort generally germinate in the fall or early spring (Harper and Wood 1957). Seedling mortality may be high (Forbes 1977), especially where competition is provided by vigorous grass stands (Harper and Wood 1957). Established vegetative plants are low-growing rosettes which compete successfully with other herbaceous plants by smothering them beneath the rosette (McEvoy 1984c). Reproductive plants flower during summer through autumn. The likelihood of a rosette flowering, as well as the number of flower heads produced, increases as rosette size increases (Van Der Meijden and Van Der Waals-Kooi 1979). Rosettes may be maintained almost indefinitely as vegetative perennials by defoliation, which keeps them from reaching flowering size (Harper 1958). Flowering plants whose flowers are removed prior to seed set often reflower that same year. Reflowering plants generally mature seed two to three months after undamaged plants; fewer seeds are produced (Cameron 1935, Harper and Wood 1957). Individual rosettes generally die after flowering. However, flowering plants may persist by producing new rosettes from the old crown and from the fleshy, laterally-spreading root system (Forbes 1977, McEvoy 1984c). Vegetative reproduction by root and crown sprouts from existing plants, together with a tendency for most seed to remain near the mother plant, often result in ragwort occurring in patches (Harper and Wood 1958, McEvoy 1984c).

Tansy ragwort is generally not very palatable to livestock. Cattle and horses may consume toxic quantities of the plant when other forage is scarce or when it is presented to them in hay (Cheeke and Shull 1985). They may also inadvertently consume it along with more palatable forage during the spring when selection between intermingled small leaves of tansy and grasses is difficult. Tansy intake of 3 to 7% of body weight is generally lethal to horses or cattle (Cheeke 1985). Sheep are less affected by tansy ragwort toxicity and display preference for the plant during the summer flowering period when other forage is mature. A lethal dose of tansy ragwort for sheep and goats is approximately 200 to 300% of body weight (Cheeke 1985). Acute tansy ragwort poisoning of sheep is rare. Sheep may be used effectively in the summer as a biological control agent to reduce tansy ragwort populations in pastures (Sharrow and Mosher 1982). While tansy ragwort is commonly seen in cattle pastures, it quickly disappears where sheep are grazed. It is not entirely clear whether sheep grazing eliminates the plant or merely so reduces its stature that tansy is no longer readily visible. Other biological agents which have proven effective in controlling tansy ragwort are the cinnabar moth (Tyria jacobaea; Lepidoptera) and the ragwort flea beetle (Longitarsus jacobaea; Coleoptera) (McEvoy 1984b). Biological control agents may reduce tansy ragwort population densities by as much as 97% (McEvoy 1984b). Chemical control of the plant may be accomplished by spring application of 2,4-D [(2,4-dichlorophenoxy)acetic acid] or by late spring-early summer application of dicamba (3,6-dichloro-2-methoxybenzoic acid) (Bedell et al. 1984). Mechanical control of established populations is effective only when root sprouting can be prevented by either removal of the entire root system or by frequent cultivation, which prevents establishment of seedlings and rootsprouts. Long-term suppression of tansy ragwort populations will be possible only where control methods are integrated into an overall program of land management which reduces the number of bare soil "safe sites" for establishment of tansy seeds, which provides strong competition from other plants, and which denies existing tansy plants the opportunity to set seed. Productive, properly grazed pasture may be the best defense against ragwort infestation (Bedell et al. 1984).

Threadleaf Groundsel

Threadleaf groundsel (Senecio douglasii var. Jamesii) (=S. longilobus) (=S. filifolius) is an evergreen, perennial subshrub. The plants have several branching stems arising from the base and a deep taproot. Stems and leaves are closely whitish floccose (with soft woolly hair) or eventually glabrate (without hairs) (Correll and Johnston 1970). The seedlings and juvenile plants are herbaceous but become woody with maturity. The plants produce yellow flowers during spring through autumn whenever growing conditions are favorable. The species is also called woolly groundsel and woolly senecio.

Threadleaf groundsel occurs from Nebraska and Wyoming southward into the western half of Texas and westward into New Mexico, northern Mexico, and Arizona (Correll and Johnston 1970, Ediger 1970). It is found in the midgrass and shortgrass prairies, southern mixed prairie, desert grasslands, Chihuahuan Desert, and pinyon-juniper vegetation types. It occurs on a wide array of soils, ranging from shallow and rocky to deep and fertile. It is most abundant on loams, clay loams, and clay soils, whereas the herbaceous, deciduous riddell groundsel (S. riddellii) (=S. spartioides) occurs on sandy soils. The species is well adapted to areas with annual precipitation of about 250 to 600 mm.

Threadleaf groundsel is rare on good condition grasslands. Its abundance on some range sites often increases with retrogression caused by overgrazing and/or drought. It is generally believed to be an increaser or invader species, depending on the particular range site. It is frequently a major component of early seral stages on areas such as road and pipeline rights-of-way, livestock bedgrounds, and sacrifice areas.

Threadleaf groundsel plants are relatively short-lived, and populations are cyclical. Germination and emergence of seedlings usually occur during autumns with above-normal precipitation (Jones et al. 1982). Seedling establishment is usually greatest when the wet autumn season follows a period of drought and/or overgrazing. Growth of the seedlings during fall and winter, while the associated grasses are dormant, allows ample time for seedling establishment before associated desirable forage plants can compete and suppress their establishment.

In some regions, threadleaf groundsel plants produce a flush of new vegetative growth during any season following significant precipitation. Winter leaf retention varies regionally; ecotypes of the northerly latitudes and higher elevations often have less winter foliage than those from southerly latitudes and lower elevations. Winter dormancy occurs in some of the colder areas (Johnson and Molyneux 1985b). Drought dormancy may also occur during extended periods of unfavorable growing conditions.

Die-offs of the species have occurred in western Texas during dry periods and also when growing conditions appeared favorable. There is some evidence that boring insect larvae in the stems and taproot may be associated with die-off of threadleaf groundsel in western Texas (W. A. McGinty, personal communication).

Palatability of threadleaf groundsel to cattle is apparently quite low. The species often increases in abundance on ranges grazed only by cattle. Sheep and goats are about 20 times more resistant to this toxic plant than are cattle (Dollahite 1972), and they readily eat it. Grazing with a mixture of cattle, sheep, and goats has been proposed as a method for controlling threadleaf groundsel in western Texas (Dollahite 1972). Ranchers who followed this practice during the 1930s and 1940s controlled the groundsel and made a profit, while many of those who continued grazing only with cattle lost money. Many had to sell their ranches because of large cattle losses (J. W. Dollahite, unpublished manuscript).

Acute poisoning on threadleaf groundsel commonly occurs when cattle are crowded into pastures where the subshrub is abundant and availability of desirable forages is limited (J. W. Dollahite, unpublished manuscript). Acute poisoning resulting in an acute hemorrhagic syndrome can occur in cattle when 0.25 to 5% of the animal's body weight of threadleaf groundsel is consumed in a short period ranging from one day to a week.

Chronic poisoning in cattle is most common and can occur (1) any time the desirable forages are dormant and/or low in phosphorus, protein, and energy due to cold temperatures and/or poor growing conditions, and (2) during or following adverse weather such as snow storms (Dollahite 1972; W. A. McGinty, personal communication). There is some evidence that PA content of threadleaf groundsel may be higher during hot, dry periods than when growing conditions are favorable (Briske and Camp 1982, Johnson and Molyneux 1985b).

Livestock losses due to threadleaf groundsel poisoning can be reduced, but probably not eliminated, by adhering to good grazing and livestock management practices and specific weed control methods. Hungry animals should never be turned into pastures where threadleaf groundsel is abundant, even though there is adequate availability of desirable forage. Stocking rates should always be carefully balanced against the availability of desirable forage. Grazing management should allow the desirable forages to increase in vigor, abundance, and variety, and a combination stocking with cattle, sheep, and goats should be used if possible.

Localized dense infestations of threadleaf groundsel may be controlled with broadcast sprays of 0.9 kg a.e./ha 2,4-D plus 0.2 kg a.e./ha picloram (4-amino-3,5,6-trichloro-2-pyridinecarboxylic acid), applied during autumn when growing conditions are favorable (Jones et al. 1982). Aerial sprays should be applied in a total volume of at least 28 liters/ha of a water-diesel fuel (11:1 ratio) emulsion carrier. Broadcast sprays applied with ground equipment should be applied in a total volume of at least 140 liters/ha. This treatment has killed 88 to 100% of the threadleaf groundsel population for one year (Jones et al. 1982; D. N. Ueckert, unpublished data). Recent work has indicated that autumn sprays of picloram at 0.28 kg/ha effectively controlled the subshrub (W. A. McGinty, personal communication), but picloram applied at rates of less than 1.1 kg/ha have not consistently achieved satisfactory control (D. N. Ueckert, unpublished data). Broadcast sprays of metsulfuron [[[[(4-methoxy-6-methyl-1,3,5-triazin-2-yl)amino]carbonyl]amino]sulfonyl]benzoic acid)] at rates as low as 17.5 g a.i./ha appear promising for control of threadleaf groundsel (W. A. McGinty, personal communication).

Low-pressure hand sprayers may be used to apply herbicides for control of low densities of threadleaf groundsel occurring along roadways, watering areas, etc. Sprays should be applied during autumn when growing conditions are favorable. Water-diesel fuel emulsion (4:1) carriers containing 0.45 kg a.e. picloram or 0.25 kg a.e. picloram plus 0.91 kg a.e. 2,4-D per 379 liters of the carrier are recommended.

Threadleaf groundsel plants normally do not die until 60 days after herbicide sprays are applied. The effects of herbicides on the palatability and toxicity of the species to livestock are not known. Therefore, it is not advisable to allow livestock to graze on herbicide-treated pastures until after the plants have died and at least partially deteriorated.

LITERATURE CITED

Bedell, T.E., R.E. Whitesides, and R.B. Hawkes. 1984. Pasture management for control of tansy ragwort. Pacific Northwest Cooperative Ext. Pub. 210, Oregon State Univ., Corvallis.

Benoit, P.M., P.C. Crisp, and B.M.G. Jones. 1975. Senecio L., pp. 404-410. In: C.A. Stace (ed), Hybridization and the flora of the British Isles. Academic Press, London.

Briske, D.D., and B.J. Camp. 1982. Water stress increases alkaloid concentrations in threadleaf groundsel (Senecio longilobus). Weed Sci. 30:106-108.

Britton, N.L., and A. Brown. 1970. An illustrated flora of the northern United States and Canada, Vol. III. Dover Pub., Inc., New York.

Cameron, E. 1935. A study of the natural control of ragwort (Senecio jacobaea L.). J. Ecol. 23:265-322.

Cheeke, P.R. 1985. Dietary additives for protection against pyrrolizidine alkaloid toxicosis in livestock, pp. 89-97. In: A.A. Seawright, M.P. Hegarty, L.F. James, and R.F. Keeler (eds), Plant toxicology. Proc. Aust.-U.S.A. Poisonous Plants Symp., Brisbane. Queensland Poisonous Plants Comm., Anim. Res. Inst., Yeerongpilly.

Cheeke, P.R., and L.R. Shull. 1985. Natural toxicants in feeds and poisonous plants. AVI Publishing Co., Inc., Westport, CN.

Correll, D.S., and M.C. Johnston. 1970. Manual of the vascular plants of Texas. Texas Res. Found., Renner, TX.

Crawford, H.S., C.L. Kucera, and J.H. Ehrenreich. 1969. Ozark range and wildlife plants. USDA For. Serv. Agr. Handbook 356.

Dayton, W.A. (ed). 1937. Range plant handbook. USDA For. Serv.

Dollahite, J.W. 1972. The use of sheep and goats to control Senecio poisoning in cattle. Southwest. Vet. 25:223-226.

Eadie, I.M., and B.D. Robinson. 1953. Control of ragwort by hormone-type weedicides. Aust. Inst. Agr. Sci. J. 19:192-196.

Ediger, R.I. 1970. Revision of section Suffruticosi of the genus Senecio (Compositae). SIDA 3:504-524.

Forbes, J.C. 1977. Population flux and mortality in a ragwort (Senecio jacobaea L.) infestation. Weed Res. 17:387-391.

Garrison, G.A., J.M. Skovlin, C.E. Poulton, and A.H. Winward. 1976. Northwest plant names and symbols for ecosystem inventory and analyses. USDA Pacific Northwest Forest and Range Exp. Sta. Gen. Tech. Rep. PNW-46.

Greene, H.E. 1937. Dispersal of Senecio jacobaea. J. Ecol. 25:569.

Harper, J.L. 1958. The ecology of ragwort (Senecio jacobaea) with special reference to control. Herbage Abstr. 28:151-157.

Harper, J.L., and W.A. Wood. 1957. Biological flora of the British Isles: Senecio jacobaea L. J. Ecol. 45:617-637.

Hart, R. 1977. Why are biennials so few? Amer. Natur. 111:792-799.

Hitchcock, A.L., A. Cronquist. M. Owenby, and J.W. Thompson. 1969. Vascular plants of the Pacific Northwest. Part 5. Univ. Washington Press, Seattle, WA.

Isaacson, D.L., and B.J. Schrumpf. 1979. Distribution of tansy ragwort in western Oregon, pp. 163-169. In: P.R. Cheeke (ed), Proc. Pyrrolizidine (Senecio) alkaloids: toxicity, metabolism, and poisonous plant control measures symp. Oregon State Univ., Corvallis.

Jago, M.V. 1969. The development of the hepatic megalocytosis of chronic pyrrolizidine alkaloid poisoning. Amer. J. Pathol. 56:405-422.

Johnson, A.E. 1979. Toxicity of tansy ragwort to cattle, pp. 129-134. In: P.R. Cheeke (ed), Proc. Pyrrolizidine (Senecio) alkaloids: toxicity, metabolism, and poisonous plant control measures symp. Oregon State University, Corvallis.

Johnson, A.E., and R.J. Molyneux. 1984. Toxicity of threadleaf groundsel (Senecio douglasii var. longilobus) to cattle. Amer. J. Vet. Res. 45:26-31.

Johnson, A.E., and R.J. Molyneux. 1985a. Delayed manifestation of pyrrolizidine alkaloidosis in cattle following their ingestion of Senecio species. No. 48. In: Proc. Sixth Ann. West. Conf. for Food Animal Vet. Med. Utah State Univ., Logan.

Johnson, A.E., and R.J. Molyneux. 1985b. Variation in toxic pyrrolizidine alkaloid content of plants, associated with site, stage of growth and environmental conditions, pp. 209-218. In: A.A. Seawright, M.P. Hegarty, L.F. James, and R.F. Keeler (eds), Plant toxicology. Proc. Aust.-U.S.A. Poisonous Plants Symp., Brisbane. Queensland Poisonous Plants Comm., Anim. Res. Inst., Yeerongpilly.

Johnson, A.E., R.J. Molyneux, and G.B. Merrill. 1985. Chemistry of toxic range plants. Variation in pyrrolizidine alkaloid content of Senecio, Amsinckia, and Crotalaria species. J. Agric. and Food Chem. 33:50-55.

Jones, R.D., D.N. Ueckert, J.T. Nelson, and J.R. Cox. 1982. Threadleaf groundsel and forage response to herbicides in the Davis Mountains. Texas Agr. Exp. Sta. Bull. 1422.

Kingsbury, J.M. 1964. Poisonous plants of the United States and Canada. Prentice-Hall Inc., Englewood Cliffs, NJ.

Kufeld, R.C. 1973. Foods eaten by the Rocky Mountain elk. J. Range Manage. 26:106-113.

McEvoy, P.B. 1984a. Dormancy and dispersal in dimorphic achenes of tansy ragwort, Senecio jacobaea L. (Compositae). Oecologia 61:160-168.

McEvoy, P.B. 1984b. Depression in ragwort (Senecio jacobaea) abundance following introduction of Tyria jacobaea and Longitarsus jacobaea on the central coast of Oregon, pp. 57-64. In: Proc. VI Int. Symp. Biol. Contr. Weeds. Vancouver, Canada.

McEvoy, P.B. 1984c. Seedling dispersion and persistence of ragwort Senecio jacobaea (Compositae) in a grassland dominated by perennial species. Oikos 42:138-143.

McLean, E.K. 1970. The toxic actions of pyrrolizidine (Senecio) alkaloids. Pharmacol. Rev. 22:429-483.

Mathews, F.P. 1933. Poisoning of cattle by species of groundsel. Texas Agr. Exp. Sta. Bull. 481.

Mattocks, A.R. 1971. Toxicity and metabolism of Senecio alkaloids, pp. 179-200. In: J.R. Harborne (ed), Phytochemical ecology. Ann. Proc. of the Phytochem. Soc. No. 8. Univ. of Reading, Reading, England.

Muntz, P.A., and D.D. Keck. 1968. A California flora. Univ. California Press, Berkeley.

Norris, J.J. 1951. The distribution and chemical control of species of Senecio, Astragalus, and Baileya in the highlands range area of west Texas. Ph.D. Diss. Texas A&M University, College Station.

Peterson, J.E., and C.C.J. Culvenor. 1983. Hepatotoxic pyrrolizidine alkaloids, pp. 637-671. In: R.F. Keeler and A.T. Tu (eds), Handbook of natural toxins, volume 1. Plant and fungal toxins. Marcel Dekker, Inc., New York.

Poole, A.L., and D. Cairns. 1940. Botanical aspects of ragwort (Senecio jacobaea L.) control. N.Z. Dept. Sci. Indust. Bull. 82.

Reagor, J.C. 1981. Overview of poison plant situation in the Trans-Pecos, pp. 1a-1f. In: A. McGinty (ed), Proc. Trans-Pecos Poison Plant Symp., Fort Stockton, TX. Texas Agr. Ext. Serv.

Sharrow, S.H., and W.D. Mosher. 1982. Sheep as a biological control agent for tansy ragwort. J. Range Manage. 35:480-482.

Sheldon, J.C., and F.M. Burrows. 1973. The dispersal effectiveness of achene-pappus units of selected Compositae in steady winds with convection. New Phytol. 72:665-675.

Smith, G.S., J.B. Watkins, T.N. Thompson, K. Rozman, and C.D. Klaassen. 1984. Oxidative and conjugative metabolism of xenobiotics of livers of cattle, sheep, swine and rats. J. Anim. Sci. 58:386-395.

Snyder, S.P. 1972. Livestock losses due to tansy ragwort poisoning. Oregon Agr. Rec. No. 255.

Thompson, A., and W. Makepeace. 1983. Short note: Longevity of buried ragwort (Senecio jacobaea L.) seed. N.Z. J. Exp. Agr. 11:89-90.

United States Department of Agriculture (USDA) Soil Conservation Service. 1982. National list of scientific plant names. SCS Tech. Pub. 159.

Van Der Meijden, E., and R.E. Van Der Waals-Kooi. 1979. Population ecology of Senecio jacobaea in a sand dune system. I. Reproductive strategy and biennial habit. J. Ecol. 67:131-153.

16

Toxicity Problems Associated with the Grazing of Oak in Intermountain and Southwestern U.S.A.

Kimball T. Harper, G.B. Ruyle,
and L.R. Rittenhouse

INTRODUCTION

Oak species (genus <u>Quercus</u> of the family Fagaceae) are all shrubs or trees. Over 75 species of oak occur in the United States and Canada (Dayton 1931). The species occur in almost all forested and shrubland habits on the continent. Two basically different groups of oak are recognized, black oaks and white oaks. Both groups have simple, lobed leaves, bear acorns, and are represented by deciduous and evergreen (or "live") species. Black oaks are recognized by leaves with bristlelike awns at the ends of lobes, dark-colored bark, and acorns that mature in two years. White oaks have smooth, prickleless lobes on leaves, light-colored, flaking bark, and mature acorns in a single year. Both groups contain low, shrubby species. In the southern midsection of the continent and in the· Southwest, large areas are dominated by shrubby oak species to the near exclusion of other species. Such species often sprout prolifically after tops are destroyed by fire, herbicides, or mechanical devices (Knipe 1983; Harper et al. 1985). Throughout the southern tier of states, these low-growing oak shrublands are known locally as shinneries (Dayton 1931).

Kingsbury (1964) reported that the three oaks most commonly involved in incidents of livestock poisoning in the United States are Gambel oak (<u>Q</u>. <u>gambelii</u>), Havard oak (<u>Q</u>. <u>havardii</u>), and white shin oak (<u>Q</u>. <u>durandii</u> var. <u>brevilobata</u>); however, the foliage, twigs, and acorns of any oak species should be considered as potentially toxic. The three species noted are all shrubs or small trees with deciduous leaves. Other oaks in the Intermountain West that are occasionally implicated in oak toxicity cases include shrub live oak (Q. turbinella), Emory oak (Q. emoryi), silverleaf oak (Q. hypoleucoides), and wavyleaf oak (Q. undulata) (Dayton 1931; Ruyle et al. 1986).

GEOGRAPHIC DISTRIBUTION

In the Southwest, oak species are often implicated in incidents of livestock poisoning. Gambel oak, the species most often involved in poisoning (Dayton 1931), occurs from northern Utah through Arizona and into northern Mexico and from southwestern Texas to southern Nevada. Throughout that region, Gambel oak occurs in an altitudinal band between the more mesic desert shrub zones and the lower elevation forests.

Havard oak is also often implicated in cases of livestock poisoning. It occurs in scattered clones on sandy plains extending from southeastern New Mexico across the panhandle of Texas and into southwestern Oklahoma. Since the species is both a good soil binder and a vigorous sprouter, it often dominates low sand dunes that have gradually accumulated on spots stabilized by the species (Marsh et al. 1919).

White shin oak is the most easterly of the oaks that commonly poison livestock. This species occurs in western Texas and Oklahoma (Preston 1976).

These troublesome oaks are all deciduous shrubs or small trees that form dense clones. Their deciduous and shrubby habits combine with the clonal habit to make them serious poisoning threats. Small size insures that much if not all the foliage will be within reach of hooved grazing animals. It is a general rule of thumb among stockmen familiar with oaks that foliage of deciduous species tends to be more palatable to ungulate grazers than foliage of evergreen species (Marsh et al. 1919). The tight clones formed by these species and the slowness with which their fallen leaves decay (Harper et al. 1985) discourages growth of understory plants. Grazing animals confined to such habitats are forced to consume primarily oak. In better watered environments, oak species achieve tree size and produce most of their foliage beyond the reach of browsing animals. More free water also encourages greater herbaceous plant growth which provides herbivores with an alternative to oak.

Shrub live oak achieves greatest abundance in Arizona, where it is the most widely distributed of 12 native oaks (Ruyle et al. 1986). Locally in that state, the species may contribute from 25 to over 40% of the forage consumed by cattle during the winter season (November through April). In a large sample of fecal material collected throughout the year and from all parts of Arizona, Ruyle et al. (1986) found that the species was least important in cattle diets during the period May through September when precipitation was most likely, and palatable grasses and forbs were abundant in the oak community. When green herbaceous growth was available, oak generally contributed less than 20% of the forage consumed by cattle. Shrub live oak was most important in cattle diets in northwestern and southeastern Arizona (Ruyle et al. 1986). The species is also common in the coast ranges of Baja and southern California. It is scattered in western New Mexico and Colorado and locally abundant in extreme southwestern Utah (Little 1976).

Emory oak and silverleaf oak were occasionally encountered in cattle fecal samples from southern Arizona (Ruyle et al. 1986). The former species is common in the Arizona-Mexico border area. Silverleaf oak is widely distributed in Arizona and fairly common (Kearney and Peebles 1969). Both species are small to moderate-sized trees with leaves that persist through the winter.

Wavyleaf oak is a small shrub of widespread distribution in the Southwest. It grows on sandy plains where it comes to dominate small dunes held in place by its dense network of roots and rhizomes. In this respect, and in its morphology, the species is reminiscent of Havard oak (Dayton 1931). The species also occupies dry, rocky mountain slopes and barren hills between 1,200 and 2,150 m. In portions of its range, wavyleaf oak forms extensive thickets in which it is the dominant component of the vegetation. The species shows considerable morphological diversity with foliage varying from deciduous to persistent through the winter. Cattle sometimes utilize the species heavily in the spring.

ENVIRONMENTAL INFLUENCES ON OAK USE

Dry spells are frequent in the Intermountain and Southwest regions of North America. In such periods, the sparse herbaceous growth associated with stands of scrub oak species (Gambel, Havard, shrub live, and wavyleaf oak) may fail almost completely. At such times, the deep rooted oaks often produce new leaves as usual and become the principal forage for large animal herbivores. Under such circumstances, livestock may consume oak to the near exclusion of other species. When oak species contribute over 50% of the forage intake of any herbivore species, animals may become ill and show symptoms of poisoning. When oak makes up over 75% of the diet, deaths can be expected (Kingsbury 1965; James et al. 1980).

GEOGRAPHIC INCIDENCE OF POISONINGS

Oak poisoning incidents are known from all parts of North America, but the vast majority of cases are reported from the American Southwest (Marsh et al. 1919). Reports of oak poisonings in California (Fowler and Richards 1965; Ruyle et al. 1986), Ohio (Sandusky et al. 1977), Oklahoma (Panciera 1978), Texas (Dollahite et al. 1966), and elsewhere in North America (Kingsbury 1964) emphasize the point that most oaks can be dangerous.

TOXIC COMPOUNDS AND ANIMAL RESPONSES

Although some have questioned whether tannins were the toxic agents in oak poisonings (Marsh et al. 1919; Kingsbury 1965), it seems clear now that tannins or their breakdown products are toxic or at least physiologically disruptive to most microorganisms and herbivores (Dollahite et al. 1962; Pigeon et al. 1962; Feeny 1970; McLeod 1974; Swain 1979). Oaks produce gallotannins, which break down to release gallic acid and pyrogallol. These compounds are known to cause death in a variety of animals if administered in sufficient quantities. Other toxins may be present but such have not been identified. Dollahite et al. (1962) found the lethal dose (LD_{50}) of pyrogallol to be 0.7 g/day/kg live weight administered orally to rabbits. Lethal doses for gallic acid and tannic acid were 2.8 g/day/kg and 3.4 g/day/kg, respectively. These chemicals cause hemorrhagic gastritis and distinctive renal damage that is identical with that observed in oak poisoning.

Tannic acid is known to hydrogen bond to proteins and the protein portion of enzymes. That reaction precipitates proteins and often renders them indigestible. The reaction with enzymes results in inactivation (McLeod 1974). Tannin also complexes with and inactivates nucleic acids. Tannin-polysaccharide complexes occur and appear to reduce polysaccharide digestibility (Swain 1979).

Experimental evidence is good that tannins in excess of 5% by weight in forage do interfere with animal performance before symptoms of toxicity appear. Tannins appear to impair recovery of dietary protein in rats (Glick and Joslyn 1970; Tamir and Alumot 1970), steers (Donnelly and Anthony 1970), and goats (Nastis and Malechek 1981). Feeny (1970) observed reduced growth of lepidopteran larvae on diets high in tannin. Nastis and Malechek (1981) observed reduced digestibility of forage when tannin exceeded 5% in the diet of goats. Ruyle et al. (1986) have suggested that oak forage may significantly reduce animal performance before any observable symptoms of toxicity appear. Experimental data strongly support that suggestion (Tamir and Alumot 1970, Nastis and Malechek 1981). There is a definite need to experimentally establish guidelines for feeding supplements capable of reducing or eliminating growth impairing effects of high tannin diets.

Tannins apparently cross cellular membranes very slowly. They readily inactivate membrane and extracellular enzymes, but have little effect on intracellular (cytoplasmic) enzymes (Herz and Kaplan 1968). Tannins and their derivatives do eventually find their way from the intestinal tract into the blood (Panciera 1978). A few enzymes resist tannins or even catalyze their breakdown. The protein portions of these enzymes often contain a carbohydrate segment and are known as glycoproteins (Strumeyer and Malin 1970).

The ease with which tannins complex with proteins, nucleic acids, and carbohydrates make them broad spectrum inhibitors of pathogens and herbivores. The tannins are stored away from cytoplasm and nuclear materials in the vacuoles or nonliving tissues (wood and bark) of plants. Pathogenic and decomposer bacteria or fungi must gain access to living cells by secreting digestive enzymes onto their surface. Tannins are potent retardants to such organisms. Likewise, in the digestive tracts of herbivores, digestion occurs extracellularly as the animal (or its intestinal microorganisms) secretes enzymes into the tract. Plant cells ruptured in the feeding process release tannins that interfere with digestion. Tannins also attack the gut lining directly and produce gastritis. On absorption, tannins and their derivatives cause serious damage in the kidneys.

Oak poisoning has been reported for a variety of animals including rabbits, deer, sheep, goats, pigs, horses, and cattle (Marsh et al. 1919; Kingsbury 1964; Dollahite et al. 1962; McLeod 1974; Panciera 1978). In the Intermountain and Southwestern regions, however, most cases of toxicity involve cattle.

Cattle poisoned by oak usually show symptoms only after several days of feeding on the plant. The first symptoms are dry, hard, dark-colored feces, sometimes containing mucus and blood. Appetites are depressed, and the animal may initially experience constipation; later, there is often diarrhea containing blood. Poisoned animals urinate frequently and have a tendency to stay near water. In later stages, the urine may be reddish. The coat becomes roughened, the animal emaciated, and the muzzle dry. Affected animals often stand with the stomach pulled up in a classic "tucked up" posture. Watery swellings may appear on the underside of the body. Symptoms may persist for three to ten days before the animal either collapses or improves (Schmutz et al. 1968; Panciera 1978).

CONDITIONS UNDER WHICH POISONINGS OCCUR

Most cases of poisoning occur in areas where oak is the overwhelming source of forage. The likelihood of poisoning is increased in years of unusually low precipitation and near failure of herbaceous growth. Since oak twigs and leaves appear to contain considerably more tannin per unit weight earlier in the growing season than after foliage is fully expanded and mature, it is a common observation that most incidents of poisoning occur in the first 30 days after the plants foliate (Marsh et al. 1919; Kingsbury 1964; Ruyle et al. 1986). The decline in percent tannin by weight between new and mature foliage is significant for Gambel oak (11.1 to 8.7%, Nastis and Malechek 1981) and Havard oak (15.1 to 4.2%, Pigeon et al. 1962). The danger is apparently heightened by frosts intense enough to kill young leaves, an occurrence that is not uncommon in the range of Gambel oak (Schmutz et al. 1968).

Kingsbury (1964) noted that oak poisonings were also common on cutover areas where there was an abundance of vigorous sprouts. This may simply be a consequence of new foliage with high levels of tannins that are not yet diluted by lignin and other secondary cell wall chemicals. It is known, however, that at least two plant species (Coleogyne ramosissima and Lespedeza cuneata) produced more foliage tannin after being cut back than occurred in uncut tissue (Provenza and Malechek 1984; Donnelly and Anthony 1970). Since some have recommended that heavy grazing of oak sprouts by goats (Knipe 1983; Harper et al. 1985) be combined with other means of controlling the plant (crushing, herbicides, fire), the possibility of enhanced tannin content following injury should be investigated.

Cattle poisonings are common in years of heavy acorn production (Fowler and Richards 1965; Sandusky et al. 1977; Panciera 1978). Calves seem more likely to be affected than mature animals. Some question reports of acorn poisoning, since acorns were an important component of the diets of some native Californian Indians. As Fowler and Richards (1965) reported, however, the Indian peoples crushed acorns and washed the pulp repeatedly with water. Acorn poisonings are also common in Europe (Long 1924).

REDUCING LOSSES

Since the oak species most frequently involved in incidents of poisoning in the Intermountain West are notoriously difficult to eliminate from rangelands (Knipe 1983; Harper et al. 1985), managers should develop programs that will minimize the probability of livestock poisoning. Fortunately, these aggressive species are only moderately palatable (Welch et al. 1983; Ruyle et al. 1986) and must be consumed in large amounts to cause death. Ranges that support a diversity of forage in addition to oak are not likely to produce cases of poisoning. Nevertheless, the oaks in question are highly tolerant of grazing and have expanded into open areas during the last century where grazing has locally eliminated palatable species. The situation is complicated by the fact that oak communities regain desirable species slowly, even under complete closure to grazing (Harper et al. 1985). Accordingly, oak ranges that must be used at times when danger of poisoning is great may require a mechanical, chemical, or fire treatment followed by reseeding of desirable species. While oak is not likely to be eliminated by these procedures, it is possible to markedly increase diversity of the forage base (Harper et al. 1985).

Wherever possible, animals should be kept off oak-dominated ranges until the foliage is at least 30 days old. By that time, understory species should be well enough developed to supply considerable forage, and tannin content of leaves will have been diluted by chemicals forming secondary cell walls. Animals should probably be removed from oak ranges after frosts that turn leaves black.

During summer and fall, animals should be removed from oak-dominated ranges when drought or grazing severely reduces availability of associated herbs. Animals should be watched closely when acorn crops are falling. Fortunately, heavy mast crops are not common in the Intermountain West.

Where animals must use oak ranges under conditions favorable for poisoning, experience has shown that feed supplements high in both energy and protein will eliminate almost all cases of poisoning. Schmutz et al. (1968) recommended at least three pounds of alfalfa hay per cow per day as a supplement. Dollahite et al. (1966) demonstrated that calcium hydroxide added to the forage base in amounts equal to 10 to 15% by weight was adequate to forestall development of symptoms of toxicity.

Poisoned animals may benefit from a mild laxative. They should be supplied adequate food and water and taken from areas where they have access to oak. Recovery may require several days (Kingsbury 1964).

ESTIMATING ECONOMIC LOSSES

Few estimates of economic losses attributable to oak poisoning appear in the literature. Dollahite et al. (1966) estimated that Havard oak caused a $10 million annual loss to the cattle industry of Texas alone. They noted that several Texas counties have reported losses of over 1,000 cattle per year to oak poisoning. Reagor (1981) concluded that oak caused more economic losses to the livestock enterprises of the state of Texas than any other plant. Although reports of oak poisoning in Utah were common in the early years of this century (Marsh et al. 1919), management of oak ranges has apparently improved, since few of the Utah ranchers and large animal veterinarians polled in connection with this study had ever seen a case of oak poisoning.

Given the negative impacts of tannins on recovery of protein and energy from diets as reported by Nastis and Malechek (1981), it seems inevitable that significant economic losses will occur before any symptoms of poisoning appear. Research quantifying reductions in weight gain attributable to dietary tannins is needed before the complete cost of oak forage to livestock producers can be determined.

LITERATURE CITED

Dayton, William A. 1931. Important western browse plants. USDA Misc. Publ. No. 101.

Dollahite, J.W., G.T. Housholder, and B.J. Camp. 1966. Effect of calcium hydroxide on the toxicity of post oak (Quercus stellata) in calves. J. Amer. Vet. Med. Assoc. 148:908-912.

Dollahite, J.W., R.F. Pigeon, and B.J. Camp. 1962. The toxicity of gallic acid, pyrogallol, tannic acid and Quercus havardii in the rabbit. Amer. J. Vet. Res. 23:1264-1266.

Donnelly, E.D., and W.B. Anthony. 1970. Effect of genotype and tannin on dry matter digestibility in sericea Lespedeza. Crop Science 10:200-202.

Feeny, P. 1970. Seasonal changes in oak leaf tannins and nutrients as a cause of spring feeding by winter moth caterpillars. Ecology 51:565-581.

Fowler, M.E., and W.P.C. Richards. 1965. Acorn poisoning in a cow and a sheep. J. Am. Vet. Med. Assoc. 147:1215-1220.

Glick, Z., and M.A. Joslyn. 1970. Effects of tannic acid and related compounds on the absorption and utilization of proteins in the rat. J. Nutrition 100:516-520.

Harper, K.T., F.J. Wagstaff, and L.M. Kunzler. 1985. Biology and management of the Gambel oak vegetative type: a literature review. USDA Forest Service General Technical Report INT-179.

Herz, F., and E. Kaplan. 1968. Effects of tannic acid on erythrocyte enzymes. Nature 217:1258-1259.

James, L.F., R.F. Keeler, A.E. Johnson, M.C. Williams, E.H. Cronin, and J.D. Olsen. 1980. Plants poisonous to live-stock in the western states. USDA Agric. Info. Bull. 415.

Kearney, T.H., and R.H. Peebles. 1969. Arizona flora, 2nd ed. University of California Press, Berkeley.

Kingsbury, J.M. 1964. Poisonous plants of the United States and Canada. Prentice-Hall, Inc., Englewood Cliffs, NJ.

Kingsbury, J.M. 1965. Deadly harvest--a guide to common poisonous plants. Holt, Rinehart and Winston, New York.

Knipe, O.D. 1983. Effects of Angora goat browsing on burned-over Arizona chaparral. Rangelands 5:252-255.

Little, E.L., Jr. 1976. Atlas of United States trees, Vol. 3. Minor western hardwoods. USDA Forest Service Misc. Pub. 1314.

Long, H.C. 1924. Plants poisonous to livestock. Cambridge University Press, Cambridge, England.

Marsh, C.D., A.B. Clawson, and H. Marsh. 1919. Oak-leaf poisoning of domestic animals. USDA Bull. 767.

McLeod, N.N. 1974. Plant tannins--their role in forage quality. Nutrition Abstracts and Reviews 44:803-815.

Nastis, A.S., and J.C. Malechek. 1981. Digestion and utilization of nutrients in oak browse by goats. J. Ani. Sci. 53:283-290.

206

Panciera, R.J. 1978. Oak poisoning in cattle, pp. 499-506. In: R.F. Keeler, K.R. Van Kampen, and L.F. James (eds), Effects of poisonous plants on livestock. Academic Press, New York.

Pigeon, R.F., B.J. Camp, and J.W. Dollahite. 1962. Oak toxicity and polyhydroxyphenol moiety of tannin isolated from Quercus havardii (Shin oak). Amer. J. Vet. Res. 23:1268-1270.

Preston, R.J. 1976. North American trees, 3rd ed. Iowa State University Press, Ames.

Provenza, F.D., and J.C. Malechek. 1984. Diet selection by domestic goats in relation to blackbrush twig chemistry. J. App. Ecol. 21:831-841.

Reagor, J.C. 1981. Overview of poisonous plant situation in the Trans-Pecos, p. 1. In: A. McGinty (ed), Proceedings of the Trans-Pecos poison plant symp. Texas Agr. Ext. Serv., College Station.

Ruyle, G.B., R.L. Grumbles, M.J. Murphy, and R.C. Cline. 1986. Oak consumption by cattle in Arizona. Rangelands 8:124-126.

Sandusky, G.E., C.J. Fosmaugh, J.B. Smith, and R. Mohan. 1977. Oak poisoning of cattle in Ohio. Amer. Vet. Med. Assoc. J. 171:627-629.

Schmutz, E.M., B.N. Freeman, and R.E. Reed. 1968. Live-stock-poisoning plants of Arizona. University of Arizona Press, Tucson.

Strumeyer, D.H., and M.J. Malin. 1970. Resistance of extracellular yeast invertase and other glycoproteins to denaturation by tannins. Biochem. J. 118:899-900.

Swain, T. 1979. Tannins and lignins, pp. 657-682. In: G.A. Rosenthal and D.H. Janzen (eds), Herbivores--their interaction with secondary plant metabolites. Academic Press, Inc., San Diego, California.

Tamir, M., and E. Alumot. 1970. Carob tannins--growth depression and levels of insoluble nitrogen in the digestive tract of rats. J. of Nutrition 100:573-580.

Welch, B.L., S.B. Monsen, and N. Shaw. 1983. Nutritive value of antelope and desert bitterbrush, Stansbury Cliffrose, and Apache-plume, pp. 173-185. In: A.R. Tiedemann and K.L. Johnson (comp) Proceedings-Research and management of bitterbrush and cliffrose in western North America. USDA Forest Service General Technical Report INT-152.

17

The Hemlocks: Poison-hemlock (Conium maculatum) and Waterhemlock (Cicuta spp.)

K.E. Panter and R.F. Keeler

INTRODUCTION

The hemlocks, poison-hemlock (Conium maculatum) and waterhemlock (Cicuta spp.), are poisonous plants growing in pastures and on ranges throughout the United States. Poison-hemlock and waterhemlock are not considered high on the list of economically important poisonous plants on a national basis; however, they do cause devastating losses to individuals.

Although both genera, Conium and Cicuta, belong to the family Umbelliferae (Apiaceae) and resemble each other in appearance and habitat, there are distinct ecological and toxicological differences between them. Umbelliferae contains over 300 genera and 3000 species worldwide (Cronquist 1981). There are some genera in the family other than Conium and Cicuta which are believed poisonous under certain conditions such as fools parsley (Aethusa cynapium), cow-parsnip (Heracleum lanatum), water-parsnip (Sium suane), and others. All members of Umbelliferae have many characteristic features in common and are often indistinguishable from one another based on casual examination. Botanists use characteristics of the fruit to distinguish features and classify species within a genus. Unfortunately, when Conium and Cicuta are at their most toxic stage, the fruiting structure is not present and identification may not always be correct. Therefore, it is important for an individual to identify those plants in question at the time of flower and fruit production to fully insure correct identification, and anticipate potential poisonous plant problems before they occur.

FIGURE 17.1
Poison-hemlock in the flower stage.

POISON-HEMLOCK

Distribution and Description

Four species of Conium have been recognized throughout the world, but only one species (C. maculatum) grows in the United States (Willis 1973). Its distribution is nationwide where adequate moisture will support its growth. The preferred common name for this species is poison-hemlock. Other common names include California fern, Nebraska fern, spotted hemlock, poison parsley, spotted parsley, poison root, bunk, cashes, cigue (Canada), herb bonnet, kill cow, St. Bennett's herb, winter fern, wade thistle, and others (Kingsbury 1964, Muenscher 1975).

Poison-hemlock (Figure 17.1) is a biennial herb which produces seed the second year of growth. The plant grows erect and averages 1 to 2.5 m in height. It may exceed heights of 3 m in favorable conditions. The stems are stout, rigid, hairless, and hollow, except at the nodes. There are numerous purple spots on the lower regions of the stem. The first year the plant grows in thick stands near the parent plant. The young plants are fine, fern-like, and usually less than 45 cm tall and the leaves are lacy and

light green.   The second year the plant grows from a
rosette.   The leaves are large, dark green, dissected,
triangular in shape, smooth without hairs, four to five
times pinnately compound, and divided oppositely by a
central stem (Figure 17.2).   The plant has a long, white
taproot, which is solid and like a parsnip. The flowers are
small and white in terminal flat-topped or umbrella-shaped
compound clusters (umbels).   The fruits are small grayish-
brown with conspicuous, wavy or knotted ribs or ridges. The
seeds generally drop close to the parent plant, but may be
spread by gusty winds, birds, water, or rodents. The seeds
have been mistaken for those of anise, the leaves for
parsley, and the roots for parsnips. All parts of the plant
are toxic (Kingsbury 1964, Muenscher 1975).

FIGURE 17.2
The leaves of poison-hemlock.

Habitat and Conditions of Growth

Poison-hemlock is native to Eurasia and northern Africa and has been naturalized in most parts of the United States and Canada. It was introduced as an ornamental for its natural beauty and has become widespread. The plant usually grows in waste places, marshy areas, ditch banks, road sides, near fences, and in uncultivated areas where adequate moisture is available (Kingsbury 1964).

Poison-hemlock is a prolific seed producer and generally dominates small areas if unchecked. It may encroach on alfalfa fields, grass pastures, and meadows, causing a serious toxicity threat to livestock, crowding out valuable forage and posing a toxicity problem in harvested silage or hay. Poison-hemlock is rapidly becoming one of the important noxious weed problems.

Toxicity

Poison-hemlock is both acutely toxic and teratogenic. Cattle (Penney 1953, Kubik et al. 1980), horses (MacDonald 1937), pigs (Buckingham 1936, Anonymous 1951, Widmer 1984), goats (Capithorne 1937), and elk (Jessup et al. 1986) have been poisoned, with the largest losses occurring in cattle and pigs. In one case, 224 hogs were poisoned from barley contaminated with poison-hemlock seed (Anonymous 1951). Forty-five of the hogs died, and the rest recovered without any permanent complications. A recent report cited 160 dairy cows and 66 heifers intoxicated by feeding green chopped hay contaminated with poison-hemlock (Kubik et al. 1980). Twenty-four cows and 26 heifers died. Milk from the cows that recovered was discarded for many days following the poisonings.

Recent studies, in which poison-hemlock was experimentally fed to cattle (Keeler 1974, Keeler and Balls 1978), horses (Keeler et al. 1980), sheep (Keeler et al. 1980, Panter et al. 1988), and pigs (Panter 1983, Panter et al. 1983, Panter et al. 1985a, 1985b), reported toxic and teratogenic effects. Poison-hemlock is most toxic to cows, less so to sheep, and least toxic to pigs. However, the teratogenic effect is most severe in cows and pigs and least in sheep. Fresh plants collected at the same site were lethal to cows, sheep, and pigs at dosages of 5.3 g/kg, 10 g/kg, and 13.7 g/kg, respectively (Keeler 1974, Keeler et al. 1980, Panter 1983, Panter et al. 1988).

The relative toxicity of the plant to different livestock species and the teratogenic effects in offspring are variable. Plant material from different locations and from different collections in the same location also vary in their toxic and teratogenic effects. Environmental conditions (moisture, temperature), season of growth (fall or spring), and soil types may alter the alkaloid composition (Cromwell 1956, Fairbairn and Suwal 1961, Panter 1983). These factors are presumed to be responsible for the variation in toxicity and teratogenicity. Plants collected in the spring season in Champaign County, Illinois, and fed free choice to pregnant gilts were toxic but not teratogenic. However, plants collected in the fall season in the same location were toxic and also teratogenic (Panter 1983). Chemical analysis of plants collected in both seasons revealed significant differences in alkaloid composition.

The clinical signs of toxicity induced by Conium are basically similar in all classes of livestock. All species show muscular weakness, loss of coordination, trembling, knuckling at the fetlock joints, slobbering, cyanotic membranes, a rapid pulse initially then slowing, dilated pupils, frequent urination and defecation, and a characteristic "mousy urine" smell on the breath of intoxicated animals. Frequent eructation in ruminants and temporary blindness in cattle and pigs from the nictitating membrane covering the eyes may occur. A preference or desire to ingest more plant after initial exposure has been reported in cows (Penney 1953), pigs (Panter 1983), goats (Capithorne 1937), and tule elk (Jessup et al. 1986).

There are eight known alkaloids in poison-hemlock. Coniine and γ-coniceine are the primary alkaloids in the plant. γ-Coniceine predominates in the early growing plant and coniine in the maturing plant and seed. Only coniine has been experimentally fed to livestock species. Commercially available coniine was administered via a stomach tube to cows, horses, and sheep. It was most toxic to cows (3.3 mg/kg), less so to mares (15.5 mg/kg), and least toxic to sheep (44.0 mg/kg) (Keeler et al. 1980). Plants, in which 98% of the total alkaloids was γ-coniciene, were teratogenic in cows and pigs (Keeler et al. 1980, Panter 1983). Toxicological and pharmacological studies have been done in cats, mice, and chicks for 3 of the conium alkaloids; coniine, γ-coniceine and N-methyl coniine. Bowman and Sanghni (1963) reported that the oral LD_{50} in mice for γ-coniceine, coniine, and N-methyl coniine is 12 mg/kg, 100 mg/kg and 204.5 mg/kg, respectively.

FIGURE 17.3
Skeletal malformations in a Hereford calf after feeding
coniine to its dam during gestation days 55 through 75.

Outbreaks of malformations in calves (Keeler 1974) and
pigs (Edmonds et al. 1972, Dyson and Wrathall 1977) from
ingestion of poison-hemlock have been reported. Skeletal
malformations in calves were induced when their mothers were
fed fresh plants and pure coniine during gestation days 55
through 75 (Keeler 1974, Keeler and Balls 1978, Keeler et
al. 1980). The malformations included excessive flexure of
the carpal joints, malalignment and twisting of the front
limbs, curvature of the spinal column, and twisted necks
(Figure 17.3). Similar skeletal malformations were induced
in pigs when their mothers were fed fresh poison-hemlock and
seed during gestation days 43 through 53 (Panter et al.
1985b). In addition, when fresh poison-hemlock was fed to
pregnant gilts during gestation days 30 to 45, cleft palate
resulted (Figure 17.4), however, no skeletal malformations
were observed (Panter et al. 1985a).

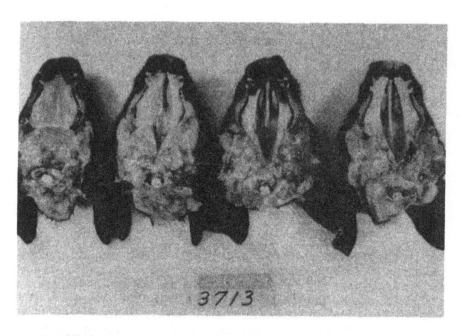

FIGURE 17.4
Cleft palates in newborn pigs after feeding fresh poison-hemlock to a pregnant gilt during gestation days 30 through 45.

Conditions of Poisoning

Usually, poisoning is most common in the early spring when poison-hemlock is often the first green forage to emerge. Concentration of alkaloids is also highest in early growth. γ-Coniceine is the predominant alkaloid during its early growth phase and is eight times more toxic than coniine. It is reduced to coniine which predominates as the plant matures. Unfortunately, the most likely time for animals to graze poison-hemlock is also the time when it is most toxic. Regrowth of poison-hemlock often occurs in the fall and may be the last green forage available to livestock in the fall. The immature fall regrowth is also toxic.

Toxicity from poison-hemlock may occur if adequate forage is not available. Hungry animals will graze plants that normally they would avoid if good quality forage was available. Poison-hemlock toxicity has occurred in cattle when adequate good quality forage was available. Cattle bedded down near or in patches of poison-hemlock have been poisoned by eating it in the early morning. Poison-hemlock is not as unpalatable as some have reported. Tule elk have been reported to graze poison-hemlock when other feed was available. Some elk would return to feed on poison-hemlock despite attempts to move them out of the poison-hemlock areas (Jessup et al. 1986).

Poison-hemlock loses its toxicity upon drying. When poison-hemlock is a contaminant in hay and allowed to dry, the risk of poisoning is greatly reduced. However, the seed remains toxic and could pose a problem in grain, hay, or if ensiled with haylage.

Management and Control

Some management guidelines can be followed to reduce chances of poison-hemlock toxicoses. One should prevent pregnant animals from grazing poison-hemlock prior to day 70 of gestation in the cow or day 60 in the pig. These are the critical stages of gestation when developmental anomalies may occur. Ranchers or producers should avoid putting animals deprived of feed in pastures where poisonous plants grow.

Poison-hemlock usually grows in dense patches in waste areas. Because the seed usually falls close to the parent plant, a stand of poison-hemlock generally remains in isolated areas. Grubbing or mowing prior to seed production can control smaller patches. Plowing or cultivation will prevent seedlings from maturing. Spraying these areas may be easily accomplished with the phenoxy herbicides, such as 2,4-D. However, poison-hemlock is a prolific seed producer, and repeated treatment in subsequent years may be needed to accomplish complete control. Poison-hemlock can be controlled in alfalfa with selective herbicides such as hexazinone.

Producers should be cautious about grazing animals in areas where poison-hemlock has been treated. It is not known whether herbicides increase the toxicity or the palatability of poison-hemlock. Therefore, ample time for sprayed plants to die and dry up should be provided before allowing animals to graze.

Economic Importance

The economic impact is difficult to assess and can only be estimated. Most deaths from poisonous plants, especially those from poison-hemlock, go undiagnosed and unreported. When considering losses, one must also consider the losses from malformed offspring and lost forage for grazing. All of these factors add to the overall loss associated with this poisonous plant.

Losses can only be accurately assessed where the cause of poisoning is known. However, there is a substantial loss to an individual when a valuable horse, a child's pony, or a registered cow in a small herd dies, or a deformed calf or litter of pigs is born. These losses are extremely difficult to express in monetary value.

We can estimate a dollar figure for the two examples of large losses in hogs and dairy cattle discussed in the toxicity section of this paper. The loss of 45 pigs approaching market weight at a current price of $52 cwt would be $4680. This is a substantial loss to the farmer. This does not include the reduced production of the 179 surviving hogs, which may cost this farmer just as much. The loss of 24 dairy cows and 26 heifers, at current prices of about $900 per cow (even at today's depressed milk industry prices), would cost this farmer $45,000 in actual death losses. This does not include losses from disposed of contaminated milk, care of those sick animals that eventually recovered, and lost production in the cows that did not die from the toxicosis. These could easily double the actual death losses.

Losses occur in almost all states; however, unless these losses can be diagnosed correctly and recorded, there is no way of accurately estimating the national loss from poison-hemlock. However, it is felt that the economic loss from this plant is substantially higher than most think. The cost of control and increased management must also be considered. Poison-hemlock contributes to the overall livestock loss from poisonous plants, estimated to be over $100 million annually in the United States (Nielsen and James 1985).

WATERHEMLOCK

Distribution and description

There are up to 20 species of Cicuta identified throughout the world, all of which are poisonous (Shishkin 1973). Most species are in North America. Nine are commonly found in the United States. Those most common to the western U.S. are C. douglasii, C. maculata, C. occidentalis, and C. vagans (Kingsbury 1964).

C. douglasii grows throughout the U.S.; C. maculata is the predominant species in the eastern U.S. and eastern Canada but is found in most of the U.S. C. occidentalis grows in the Rocky Mountain areas from the Black Hills and Washington south into Nevada and New Mexico. C. vagans grows throughout the Pacific Northwest and Canada (Kingsbury 1964, Muenscher 1975). Other species with wide geographical distribution include C. bulbifera, found across the northern U.S., southern Canada, and southeast to Virginia, and C. curtissii, distributed in the southeastern U.S. on the coastal plains from Virginia to Louisiana. Other species with more localized distribution include C. bolanderi and C. californica in central to western California and C. mackenziana, found in southeastern Alaska north to the Brooks range (Kingsbury 1964, Muenscher 1975). Species of Cicuta are often confused because many of the taxonomic characteristics are common between species.

Cicuta is also known by a large number of common names which have not been distinguished by species. The most common name used is waterhemlock and is descriptive of where the plant is most commonly found. Other names include spotted waterhemlock, spotted cowbane, beaver poison, childrens bane, death-of-man, musquash poison, musquash root, poison parsnip, snakeroot, snakeweed, false parsley, spotted parsley, wild carrot, mock-eel root, fever root, and poison-hemlock (Kingsbury 1964, Muenscher 1975). As can be seen from some of these names, they attempt to describe the observations of local people in different regions who have had experience with waterhemlock.

Botanists use characteristics of the fruit as distinguishing features; leaf structure is also used to classify species of waterhemlock (Kingsbury 1964). Unfortunately, when the plant is most toxic, the fruiting structure is not present, and positive identification may be incorrect. A few species are distinctive and are distinguished by morphology, habitat, and geographical distribution.

Cicuta spp. are considered perennial herbs; however, in the true sense, they are not a perennial. A stand of waterhemlock tends to persist year after year in the same location, which led early botanists to this conclusion. Before the root system dies in the fall, it produces a lateral tuber which winters over and initiates a new plant the following spring. Waterhemlock spreads primarily by seed, which are produced in abundance in late summer or early fall. If moisture and temperature allow, the seeds germinate and develop into small, fern-like plants the first

FIGURE 17.5
Waterhemlock.

year. The plants will blossom and produce seed the second
year (later, if water or nutritional reserves are limited).
Flowering occurs in late spring or early summer, after which
the upper parts of the plant become dried and desiccated.
Seeds are dense and generally drop close to the parent
plant, although they may be spread by animals, gusty winds,
and water.

   Cicuta grows to about 1 to 2 m (Figure 17.5) with
thickened tubers possessing long slender parsnip-like roots
radiating out from the main tubers (Figure 17.6). The
tubers exhibit several chambers or cavities separated by
cross-partitions as seen in a vertical cut through the tuber
at the base of the stem (Fig 17.7). The cut surface of the
stem and tuber exude a yellowish, thick, oily liquid with a
parsnip-like odor. The roots extending from the tuber or
lower thickened portion of the stem may be solid, whitish
and fleshy, resembling a parsnip. The stems are hollow

FIGURE 17.6
The tuber and root of waterhemlock (the most toxic portion
of the plant).

FIGURE 17.7
The cut surface of the waterhemlock tuber showing the thin
diaphragm-like pith layers which are distinguishing
features.

(except at the nodes), hairless, and may have purple stripes or spots on them. The stem is distinctly jointed.   Leaf stalks alternate up the stem partially clasping the main stem.   Leaves alternate two to three pinnately divided and may be 30 to 60 cm long. The leaflets are narrow, 2.5 to 10 cm long, the margins serrated, and the major veins ending in the notches between the teeth.   Flowers are small, white, and in terminal flat-topped or compound umbrella-shaped (umbellate) clusters. The fruits are small, with prominent ribs, and encased in a hard brownish shell (Kingsbury 1964, Muenscher 1975). In all Cicuta species except C. bulbifera and C. californica, the primary veins of the leaves are directed toward the notches of the serrations rather than the points or tips.   In most genera of Umbelliferae, the veins approach the tip or point of the serration (Bomhard 1936).

Habitat and Growth Conditions

Cicuta grows in wet meadows and marshes and along creeks, rivers, and ditches. It requires an abundance of moisture and is most often growing directly in water. The high water requirement limits its range and so it does not cover large areas of grazing land. All species of Cicuta grow in similar habitats.

Toxicity

Waterhemlock is considered to be the most violently poisonous plant known. It is toxic to all classes of livestock and to humans (Marsh and Clawson 1914, Fleming and Peterson 1920, Skidmore 1954, Starreveld and Hope 1975). Cicutoxin, the toxic principle, is a long chain highly unsaturated alcohol (Anet et al. 1953).

The toxin acts directly on the central nervous system, causing violent convulsions which are distinctive and severe.   Clinical signs of poisoning can appear within 15 minutes after ingestion of a lethal dose and begin with excessive salivation, nervousness, stumbling, and incoordination.   This is quickly followed by tremors, muscular weakness, and convulsive seizures interspersed intermittently by periods of relaxation.   Each successive convulsive period is shorter and causes further weakness, exhaustion, and death due to suffocation from a final paralytic seizure (Marsh and Clawson 1914, Fleming and Peterson 1920, Kingsbury 1964).

The convulsions are extreme and violent, often resulting in lacerations of the tongue or injury to the head or limbs; the head may be thrown rigidly back; legs become stiffened; eyes and mouth tightly closed; the animal laterally recumbent; and legs may flex as though running. There may be chewing and grinding of the teeth, the pupils of the eyes dilated, and temperature may be elevated a few degrees. Death may occur anywhere from 15 minutes to eight hours after ingestion of a lethal dose. Bloat in ruminants is common. To date, post mortem examination is not pathognomonic, although irritation of mucous membranes lining the stomach or rumen has been noted (Kingsbury 1964). Because of waterhemlock's rapid and violent mode of action, the veterinarian is usually unable to respond soon enough to help the intoxicated animal. In addition, there is no adequate treatment for this intoxication.

Three ewes (K.E. Panter and D.C. Baker, unpublished data) weighing 70 kg each were fed a single oral dose of ground, fresh waterhemlock root. The first ewe was fed 450 g and died in less than 2 hours, showing the signs described above. The second ewe, fed 100 g of fresh root, experienced slight nervousness and tremors but no convulsions or death. The third ewe, fed 200 g of fresh root, showed the clinical signs described above. In addition, there occurred frequent urination and defecation and convulsive seizures lasting one to two minutes each, followed by a period of relaxation with labored breathing, loss of muscular control, and sternal or lateral recumbency. These relaxed periods lasted eight to ten minutes in between eight seizures which occurred over a two-hour period. After two hours, no more seizures occurred, but she was exhausted and unable to stand until 12 hours later. She had muscular weakness and an unsteady gait for the next three days. Three days after feeding the plant, a blood sample was taken for blood chemistry. The only remarkable findings were lactic dehydrogenase (LDH) 3,280 IU/liter, serum glutamic oxalocetic transaminasae (SGOT) 3,060 IU/liter, and creatinine phosphokinase (CPK) 6,570 IU/liter compared to control values of 165, 55, and 10 IU/liter, respectively, for LDH, SGOT, and CPK. These values indicate muscle damage and may have resulted from the convulsive seizures.

Hamsters were fed different plant parts of waterhemlock to determine relative toxicity (K.E. Panter and D.C. Baker unpublished data). Fresh tuber was over two times more toxic than the white parsnip-like roots growing from the tuber; both were lethal. The tubers have been reported to be toxic at all stages of growth, although reports suggest the most dangerous period of toxicity is winter and early spring (Marsh and Clawson 1914). Drying appears to reduce the toxicity of tubers. Dry tubers fed to hamsters at 2.5 g/kg caused no signs of toxicity, which is contrary to earlier reports (Fleming and Peterson 1920). The difference between tuber and root toxicity has not been previously reported.

There are reports of the green foliage being toxic to livestock, especially in the early spring (Fleming and Peterson 1920). As the above-ground foliage grows and matures, it becomes less toxic. Seed, dry stems, and dry leaves were fed at 2.5 g/kg to hamsters; there were no signs of toxicity (K.E. Panter and D.C. Baker unpublished data). Mature foliage is reported to be completely harmless to livestock (Fleming and Peterson 1920). Further studies are needed to unequivocally determine the relative toxicity of waterhemlock.

Many cases of human poisoning and death have been reported from ingesting waterhemlock (Starreveld and Hope 1975). The tuberous roots are appealing to children. The roots are often mistaken for those of wild parsnip (Postinaca satina) or the tubers for wild artichoke (Helianthus annuus) (Kingsbury 1964). Treatment in humans with hemodialysis, hemoperfusion, forced diuresis, artificial ventilation and pancuronium bromide (to control convulsions) has been successful (Knutsen and Paszkowski 1984).

Conditions of Poisoning

Poisoning from waterhemlock usually occurs in early spring when the first growth of waterhemlock may be some of the first green forage to emerge. The early growth of the plant is toxic, but the tubers and the roots are the most dangerous part of the plant. Death occurs when animals pull the tuber out of the wet, moist ground and ingest it. One tuber has been reported to kill an adult cow. As the plant matures and the ground firms up, the likelihood of waterhemlock toxicoses diminishes.

Management and Control

Prevention of livestock losses from waterhemlock can be accomplished by following a few management steps. First, determine that waterhemlock is present. Second, control the plant by digging or spraying. Waterhemlock often grows in isolated clumps and can easily be dug with a shovel. One should be certain that exposed tubers are not left out in the open for access by livestock. Drying and burning is the safest way of destroying tubers. Waterhemlock is easily controlled with phenoxy herbicides. Wetting the plants when in full flower with 2,4-D at a rate of 2.25 kg/ha will give good control. Picloram at 0.56 kg/ha will also control waterhemlock (Humberg et al. 1981).

Economic Importance

Economic losses from waterhemlock are significant to the livestock industry every year. A specific dollar value, however, is difficult to estimate. Losses are generally isolated, are usually undiagnosed, and may involve only a few animals in a given herd, but over a large area the losses are larger than most believe. In 1897 it was estimated that 100 cattle died each year in Oregon from waterhemlock (Hendrick 1897). At today's cattle prices we could estimate that the loss may be in excess of $50,000 in Oregon alone. This loss could be similar in all 17 of the Western States and given an average value of $400 per animal, the loss could exceed $650,000 in the Western U.S. annually. Waterhemlock grows throughout the U.S. and is a potential problem whenever livestock graze areas where it grows. These projected losses plus the loss to human life warrant being aware of waterhemlock and its potential dangers.

SUMMARY

Poison-hemlock and waterhemlock are similar in habitat and patterns of growth and reproduction. They grow where adequate moisture is available. They can be controlled by similar methods, i.e., grubbing, digging, or spraying with phenoxy herbicides. Both genera are toxic, although the clinical effects of toxicity in livestock are different. Poison-hemlock causes muscular weakness, trembling, collapse, and death. It is teratogenic if ingested by pregnant livestock. All parts of poison-hemlock are toxic. Waterhemlock causes violent convulsions and death. The roots, tubers, and early spring growth of the waterhemlock are toxic.

LITERATURE CITED

Anet, E.F.L.J., B. Lythgoe, M.H. Silk, and S. Trippett. 1953. Oenanthotoxin and cicutoxin. Isolation and structures. J. Chem. Soc. 66:309-322.

Anonymous. 1951. Unusual case of hemlock poisoning in swine. California Vet. 2:26.

Bomhard, M.L. 1936. Leaf venation as a means of distinguishing Cicuta from Angelica. J. Wash. Acad. Sci. 26:102.

Bowman, W.C., and I.S. Sanghni. 1963. Pharmacological actions of hemlock (Conium maculatum) alkaloids. J. Pharm. and Pharmacology 15:1-25.

Buckingham, J.L. 1936. Poisoning in a pig by hemlock (Conium maculatum). Vet. J. 92:301-302.

Capithorne, B. 1937. Suspected poisoning of goats by hemlock (Conium maculatum). The Vet. Record 49:1018-1019.

Cromwell, B.T. 1956. The separation, micro-estimation and distribution of the alkaloids of hemlock (Conium maculatum L.). Biochem J. 64:259.

Cronquist, A. 1981. An integrated classification of flowering plants. Columbia University Press, New York. pp. 848-849

Dyson, D.A., and A.E. Wrathall. 1977. Congenital deformities in pigs possibly associated with exposure to hemlock (Conium maculatum). Vet. Rec. 10:241-242.

Edmonds, L.D., L.A. Selby, and A.A. Case. 1972. Poisoning and congenital malformations associated with consumption of poison hemlock by sows. J. Amer. Vet. Med. Assoc. 160:1319-1324.

Fairbairn, J.W., and P.N. Suwal. 1961. The alkaloids of hemlock (Conium maculatum L.) - II. Evidence for a rapid turnover of the major alkaloids. Phytochemistry 1:38-46.

Fleming, C.E., and N.F. Peterson. 1920. The poison parsnip or water hemlock (Cicuta occidentalis). University of Nevada Ag. Exp. Stat. Bull. 100.

Hendrick, U.P. 1897. A plant that poisons cattle. Cicuta. Oregon Exp. Sta. Bull. 46.

Humberg, N.E., H.P. Alley, and R.E. Bore. 1981. Research in Weed Science. University of Wyoming Ag. Exp. Stat. Bull. 163.

Jessup, D.A., H.J. Boermans, and N.D. Kock. 1986. Toxicosis in tule elk caused by ingestion of poison hemlock. J. Amer. Vet. Med. Assoc. 189(9):1173-1175.

Keeler, R.F. 1974. Coniine, a teratogenic principal from Conium maculatum producing congenital malformations in calves. Clin. Toxicol. 7(2):195-206.

Keeler, R.F., and L.D. Balls. 1978. Teratogenic effects in cattle of Conium maculatum and conium alkaloids and analogs. Clin. Toxicol. 12(1):49-64.

Keeler, R.F., L.D. Balls, J.L. Shupe, and M.W. Crowe. 1980. Teratogenicity and toxicity of coniine in cows, ewes, and mares. Cornell Vet. 70(1):19-26.

Kingsbury, J.M. 1964. Poisonous plants of the United States and Canada. Prentice-Hall, Inc., Englewood Cliffs, NJ. pp. 372-383.

Knutsen, O.H., and P. Paszkowski. 1984. New aspects in the treatment of water hemlock poisoning. Clin. Tox. 22:157-166.

Kubik, M., J. Refholec, and Z. Zachoval. 1980. Outbreak of hemlock poisoning in cattle. Veterinarstvi 30(4):157-158.

MacDonald, H. 1937. Hemlock poisoning in horses. The Vet. Record 49:1211-1212.

Marsh, C.D., and A.B. Clawson. 1914. Cicuta or water hemlock. USDA Bull. 69.

Muenscher, W.C. 1975. Poisonous plants of the United States, pp. 179-182. Collier Books, MacMillan Publishing Co., New York.

Nielsen, D.B., and L.F. James. 1985. Economic aspects of plant poisonings in livestock, pp. 24-31. In: A.A. Seawright, M.P. Hegarty, L.F. James and R.F. Keeler (eds), Plant toxicology. Proc. Aust.-U.S.A. Poisonous Plants Symp., Brisbane. Queensland Poisonous Plants Comm., Anim. Res. Inst., Yeerongpilly.

Panter, K.E. 1983. Toxicity and teratogenicity of Conium maculatum in swine and hamsters. Ph.D. Thesis. University of Illinois, Urbana, IL.

Panter, K.E., T.D. Bunch, and R.F. Keeler. 1988. Maternal and fetal toxicity of poison-hemlock (Conium maculatum) in sheep. Amer. J. Vet. Res. (In press)

Panter, K.E., R.F. Keeler, and W.B. Buck. 1985a. Induction of cleft palate in newborn pigs by maternal ingestion of poison hemlock (Conium maculatum). Amer. J. Vet. Res. 46(6):1368-1371.

Panter, K.E., R.F. Keeler, and W.B. Buck. 1985b. Congenital skeletal malformations induced by maternal ingestion of Conium maculatum (poison hemlock) in newborn pigs. Amer. J. Vet. Res. 46(10):2064-4066.

Panter, K.E., R.F. Keeler, W.B. Buck, and J.L. Shupe. 1983. Toxicity and teratogenicity of Conium maculatum in swine. Toxicon. Suppl. 3:333-336.

Penney, R.H.C. 1953. Hemlock poisoning in cattle. The Vet. Record 65:669-670.

Shishkin, B.K. 1973. Flora of the USSR, Umbelliflorae. Volume XVI. U.S. Dept. Commerce, National Technical Information Service, Springfield, VA. pp. 271-272.

Skidmore, L.V. 1954. Water hemlock (Cicuta maculata L.) poisoning in swine. Vet. J. 89:76.

Starreveld, E., and E. Hope. 1975. Cicutoxin poisoning (water hemlock). Neurology 25:730-734.

Widmer, W.R. 1984. Poison hemlock toxicosis in swine. Vet. Med. 79:405-408.

Willis, J.C. 1973. A dictionary of the flowering plants and ferns, p. 286. Revised by K.K. Airy Shaw. Cambridge University Press, New York.

18

Rain Lily (Cooperia pedunculata) Caused Photosensitization in Cattle and Deer in Texas

J.L. Schuster, B.S. Rector, L.D. Rowe, E.M. Bailey, Jr.,
J.C. Reagor, and S.L. Hatch

DISCOVERY OF RAIN LILY INDUCED PHOTOSENSITIZATION

Photosensitization (PS) is the syndrome which develops when an animal which is hyperreactive to sunlight is exposed to the sun. This syndrome consists of dermatitis and/or conjunctivitis, and in some cases hyperaesthesia, elicited only after exposure to light. Abnormal sensitivity to sunlight is produced when a photodynamic agent present in the skin is activated by light energy in sufficient quantities to cause cellular injury and tissue inflammation (Clare 1952).

Photosensitization of cattle was first reported in southcentral Texas in 1913 (Sperry et al. 1955). Studies during the 1950s (Hoffman 1954, Stroud 1955) in southcentral Texas cataloged epidemiologic factors associated with the condition and attempted unsuccessfully to establish its cause. The disease was believed to be due to the consumption of a plant containing a photodynamic agent. Cattle, sheep, goats, horses, and white-tailed deer were known to be affected.

There have been increasingly frequent occurrences of the disease during the past 30 years in DeWitt and surrounding counties in southcentral Texas. An estimated 15,000 cattle and 90% of the white-tailed deer of DeWitt County were affected in 1983 and 1984. Following severe outbreaks in 1982-83 and 1984-85, a task force was organized by the Texas Agricultural Extension Service to investigate the cause of the disease and develop control and management practices.

Systematic studies of ranches with known PS problems were conducted. Plant surveys, diet analyses, animal necropsies, blood analyses, and histological examinations of

227

animal tissue were used in the search for the causative agent. These early field investigations failed to identify known causes of PS as the source of the problem. Bishop's weed (<u>Ammi</u> <u>majus</u>), a plant containing furanocoumarin, was proposed as a potential cause of the problem (Dollahite et al. 1978) but was not found in sufficient quantities to be the causative agent. It was determined through serum biochemical profiles and gross and histopathological examinations of natural cases that liver damage was not a feature of the disorder in cattle or deer; thus, secondary photosensitization was ruled out.

Plant samples from pastures with affected animals were subjected to a microbiological assay for phototoxic activity (Daniels 1965). A large number of plant samples obtained from pastures where PS was documented during three outbreaks occurring between February 1983 and May 1985 were screened. This process led to the identification of potent phototoxic activity associated with dead leaf tips of rain lily (<u>Cooperia</u> <u>pedunculata</u>). Subsequent feedings of this material to white mice produced classical PS responses following long-wave ultraviolet (UVA) irradiation (Rowe et al., in press). Only dead leaf material produced the photosensitization; no other plant parts appeared to be involved. Green leaf material (fresh and dried) was not phototoxic in the microbiological assay despite repeated testing. Plant material of rain lily, consisting primarily of green leaves, was fed to sheep exposed to natural sunlight without producing evidence of PS. We have not been able to get mice to eat green leaf material, so we cannot say that it will not produce PS in that species if consumed.

The availability and phototoxic activity of dead leaf material of rain lily corresponds with outbreaks of photosensitization in cattle in DeWitt County during 1985 and 1986. Rain lily plant material collected during peak phototoxic activity produced PS in calves when they were fed between 0.6% (6.5 g/kg) and 1.5% (40 g/kg) of their body weight (Casteel et al., in press). This evidence strongly suggests that rain lily is involved with the etiology of recurring photosensitization of cattle and deer in southcentral Texas. The chemical nature of the causative agent is not yet known. Mycotoxins are suspected because of the environmental conditions characteristically present prior to outbreaks of PS. Peak PS activity historically has occurred two to ten days after rains when sufficient dead leaves of rain lily are available for animal consumption.

DESCRIPTION AND DISTRIBUTION

Rain lily is a member of the Amaryllidaceae family. It grows from a deep perennial bulb the size of a small onion (Figure 18.1). Common names are rain lily, giant rain lily, prairie rain lily, and white rain lily. Its leaves are fleshy, linear, 15 to 30 cm long and 0.6 to 1.2 cm wide, glabrous, and succulent. The plant blooms in the spring, summer, and fall after sufficient rains wet the root zone. The unbranched flower stalk is 12 to 23 cm tall, supporting a solitary white, six-petaled, trumpet-shaped flower measuring up to 5 cm wide. It occurs in south and central Texas with a documented distribution in 32 counties. Rain lily often dominates the aspect of old fields and native pastures in DeWitt County in early spring and late summer.

The peak flowering period occurs in late spring and early summer. Flowering continues from March to October when soil moisture is available. Leaves mature in late spring, and the entire plant becomes desiccated during hot summer months. Leaves again emerge during the fall if moisture is available. The plants grow until killed by frosts in late fall. The plants remain dormant during cold winters, but the cycle may begin again during warm periods of mild winters. Rain lily is an opportunistic species, putting out leaves throughout the year when moisture and temperature allow. Green rain lily leaves have been observed during each month of the year.

CLINICAL SIGNS

Photosensitization of livestock and deer occurs during the time dead leafy material is available. Rain lily appears to be sought out by white-tailed deer, but cattle graze it only by chance or when overstocked. In early stages of PS, light-pigmented areas of animals become yellow or red in color. The animals become extremely sensitive to sunlight and react to the sensitivity. Other signs include lacrimation, tail switching, and scratching, rubbing, kicking, or biting at affected parts. Unpigmented areas of skin may blister, and the skin and hair may slough. Animals become irritable, lame, lose their appetites, and may develop keratitis resulting in blindness. Severe cases may result in death; however, most adult animals recover completely. Suckling calves are often kept from nursing by cows with affected teats and udders.

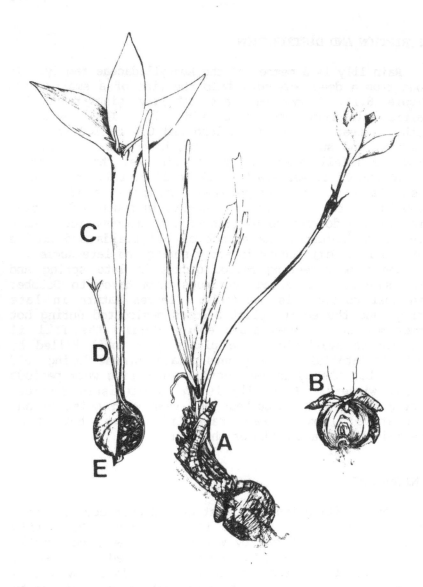

FIGURE 18.1
Rain lily (<u>Cooperia pedunculata</u>): (A) plant habit (bulb,
fresh leaves, peduncle, and flower); (B) longitudinal
section of bulb; (C-E) flower; (C) tepals and anthers at
apex; (D) style and stigma dislodged at apex from the
floral tubes of the pistil; (E) inferior ovary.

Calves force-fed rain lily aerial plant parts including senescent and dead leaves (0.6% to 1.5% of body weight) exhibited classical symptoms of PS. These clinical signs included increased lacrimation (tearing), photophobia (rapid blinking of eyes in sunlight), and erythema (reddening) of the nose, ears, and other hairless areas of the body. Subsequent clinical signs included slight corneal opacity and sloughing of skin over most of the body. One calf became debilitated and died.

CONDITIONS UNDER WHICH POISONING OCCURS

Poisoning occurs after animals ingest dead leaf materials of rain lily. Stalks, flowers, roots, and bulbs have not shown phototoxicity in the Candida assay. Most current outbreaks occur either in the late fall and winter (November to January) or after dry weather in late spring causes significant senescence of aboveground parts, and animals have access to dry material two to ten days after rains. The unique combination of available dead material and environmental conditions causes us to suspect microbial involvement in the PS process. Efforts are being made to isolate and purify the phototoxins and identify fungi which might be associated with the disease.

MANAGEMENT RECOMMENDATIONS

Prevention is the only tool currently available for management of the condition. Animals should not be allowed to graze pastures heavily infested with rain lily when large amounts of dead leaf material are available. At first signs of PS, affected animals should be moved to rain lily-free pastures or fed in pens. Animals may safely utilize the rain lily-infested pasture only when insignificant amounts of dead rain lily are available for consumption.

Development of rain lily-free pasture is recommended. Herbicidal control methods are being tested, but exact methods and specific herbicides are not known at this time.

Treatment recommendations for the disease follow those for classical primary photosensitization. Affected animals should either be removed from rain lily-infested pastures, placed in a shaded area, and given dry feed, or moved to pastures where rain lily does not occur. Since sunburn is the result, animals should be kept in shaded areas until recovery. Any secondary infections should be treated.

232

LITERATURE CITED

Casteel, S.W., L.D. Rowe, E.M. Bailey, J.C. Reagor, R.A. Fisk, and W.L. Schwartz. (In press). Photosensitization in cattle of South Texas: Experimentally induced photosensitization in cattle by Cooperia pedunculata. Amer. J. Vet. Res.

Clare, N.T. 1952. Photosensitization in diseases of domestic animals. Commonwealth Bureau of Animal Health Review Series No. 3. Farnham Royal, Bucks, England.

Daniels, F. 1965. A simple microbiological method for demonstrating phototoxic compounds. J. Invest. Dermatol. 44:259-263.

Dollahite, J.W., R.L. Younger, and G.O. Hoffman. 1978. Photosensitization in cattle and sheep caused by feeding Ammi majus (Greater Ammi; Bishop's Weed). Amer. J. Vet. Res. 39:193-197.

Hoffman, G.O. 1954. A study of photosensitization in cattle in Southeast Texas during 1953. M.S. Thesis, Texas A&M Univ., College Station.

Rowe, L.D., J.O. Norman, D.E. Corrier, S.W. Casteel, B.S. Rector, E.M. Bailey, J.L. Schuster, and J.C. Reagor. (In press). Photosensitization of cattle in South Texas: Identification of phototoxic activity association with Cooperia pedunculata. Amer. J. Vet. Res.

Sperry, O.E., R.D. Turk, G.O. Hoffman, and F.B. Stroud. 1955. Photosensitization of cattle in Texas. Tex. Agr. Exp. Sta. Bull. 812.

Stroud, F.B. 1955. A study of photosensitization of cattle in Southeast Texas during 1954. M.S. Thesis, Texas A&M Univ., College Station.

19

Nitrate Intoxication of Ruminant Livestock

James A. Pfister

INTRODUCTION

Nitrate intoxication in livestock from some roughages has been of concern since Bradley et al. (1940) identified the toxic compound in oat hay as nitrate (NO_3). Subsequent work worldwide indicated that nitrates were a potentially serious hazard to livestock (Wright and Davison 1964). Nitrates are ubiquitous in feed, food, and water, and are essentially nontoxic. Nitrate becomes toxic when reduced to nitrite (NO_2). Much current research on nitrate toxicity is centered on nitrite as a precursor of potential carcinogens in human foods (Archer 1982). This review will deal primarily with nitrate toxicity in ruminants consuming grazed or harvested forages. Nitrate toxicity has been demonstrated in cattle, horses, sheep, swine (Deeb and Sloan 1975), goats (Prasad 1983), reindeer (Nordkvist et al. 1984), and water buffalo (Prasad et al. 1984b).

TABLE 19.1
Conversion of factors for various methods of expressing nitrate and nitrite content of forages[1]/

Nitrate (NO_3) = Nitrate-nitrogen (NO_3-N) X 4.4
Nitrate (NO_2) = Nitrite-nitrogen (NO_2-N) X 3.3
Nitrate (NO_3) = Potassium nitrate (KNO_3) X 0.61
Parts per million (ppm) = % NO_x X 10,000
ppm = mg/kg = mg/liter

---

[1]/Adapted from Knight (1985) and Deeb and Sloan (1975)

TABLE 19.2.
Nitrate content (% of dry matter) of forages at various phenological stages or harvest date.

Species	Reference	Phenological stage or harvest date	High NO_3-N	Mean NO_3-N
Agropyron desertorum	Lawrence et al. 1981	May	-	.03
Agropyron smithii	White & Halvorson 1980	Vegetative	.03	.015
Agropyron smithii	Houston et al. 1973	July	-	.009
Avena sativa	Knott 1971	ND[1]	1.13	-
Bouteloua gracilis	Houston et al. 1973	July	-	.017
Brachiaria mutica	Prasad 1983	July	.55	.48
Bromus inermis	George et al. 1973	June 30-July 2	.04	.02
Chenopodium leptophyllum	Houston et al. 1973	July	-	.113
Cynodon dactylon	Hojjati et al. 1972	August	-	.004
Dactylis glomerata	George et al. 1973	May	.02	.161
Echinochloa crus-galli	Knott 1971	ND	.41	-
Elymus angustus	Lawrence et al. 1981	May	-	.03
Elymus angustus	Lawrence et al. 1981	June	-	.07
Elymus junceus	Lawrence et al. 1981	May	-	.03
Elymus junceus	Lawrence et al. 1981	June	-	.08
Festuca arundinacea	Hojjati et al. 1972	August	-	.005
Galenia pubescens	Williams 1979	ND	.35	.10
Hordeum vulgare	O'Hara & Fraser 1975	Apr. to Oct.	.35	.24
Lepidium densiflorum	Houston et al. 1973	July	-	.005
Lolium multiflorum	O'Hara & Fraser 1975	Apr. to Oct.	1.25	.29
Malva parviflora	Knott 1971	ND	.84	-

Species	Reference	Date		
Nicotiana megalosiphon	Knott 1971	ND	.60	-
Pennisetum clandestinum	O'Hara & Fraser 1975	Apr. to Oct.	.14	.11
Pennisetum spicatum	O'Hara & Fraser 1975	Apr. to Oct.	.14	.08
Phalaris sp.	O'Hara & Fraser 1975	Apr. to Oct.	.28	-
Phleum pratense	George et al. 1973	June	.014	.01
Portulaca oleracea	Knott 1971	ND	2.6	-
Salvia reflexa	Knott 1971	ND	.70	-
Secale cereale	Hojjati et al. 1972	Apr.-May	-	.01
Silybus marianum	Knott 1971	ND	-	2.7
Solanum nigrum	Knott 1971	ND	.58	-
Stipa viridula	White & Halvorson 1980	Vegetative	.08	.02
Stipa comata	Houston et al. 1973	July	-	.024
Trianthema monogyna	Gupta et al. 1983	ND	-	.25
Tribulus terrestris	Knott 1971	ND	.59	-
Urochloa panicoides	Knott 1971	ND	.69	-
Zea mays	O'Hara & Fraser 1975	Apr. to Oct.	.03	-
Zea mays	Knott 1971	ND	.83	-

[1]ND=No date given

Catastrophic death losses and numerous effects of chronic ingestion of nitrates on production have been reported (Kingsbury 1964). For earlier excellent reviews, see Wright and Davison (1964) and Deeb and Sloan (1975). Conversion factors for various methods of expressing nitrate levels in forages are given in Table 19.1.

NITRATE ACCUMULATION IN PLANTS

Numerous plants are known to accumulate nitrates at potentially toxic levels. A number of pasture and forage crop species [e.g., corn (Zea Mays), kikuyugrass (Pennisetum clandestinum), Johnson grass (Sorghum halepense)] and common hays [e.g., oats (Avena sativa), alfalfa (Medicago sativa)] may accumulate nitrates under certain circumstances. Table 19.2 gives nitrate levels for numerous forage species. Kingsbury (1964) and Muenscher (1951) have given detailed lists of potentially dangerous species. Harker and Kamau (1961) reported tests for nitrates in 26 African grasses, only one of which [elephant grass (Pennisetum purpureum)] was considered dangerous. Knott (1971) and O'Hara and Fraser (1975) have listed Australian and New Zealand forage crops likely to be hazardous. Recent recommendations for pen feeding potentially toxic forages are given in Table 19.3. Silages and water may also have toxic nitrate levels (Deeb and Sloan 1975) but will not be reviewed here.

The principal inorganic nitrogen source for plants is nitrate (Wright and Davison 1964). Generally, only small quantities of nitrates accumulate because they are rapidly reduced and combined with carbohydrates to form amino acids (O'Hara and Fraser 1975). However, excess nitrates can accumulate in plants under certain environmental conditions. The ability to accumulate nitrates is greatest in some plant families. For example, the Amaranthaceae, Chenopodiaceae, Compositae, Cruciferae, Solanaceae, and Poaceae families are considered to be potentially troublesome (Wright and Davison 1964). Perennial forage grasses generally contain low levels of nitrates (Carey et al. 1952; Murphy and Smith 1967; Lawrence et al. 1981), although in some areas they can accumulate dangerously high nitrate levels (Wright and Davison 1964).

As plants mature, nitrate content rises until the prebloom stage, peaks, and then begins to decline (Wright and Davison 1964). The nitrate concentration has also been shown to increase during the first ten days of regrowth (Sorensen 1971). Changing proportions of leaf, stem, and

TABLE 19.3
Recommendations for feeding forages containing nitrates[1]

NO$_3$-N (%) [2]	NO$_3$ (%) [2]	NO$_3$-N (ppm)	Feeding level
<0.10	<0.44	1000	Safe at all levels
0.10-0.15	0.44-0.66	1000-1500	Safe at all levels to nonpregnant animals. Limit use to 50% of diet dry matter (DM) for pregnant animals.
0.15-0.20	0.66-0.88	1500-2000	Safe if less than 50% of diet DM
0.20-0.35	0.88-1.54	2000-3500	Safe if less than 35% of diet DM
0.35-0.40	1.54-1.76	3500-4000	Safe for nonpregnant animals at less than 25% of diet DM. Not for pregnant animals.
>0.40	>1.76	>4000	Potentially toxic to ruminants.

[1] Adapted from Knight (1985)
[2] % of dry matter

fruit and decreased nitrogen levels in soil as the plant matures may account for the reduction in nitrates (Wright and Davison 1964). Stems are generally highest in nitrate concentration, followed by leaves and flowering structures. Nitrate levels tend to be lower in leaves because of relatively higher nitrate reductase levels.

The activity of the plant enzyme nitrate reductase is a key factor in nitrate accumulation (Walters and Walker 1979). Nitrate reductase reduces nitrate to nitrite before further reduction to amino nitrogen. Nitrate reductase activity is influenced by many factors, including genetic control, intensity of light, hydration, and nitrate substrate (Wright and Davison 1964). Light intensity may be an especially important factor in excess nitrate accumulation as this enzyme has very low activity in dark or shaded conditions (Hageman et al. 1961).

Numerous environmental factors influence nitrate levels in forage plants. One of the most common is a short drought (Wright and Davison 1964). Nitrate often accumulates in the soil during drought because it is not taken up by plants or leached. Once established plants begin growth after moisture-induced dormancy, high levels of nitrate are available for uptake (Harris and Rhodes 1969, Knott 1971, Gibson 1982). Ray and Sisson (1986) also found that kleingrass (Panicum coloratum) accumulated toxic nitrate levels during drought due to lower nitrate reductase activity. The combination of nitrate uptake at night and lack of reduction of nitrate by nitrate reductase during daylight hours, due to wilting and water loss, is an ideal combination for nitrate accumulation in plant tissues. If sufficient soil moisture is present for nitrate uptake during darkness when nitrate reductase activity is low, yet wilting occurs during hot, dry daylight hours, lowering nitrate reductase activity, nitrate accumulations in plants are likely. Thus a moderate drought is more likely to provide the necessary moisture conditions for nitrate accumulation than is a severe drought.

Cloudy weather or shading may enhance nitrate accumulation by reducing photosynthesis and decreasing amounts of carbohydrate intermediates available for carbon skeletons and assimilation of nitrogen (Wright and Davison 1964). Lower light levels often act in conjunction with drought conditions to produce very dangerous levels of nitrate in plants.

Levels of various nutrients have been implicated in nitrate accumulation (Wright and Davison 1964, Sorensen 1971). Molybdenum is an essential constituent of nitrate reductase, and molybdenum deficiency can interfere with the reduction of nitrate in plants (Walters and Walker 1979). Manganese deficiency (Wright and Davison 1964), excess potassium (Sorensen 1971), and high and low phosphorus levels (Wright and Davison 1964) have been related to nitrate accumulations.

Fertilizer application of nitrogen has perhaps received more research attention than any other nutrient. Numerous studies have shown that nitrogen fertilization can lead to high nitrate levels in forages (Wright and Davison 1964, Deeb and Sloan 1975). Nitrate levels above 2,000 ppm (0.2% NO_3-N) have generally been considered the minimum toxic level in forages by researchers working with fertilized and unfertilized forages (Kingsbury 1964; see Table 19.3 for recent recommendations). Lawrence et al. (1980, 1981) studied the effect of fertilizing ryegrass (Elymus angustus

and E. junceus) and crested wheatgrass (Agropyron desertorum) and found that the nitrate content in all species
increased with levels of fertilization. High levels of
nitrates were noted in all three grasses with application of
400 kg N/ha. The ryegrasses accumulated high nitrate levels
more readily than did crested wheatgrass. Crested wheatgrass had high nitrate levels for only a short period of
time during early June. Lawrence et al. (1968) also
reported that intermediate wheatgrass (Agropyron inter-
medium) had high levels of nitrates when fertilized with 225
kg N/ha.

Malhi et al. (1986) reported that fertilization of
bromegrass (Bromus inermis) produced high levels of nitrate
only during the first year of fertilization. Vegetative or
floral tillers of western wheatgrass (Agropyron smithii) did
not accumulate high levels of nitrate when fertilized with
640 kg N/ha (Houston et al. 1973; White and Halvorson 1980).
White and Halvorson (1980) also found that green needlegrass
(Stipa viridula) accumulated high nitrate levels when
fertilized above 320 kg N/ha. Massive fertilization (up to
672 kg N/ha) of shortgrass prairie did not result in high
nitrate accumulations in perennial grasses or forbs, but
dangerous levels were noted in two annual forbs (Houston et
al. 1973). Table 19.4 gives nitrate values for numerous
forage species at various levels of nitrogen fertilization.
Fertilizing rangelands is generally not economically
feasible, and rates of fertilization used on forage crops
are low enough that nitrate accumulation is generally not a
problem (O'Donovan and Conway 1968).

Herbicidal applications can influence nitrate levels
(Williams and James 1983). Stahler and Whitehead (1950)
found greatly increased nitrate levels in sugar beet leaves
due to 2,4-D application. Frank and Grigsby (1957) reported
variable results in 14 plant species after 2,4-D treatments.
Several plants, such as pigweed (Amaranthus retroflexus),
had higher nitrate concentrations while others had lower
levels. Some herbicide treatments have increased
palatability of toxic plants, but no information was found
on interactions between herbicides, nitrate levels, and
forage acceptance by livestock.

Nitrate Test

A simple field procedure called the diphenylamine blue
test to indicate the presence of high nitrate levels in
forages was detailed by Helwig and Setchell (1960). The

TABLE 19.4
Nitrate levels (% of dry matter) in forage fertilized at various rates as measured on various dates or phenological stages

Species	Reference	Date of Harvest or Phenological stage	N Application (kg/ha)	NO_3-N
Agropyron desertorum	Lawrence et al. 1981	June	400	.21
Agropyron smithii	White & Halvorson 1980	vegetative	160	.04
Agropyron smithii	White & Halvorson 1980	heading	160	.03
Agropyron smithii	White & Halvorson 1980	heading	640	.06
Agropyron smithii	White & Halvorson 1980	vegetative	640	.07
Agropyron smithii	Houston et al. 1973	July	448	.080
Agropyron smithii	Houston et al. 1973	July	224	.056
Agropyron smithii	Houston et al. 1973	July	672	.101
Avena elatior	Nikolic & Cmiljanic 1986	early heading	250	.134
Avena elatior	Nikolic & Cmiljanic 1986	immature	250	.264
Bouteloua gracilis	Houston et al. 1973	July	672	.031
Bouteloua gracilis	Houston et al. 1973	July	448	.026
Bouteloua gracilis	Houston et al. 1973	July	224	.018
Bromus inermis	MacLeod & MacLeod 1974	early heading	896	.43
Bromus inermis	MacLeod & MacLeod 1974	early heading	112	.06
Bromus inermis	George et al. 1973	May	336	.30
Bromus inermis	George et al. 1973	May	168	.09
Chenopodium leptophyllum	Houston et al. 1973	July	672	.551
Chenopodium leptophyllum	Houston et al. 1973	July	448	.465
Chenopodium leptophyllum	Houston et al. 1973	July	224	.487
Cynodon dactylon	Hojjati et al. 1972	August	300	.030
Cynodon dactylon	Hojjati et al. 1972	August	200	.005
Dactylis glomerata	Nikolic & Cmiljanic 1986	early heading	250	.317
Dactylis glomerata	Nikolic & Cmiljanic 1986	immature	250	.206

Species	Reference	Stage		
Dactylis glomerata	George et al. 1973	May	336	.40
Dactylis glomerata	George et al. 1973	May	168	.20
Elymus angustus	Lawrence et al. 1981	June	400	.46
Elymus junceus	Lawrence et al. 1981	June	400	.43
Festuca arundinacea	Nikolic & Cmiljanic 1986	immature	250	.007
Festuca arundinacea	Nikolic & Cmiljanic 1986	early heading	250	.104
Festuca arundinacea	Stritzke & McMurphy 1982	March	100	.75
Festuca arundinacea	Hojjati et al 1972	August	300	.15
Festuca arundinacea	Hojjati et al 1972	August	200	.02
Festuca pratensis	Nikolic & Cmiljanic 1986	immature	250	.198
Festuca pratensis	Nikolic & Cmiljanic 1986	early heading	250	.087
Lepidium densiflorium	Houston et al. 1973	July	672	.207
Lepidium densiflorium	Houston et al. 1973	July	448	.176
Lepidium densiflorium	Houston et al. 1973	July	224	.124
Phleum pratense	MacLeod & MacLeod 1974	early heading	896	.37
Phleum pratense	MacLeod & MacLeod 1974	early heading	112	.09
Phleum pratense	Nikolic & Cmiljanic 1986	early heading	250	.079
Phleum pratense	Nikolic & Cmiljanic 1986	immature	250	.307
Phleum pratense	George et al. 1973	June	336	.03
Phleum pratense	George et al. 1973	June	168	.04
Secale cereale	Hojjati et al. 1972	April-May	400	.08
Secale cereale	Hojjati et al. 1972	April-May	200	.05
Sorghuam sudanese	Gillingham et al. 1969	July-August	337	.18
Sorghuam sudanese	Gillingham et al. 1969	July-August	169	.11
Sorghuam sudanese	Gillingham et al. 1969	July-August	84	.06
Stipa comata	Houston et al. 1973	July	672	.131
Stipa comata	Houston et al. 1973	July	448	.111
Stipa comata	Houston et al. 1973	July	224	.062
Stipa viridula	White & Halvorson 1980	heading	640	.14
Stipa viridula	White & Halvorson 1980	heading	160	.06
Stipa viridula	White & Halvorson 1980	vegetative	640	.05
Stipa viridula	White & Halvorson 1980	vegetative	160	.04

reagent consists of diphenylamine and sulfuric acid. A representative plant tissue or blood or urine sample (taken within 24 hours of death) can be treated with a few drops of the reagent; high levels of nitrate will yield a blue color almost immediately (Helwig and Setchell 1960; Housholder et al. 1966). This test has also been used for other body fluids after death (Dollahite and Holt 1970).

Many methods are available for laboratory analysis of forage nitrates. McMahon and Casper (1984) reported wide variation in a controlled study where 16 laboratories analyzed four forage samples for KNO_3. Coefficients of variation for the samples were between 13 and 55%, and recovery of nitrate from a spiked sample averaged 105%. McMahon and Casper (1984) indicated that the results were unacceptable and that if feeding decisions were based on nitrate levels, some analyses may have resulted in feeding toxic material to livestock.

NITRATE TOXICITY IN RUMINANT LIVESTOCK

Nitrate (NO_3) is not toxic (Wolff and Wasserman 1972; Archer 1982), but becomes toxic when it is reduced to nitrite (NO_2) or another intermediate, hydroxylamine (NH_2OH), before being reduced to ammonia in the rumen (Smith and Layne 1969). Accumulation of nitrite occurs in the rumen when nitrate ingestion and reduction to nitrite exceeds that of nitrite reduction to ammonia due to different levels of reductive enzymes (Allison 1978).

Nitrate can be absorbed directly into the bloodstream, but evidence indicates that circulating nitrate cannot be reduced to nitrite in the bloodstream (Wang et al. 1961; WInter 1962). Significant amounts of nitrate (up to 27%) are excreted in the urine within a few hours of dosing (Wang et al. 1961; Setchell and Williams 1962), although some can be recycled from the bloodstream into the gut by salivary and gastrointestinal secretions (Deeb and Sloan 1975). Approximately 25% of fed nitrate was cleared from the rumen of sheep adapted to high nitrate diets within three hours after feeding (Alaboudi and Jones 1985).

Nitrite toxicity occurs when nitrite oxidizes the ferrous iron of hemoglobin to ferric iron, producing a chocolate-colored pigment called methemoglobin that cannot carry oxygen to body tissues. Clinical signs of nitrate toxicosis appear when methemoglobin concentrations reach 40% (Deeb and Sloan 1975). Small amounts of methemoglobin can be reconverted to hemoglobin by NADPH reductase (Vertregt

1977). When the capacity for reconversion is exceeded and concentrations of methemoglobin reach 70-80%, then death from methemoglobinemia usually results. However, functional hemoglobin or plasma nitrate levels may give a better indication of toxicity than blood methemoglobin level (Diven et al. 1962; Dollahite and Holt 1970).

Ruminants are particularly susceptible to nitrite poisoning because rumen microorganisms are responsible for essentially all reduction of nitrate to nitrite (Lewis 1951). Monogastric animals with significant microfloral fermentation in the lower gastrointestinal tract (e.g., horses with cecums) are also more susceptible to nitrite poisoning than are monogastric animals such as pigs. Ruminal pH is also an important factor in nitrate reduction by rumen microflora. Tillman et al. (1965b) found that optimum pH for nitrate reduction is 6.5 and 5.6 for nitrite. When pH favored nitrate reduction, nitrite appeared rapidly in the blood of sheep (Tillman et al. 1965b).

Species susceptibility to methemoglobinemia is also related to the capacity to reduce methemoglobin (O'Hara and Fraser 1975). Sheep are less susceptible to nitrate intoxications than are cattle (Harris and Rhodes 1969, Nielson 1974). Smith and Beutler (1966) studied the rate of methemoglobin formation in man, goats, sheep, horses, cattle, and pigs. The rate of methemoglobin reduction in erythrocytes of ruminants was highest in sheep and lowest in cattle. Hemoglobin from ruminants was more easily oxidized to methemoglobin than was hemoglobin from nonruminants. Swine showed the slowest rate of formation of methemoglobin of all species studied.

Ruminants appear to adapt to continuing ingestion of sublethal levels of nitrate through the compensatory erythropoietic response (Diven et al. 1964) of increasing hemoglobin levels, hematocrits, and blood volumes (Jainudeen et al. 1964). Coombe and Hood (1980) demonstrated that dairy cows grazing herbage fertilized with 750 kg N/ha could tolerate much higher levels of blood nitrite as a result of increasing functional hemoglobin levels. Methemoglobin-induced hypoxia appears to result in an erythropoietic response such that animals can tolerate relatively high levels of dietary nitrate (Deeb and Sloan 1975).

Nitrate toxicity is influenced greatly by the diet of livestock. Adequate levels of readily fermentable carbohydrates increase nitrite disappearance in the rumen (Wright and Davison 1964; Emerick et al. 1965; Knight 1985). Burrows et al. (1987) found that corn supplement protected against nitrate intoxication by reducing intraruminal nitrite. Fasted animals are more susceptible to nitrate intoxication than are fully fed animals (O'Hara and Fraser 1975). Kretschmer (1958) reported that fasted animals eat more rapidly and may become intoxicated on forages grazed safely by fed animals.

Geurink et al. (1979) have conducted 40 feeding experiments with high nitrate forages. Their work indicated that wide differences existed among and within forages in methemoglobin formation at equal nitrate intakes. Rate of nitrate ingestion was a significant factor in these studies. When total nitrate intake was equal, less hemoglobin was converted to methemoglobin as rate of nitrate intake slowed (Geurink et al. 1979). Dollahite and Holt (1970) reported that a calf ingested 1,134 mg NO_3-N/kg body weight in hay over 24 hours with no signs of toxicosis, yet another calf force-fed 320 mg NO_3-N/kg died within four hours.

Physical form of the feed can also influence toxicity. Nitrates are readily reduced in silages and moist hays (Deeb and Sloan 1975). Rate of methemoglobin formation was higher in cows fed dried hay compared with cows offered fresh mowed hay at the same total nitrate intake and at the same rate of ingestion (Geurink et al. 1979). Subsequent in vitro studies showed that the dried hay released 80% of total nitrate into distilled water within 20 minutes compared with only 30% from the fresh cut grass. Geurink et al. (1979) speculated that cell membranes of preserved (dried) roughage are more permeable than are membranes in fresh roughage.

Animals can adapt to relatively high nitrate levels in their diets if the levels of nitrate are increased gradually. This adaptation is due to an increase in the rate of nitrite reduction in the rumen (Sinclair and Jones 1964; Farra and Satter 1971; Kemp et al. 1977). Alaboudi and Jones (1985) found that nitrate and nitrite reduction were three and five times higher, respectively, in sheep already adapted to high nitrate diets. After withdrawing the nitrate from the diet of adapted sheep, reducing activities in the rumen fell to initial (pre-adaptation) levels within three weeks (Alaboudi and Jones 1985). Allison and Reddy (1984) and Reddy and Allison (1982) reported that nitrite reduction was many times greater in sheep adapted to nitrate diets compared with sheep receiving no nitrate. These studies by Allison have also demonstrated that much of the adaptation by rumen microorganisms occurs within four hours after ingestion of nitrate feed.

Although rumen microfloral populations can adapt to high nitrate levels, nitrate can be very toxic to nonadapted microorganisms. Nitrates are commonly used as preservatives in human food because of inhibition of microbial growth (Allison 1978). The incorporation of 10 mM of nitrite to bacterial colonies produced an 80% reduction in colony counts in rumen fluid of nonadapted sheep (Marinho 1986). Marinho (1986) tested six major groups of rumen bacteria for susceptibility to nitrate and nitrite and found that only four grew in the presence of nitrite. Little information is available on toxicity of nitrate to rumen protozoa. Marinho (1986) was able to detect no significant change in protozoal populations in rumen fluid from sheep fed nitrates and suggested that protozoa are important in the reduction of nitrite in the rumen.

Experimentally-induced nitrite intoxications have taken several forms and include massive doses placed directly into the rumen, intraperitoneal administration, oral doses in feed or water, or injections directly into the bloodstream. Plasma nitrite concentration in sheep was highest within two hours of administrating a single lethal intraruminal dose of 20 g potassium nitrite (Sinclair and Jones 1967). The same dosage given in the daily feed produced no effect. Diven et al. (1962) gave sodium nitrate doses of 0.1 to 0.5 g/kg body weight intraruminally to sheep and found a direct relationship between magnitude of the dose and the response. Blood nitrite levels were maximal at four to six hours post-treatment; signs of nitrite poisoning appeared within six to eight hours after treatment. Generally, peak levels of nitrite and methemoglobin in blood are reached approximately four to eight hours after animals are given large single doses of nitrite (Deeb and Sloan 1975). It appears that large animal to animal variation can be expected from the same dose of nitrate or nitrite (Winter 1962; Diven et al. 1962).

Because experiments with induced nitrite intoxications have shown such wide variation in responses, there is little agreement as to what is considered a toxic dose (LD_{50}) in cattle (Dollahite and Holt 1970). Estimates have ranged from 45 to 226 mg NO_3-N/kg body weight (O'Hara and Fraser 1975). Bradley et al. (1940) reported LD_{50} for nitrate (NO_3-N) to be approximately 74 mg/kg body weight for cattle. The trial upon which this figure is based was conducted by drenching cattle with single large doses. However, oat hay with much higher levels of nitrate, fed to cattle and sheep, was not lethal. In spite of this, Bradley et al. (1940) gave the lower limit for ingestion of toxic oat hay based on the drenching results, and this value in the past has been widely accepted. Crawford et al. (1966) estimated LD_{50} at 226 mg NO_3-N/kg when fed to cattle over a 24-hour period, but this figure may be too high because of the reported difficulty in getting the cattle to consume the experimental rations with added nitrate salts. Wright and Davison (1964) reported LD_{50} for cattle at 160 to 224 mg NO_3-N/kg when fed with the roughage. LD_{50} for sheep for a single oral dose is reported to be 70 mg NO_3-N/kg body weight (Deeb and Sloan 1975). When administered as a drench, the mass action of large doses of nitrate appears to favor a shift in the reduction equilibrium towards nitrite. Forage nitrate is released more slowly, and a closer balance can be maintained in reduction of nitrite to ammonia (O'Hara and Fraser 1975). Geurink et al. (1982), in summarizing results from feeding nearly 300 animals nitrate forage, have graphically shown the relationship between dry matter intake, rate and amount of nitrate ingestion, formation of methemoglobin, and clinical signs of toxicity.

Although adaptability to a high nitrate diet is commonly reported, Dollahite and Holt (1970) noted increased susceptibility to nitrate poisoning with repeated feeding. Apparently when animals are adjusted to low levels of nitrate in the feed, a sudden increase in nitrate level may be associated with an increased rate of reduction (Burrows et al. 1987). A more rapid reduction of nitrate may exceed nitrite reduction to ammonia, and intoxication may result. Kemp et al. (1977) demonstrated that the length of the period in which equal doses of high-nitrate hay are fed may have considerable effects on methemoglobin levels in blood. Much less nitrite was formed after the first dose of nitrate than after giving nitrate in equal doses on successive days. Kemp et al. (1977) also noted that conclusions on acceptable nitrate levels from single-day experiments may vary greatly from those of longer experiments.

Discrepancies in reported toxic doses of nitrate for livestock are not surprising considering the influence of such variables as biological variation in experimental animals, route of administration, dosage and duration interactions, length of adaptation to the nitrate diet, and physical and temporal dietary differences. Further, if laboratory estimates of feed nitrate are as inaccurate as suggested by McMahon and Casper (1984), this may also account for some of the variation in toxic doses reported in the literature. As a general rule, nonpregnant cattle can safely graze ad libitum forage with a nitrate content of 0.34 to 0.46% NO_3-N on a dry weight basis (Geurink et al. 1982).

Acute and Chronic Toxicity

Nitrate toxicity may be classified as acute or chronic (Wright and Davison 1964). In acute intoxications, the affected animal may die within hours of ingestion of a lethal dose or collapse and recover spontaneously. The primary response to acute nitrate intoxication is hypoxia, resulting in death from lack of oxygen due to methemoglobinemia (Deeb and Sloan 1975). Ruminants develop a brownish discoloration of nonpigmented skin and vaginal membranes. Geurink et al. (1982) indicate that discoloration of the vaginal membranes in females is a reliable criterion for detecting nitrate poisoning before other clinical signs are visible. Deeb and Sloan (1975) list a staggered gait, accelerated pulse, frequent urination, and labored breathing, followed by collapse, as visible symptoms of progressive intoxication. Lethal intoxications usually result in coma and death within two to three hours after symptoms appear (Deeb and Sloan 1975).

Chronic nitrate toxicity has been associated with a variety of problems, including reduction in weight gains, decreased milk production, abortion, vitamin A deficiency, and hypothyroidism (Wright and Davison 1964; Deeb and Sloan 1975). Considerable controversy surrounds these reported problems. Attempts to substantiate or refute effects of chronic nitrate ingestion have been variable (Deeb and Sloan 1975).

Depressed appetite and growth rate have been attributed to chronic nitrate ingestion (O'Hara and Fraser 1975). Pfander et al. (1964) reported reduced weight gains in cattle, sheep, and swine. Weichenthal et al. (1963) noted reduced gains in yearling steers of 0.2 kg/day when sodium nitrate was fed in the ration. Other reports have failed to demonstrate any effect on weight gains (Sokolowski et al. 1961; Cline et al. 1962; Smith et al. 1962; Crawford et al. 1966).

Effects of nitrates on milk production have been variable. Wright and Davison (1964) found that nitrate did not lower milk production until feed consumption was reduced or cows were near collapse. Milk production was not reduced in other studies (Davison et al. 1963; Crawford et al. 1966; Farra and Satter 1971). However, Nielson (1974) reported lower milk yields from cows on high nitrate forages.

Sublethal levels of nitrate have been reported to cause abortion (Deeb and Sloan 1975). Pfander et al. (1964) observed abortions in cows eating high nitrate silage. Cattle grazing of weeds with high nitrate content was implicated in abortions in Wisconsin (Simon et al. 1958, 1959a, b; Sund et al. 1960). Simon et al. (1959a) produced abortions in pregnant heifers with nitrate placed in the rumen. Davison et al. (1964) fed nitrate to 45 heifers before breeding and through gestation. Several abortions were noted, but estrous cycles, placentas, reproductive tracts, birth weights, and gains of calves were all normal. Winter and Hokanson (1964) found no detrimental effects of nitrate feeding on pregnancy in 15 heifers. Analysis of nitrate content of aqueous humor from 227 fetuses, stillborn, or weak calves indicated a high nitrate level in 83 cases (Johnson et al. 1984). Blood nitrate levels were high in 54 of the 83 cases in this study. A correctly balanced ration may alleviate nitrate effects on reproduction in cattle (Knight 1985).

Reproduction in sheep is apparently not greatly affected by nitrate consumption. Eppson et al. (1960) observed no abortions in ewes fed oat hay with 0.11% NO_3-N. Similar results were found by Setchell and Williams (1962). Sinclair and Jones (1964) could not produce abortions in sheep fed 0.21% NO_3-N but did note a reduction in lamb birth weights in the nitrate treated groups. Feeding ewes high levels of nitrate (0.77% NO_3-N in forage) produced abortions in three of seven ewes, but lower levels of nitrate produced no abortions (Davison et al. 1965). Subtoxic amounts of Tribulus terrestris fed to pregnant ewes during the last trimester of gestation resulted in high lamb mortality and an impaired ability of newborns to seek a teat (Broadmeadow et al. 1984).

The mechanism of abortion from nitrate has not been extensively studied. Work with guinea pigs suggested that acute nitrite toxicosis produced fetal deaths from hypoxia due to maternal methemoglobinemia (Sinha and Sleight 1971). Malestein et al. (1980) hypothesized that fetal death in nitrate toxicosis results from insufficient oxygen transfer to the fetus. Doses of 9 to 12 mg NO_2/kg body weight intravenously or 30 mg NO_2/kg body weight orally to pregnant cows resulted in much higher methemoglobin levels in maternal blood than in fetal blood. Results from this study demonstrated that oxygen capacity of the maternal blood decreases after nitrite treatment, and in pregnant animals, the fetus will be affected through lower oxygen transfer. When oxygen transfer to the fetal blood is reduced, intrauterine death may result (Malestein et al. 1980).

Nitrates have been postulated to impair thyroid function (Wright and Davison 1964). In rats, high levels of dietary nitrate apparently compete with iodine in the thyroid (Lee 1970) and interfere with thyroxine synthesis (Deeb and Sloan 1975). Data from rats has been used to explain poor performance of livestock consuming low amounts of nitrate (Wright and Davison 1964). Nitrates appear to have no effect on thyroid function or weight in cattle (Jainudeen et al. 1965), dogs (Kelley et al. 1974), and sheep (Arora et al. 1968). Long-term experiments with cattle using sublethal doses of nitrate had no effect on thyroid weight or histology (Wright and Davison 1964). Prasad (1983) observed slight hyperplasia of the thyroids of goats fed Brachiaria mutica. Deeb and Sloan (1975) concluded that if nitrates cause thyroid dysfunction, the effect is short-lived. Iodine stimulates thyroid activity, and dietary iodine level was shown to be very important for nitrate utilization in rat and sheep diets (Bloomfield et al. 1961). Knight (1985) concluded that hypothyroidism was unlikely to be a problem for ruminants on diets with high nitrate content and adequate iodine.

If dietary nitrate impaired thyroxine production by the thyroid, reduced vitamin A activity could result. An active thyroid increases intestinal absorption of carotene and subsequent conversion of carotene to vitamin A in rats (Johnson and Baumann 1947). Nitrate has been implicated as a possible factor in vitamin A deficiency in livestock (Smith et al. 1961; Hale et al. 1961). However, Cline et al. (1963) failed to affect the vitamin A status of lambs using a thyroid inhibitor in conjunction with high nitrate feed. It appears unlikely that long-term feeding of chronic levels of nitrate will produce a thyroid dysfunction resulting in vitamin A deficiency (Jainudeen et al. 1965).

Ingestion of nitrate may result in vitamin A deficiencies in livestock by interfering with carotene and vitamin A metabolism (McIlwain and Schipper 1963). Studies with rats indicated that nitrite can oxidize vitamin A or carotene in the gastrointestinal tract (Wright and Davison 1964; Deeb and Sloan 1975). Davison and Seo (1963) found that nitrate did not destroy carotene in rumen fluid. Several studies have found that nitrate had no influence on liver storage of vitamin A (Cline et al. 1962; Smith et al. 1962; Weichenthal et al. 1963; O'Donovan and Conway 1968). Vitamin A storage in the liver was not affected in sheep and cattle even though nitrate intoxication induced methemoglobinemia (Tillman et al. 1965a; Cunningham et al. 1968). Deeb and Sloan (1975) concluded that nitrates do not interfere with liver storage of vitamin A in ruminants. Nitrites may destroy vitamin A in ruminants under some circumstances, but acute nitrite toxicity would likely be of greater concern (Deeb and Sloan 1975).

Comparison to Other Toxicoses

Symptoms of nitrite intoxication may be confused with cyanosis (carbon dioxide poisoning) due to discoloration of blood and tissues (Wright and Davison 1964). In acute cyanosis, the blood turns dark red to brownish because of a lack of oxyhemoglobin and is distinguished from cyanide poisoning (hydrocyanic or prussic acid poisoning) where the blood turns cherry red because of an abundance of oxyhemoglobin. Nitrite and cyanide have different modes of toxicological action. Nitrite prevents oxygen transport to body tissues due to formation of methemoglobin, while cyanide prevents use of oxygen by the cytochrome system during oxidative phosphorylation. Nitrite may actually alleviate cyanide poisoning (Cran 1985). Sodium nitrite or sodium thiosulfate can be administered alone or together to produce low-level methemoglobinemia (Burrows 1981). Methemoglobin has a strong affinity for cyanide, and will bind the cyanide molecule and prevent inhibition of the cytochrome oxidation system (Wright and Davison 1964). Nitrites have also been used as a treatment to fix free hydrogen cyanide in the rumen in cases of cyanide poisoning (Webber et al. 1985). Burrows (1981) reported that the conventional dose of nitrite and thiosulfate (6.7 and 67 mg/kg, respectively) was less effective than larger doses (22 and 660 mg/kg nitrite and thiosulfate, respectively) for treatment of cyanide intoxication in sheep. It was also concluded that sodium thiosulfate without nitrate was an effective antidote for cyanide poisoning.

Treatment

The standard treatment for acute nitrite toxicity is intravenous administration of methylene blue solution (Deeb and Sloan 1975). Methylene blue apparently accepts electrons for NADPH reductase in the blood and accelerates the reconversion of methemoglobin to functional hemoglobin (Burrows 1984). It was recommended that the dosage of methylene blue be increased from 2 to 4 mg/kg to 15 mg/kg in severe methemoglobinemia. Therapeutic doses of ascorbic acid (vitamin C) have not been successful in reducing methemoglobin (Tai-hsiumg and Yu 1965; Prasad et al. 1984a). Dutch researchers (Korzeniowski et al. 1980, 1981) have reported considerable success using sodium tungstate to prevent nitrate toxicity. In vitro experiments indicated that tungsten inhibited nitrate reductase activity in rumen microbes. This inhibitory effect could be overcome by high levels of molybdenum (Korzeniowski et al. 1980). Later work (Korzeniowski et al. 1981) confirmed the effectiveness of tungsten in depressing nitrite formation in vivo. Antibiotics such as chlortetracycline may also give limited protection from nitrate toxicosis (Emerick and Embry 1961).

SUMMARY

Nitrates can be a serious health hazard to ruminants grazing pastures or offered harvested feed. Numerous factors may favor accumulation of nitrates by plants. The most common factors favoring accumulation of nitrates in plants are high concentrations of nitrate in soil, low light intensity, and moderate drought. During drought, enough soil moisture may be present to allow uptake of nitrite at night when activity of nitrate reductase is low. Hot, dry days may cause water loss and wilting, thus nitrate accumulated during darkness may not be reduced because of lowered nitrate reductase activity. Cloudy weather may also inhibit reduction of nitrate to nitrite so that toxic quantities of nitrate accumulate. Application of nitrogen fertilizers may increase nitrate levels in some forages to toxic levels. Livestock producers should be aware of plant species that have a tendency to accumulate nitrates to prevent ingestion or mechanical harvest of these forage crops when risk of nitrate accumulation is high. Field tests can give qualitative indications if forage has potentially dangerous nitrate levels. Nonpregnant cattle may generally graze forage of 0.34 to 0.46% NO_3-N (dry matter basis) with no risk. Harvested forage fed to livestock may be more toxic due to a more rapid rate of nitrate ingestion. Suspected nitrate toxicity should be verified by a veterinarian after collection of plant material and body fluids of affected animals.

LITERATURE CITED

Alaboudi, A.R., and G.A. Jones. 1985. Effect of acclimation to high nitrate intakes on some rumen fermentation parameters in sheep. Can J. Anim. Sci. 65:841.

Allison, M.J. 1978. The role of ruminal microbes in the metabolism of toxic constituents from plants, pp. 101-118. In: R.F. Keeler, K.R. Van Kampen, and L.F. James (eds), Effects of poisonous plants on livestock. Academic Press, New York.

Allison, M.J., and C.A. Reddy. 1984. Adaptations of gastrointestinal bacteria in response to changes in dietary oxalate and nitrite, pp. 248. In: M.J. Klug, and C.A. Reddy (eds), Proc. 3rd Int. Symp. Microb. Ecol. Michigan State Univ., East Lansing.

Archer, M.C. 1982. Hazards of nitrate, nitrite and N-nitroso compounds in human nutrition, pp. 327-381. In: J.H. Hathcock (ed), Nutritional toxicology, Vol. 1. Academic Press, New York.

Arora, S.P., E.E. Hatfield, U.S. Garrigus, F.E. Romack, and H. Motyka. 1968. Effect of adaptation to dietary nitrate on thyroxine secretion rate and growth in lambs. J. Anim. Sci. 27:1445.

Bloomfield, R.A., C.W. Welsch, G.B. Garner, and M.E. Muhrer. 1961. Effect of dietary nitrate on thyroid function. Science 134:1690.

Bradley, W.B., H.F. Eppson, and O.A. Beath. 1940. Livestock poisoning by oat hay and other plants containing nitrate. Wyo. Agr. Exp. Sta. Bull. 241.

Broadmeadow, A.C.C., D. Cobon, P.S. Hopkins, and B.M. O'Sullivan. 1984. Effects of plant toxins on peri-natal viability in sheep, pp. 216-219. In: D.R. Lindsay and D.T. Pearce (eds), Reproduction in sheep. Cambridge University Press, New York.

Burrows, G.E. 1981. Cyanide intoxication in sheep therapeutics. Vet. Hum. Toxicol. 23:22.

Burrows, G.E. 1984. Methylene blue: effects and disposition in sheep. J. Vet. Pharmacol. Therap. 7:225.

Burrows, G.E., G.W. Horn, R.W. McNew, L.I. Croy, R.D. Keeton, and J. Kyle. 1987. The prophylactic effect of corn supplementation on experimental nitrate intoxication in cattle. J. Anim. Sci. 64:1682.

Carey, V., H.L. Mitchell, and K. Anderson. 1952. Effect of nitrogen fertilizer on the chemical composition of bromegrass. Agron. J. 44:467.

Cline, T.R., E.E. Hatfield, and U.S. Garrigus. 1962. Effects of potassium nitrate, alpha-tocopherol, thyroid treatments, and vitamin A on weight gain and liver storage of vitamin A in lambs. (Abstract). J. Anim. Sci. 21:991.

Cline, T.R., E.E. Hatfield, and U.S. Garrigus. 1963. Effects of potassium nitrate, alpha-tocopherol, thyroid treatments, and vitamin A on weight gain and liver storage of vitamin A in fattening lambs. J. Anim. Sci. 22:911.

Coombe, N.B., and A.E.M. Hood. 1980. Fertilizer nitrogen effects on dairy cow health and performance. Fert. Res. 1:157.

Cran, H.R. 1985. Suspected hydrocyanic acid poisoning in cattle. Vet. Rec. 116:349.

Crawford, R.F., W.K. Kennedy, and K.L. Davison. 1966. Factors influencing the toxicity of forages that contain nitrate when fed to cattle. Cornell Vet. 56:1.

Cunningham, G.N., M.B. Wise, and E.R. Barrick. 1968. Influence of nitrite and hydroxylamine on performance and vitamin A and carotene metabolism of ruminants. J. Anim. Sci. 27:1067.

Davison, K.L., W. Hansel, L. Krook, K. McEntee, and M.J. Wright. 1964. Nitrate toxicity in dairy heifers. I. Effects on reproduction, growth, lactation, and vitamin A nutrition. J. Dairy Sci. 47:1065.

Davison, K.L., W. Hansel, K. McEntee, and M.J. Wright. 1963. Reproduction, growth and lactation of heifers fed nitrate. (Abstract). J. Anim. Sci. 22:835.

Davison, K.L., K. McEntee, and M.J. Wright, 1965. Responses in pregnant ewes fed forages containing various levels of nitrate. J. Dairy Sci. 48:968.

Davison, K.L., and J. Seo. 1963. Influence of nitrate upon carotene destruction during in vitro fermentation with rumen liquor. J. Dairy Sci. 46:862.

Deeb, B.S., and K.W. Sloan. 1975. Nitrates, nitrites and health. University of Illinois, Urbana-Champaign, Bull. 750.

Diven, R.H., R.E. Reed, and W.J. Pistor. 1964. The physiology of nitrite poisoning in sheep. Ann. N.Y. Acad. Sci. 111:638.

Diven, R.H., R.E. Reed, R.J. Trautman, W.J. Pistor, and R.E. Watts. 1962. Experimentally induced nitrite poisoning in sheep. Amer. J. Vet. Res. 23:494.

Dollahite, J.W., and E.C. Holt. 1970. Nitrate poisoning. S. Afr. Med. J. 44:171.

Emerick, R.J., and L.B. Embry. 1961. Effect of chlortetracycline on methemoglobinemia resulting from the ingestion of sodium nitrate by ruminants. J. Anim. Sci. 20:844.

Emerick, R.J., L.B. Embry, and R.W. Seerley. 1965. Rate of formation and reduction of nitrite-induced methemoglobin in vitro and in vivo as influenced by diet of sheep and age of swine. J. Anim. Sci. 24:221.

Eppson, H.F., M.W. Glenn, W.W. Ellis, and C.S. Gilbert. 1960. Nitrate in the diet of pregnant ewes. J. Amer. Vet. Med. Assoc. 137:611.

Farra, P.A., and L.D. Satter. 1971. Manipulation of the ruminal fermentation. III. Effect of nitrate on ruminal volatile fatty acid production and milk composition. J. Dairy Sci. 54:1018.

Frank, P.A., and B.H. Grigsby. 1957. Effects of herbicidal sprays on nitrate accumulation in certain weed species. Weeds 5:206.

George, J.R., C.L. Rhykerd, C.H. Noller, J.E. Dillon, and J.C. Burns. 1973. Effect of N fertilization on dry matter yield, total N, N recovery, and nitrate N concentration of three cool-season forage grass species. Agron. J. 65:211.

Geurink, J.H., A. Malestein, A. Kemp, A. Korzeniowski, and A. Th. Van't Klooster. 1982. Nitrate poisoning in cattle. 7. Prevention. Neth. J. Agric. Sci. 30:105.

Geurink, J.A., A. Malestein, A. Kemp, and A. Th. Van't Klooster. 1979. Nitrate poisoning in cattle. 3. The relationship between nitrate intake with hay or fresh roughage and the speed of intake on the formation of methemoglobin. Neth. J. Agric. Sci. 27:268.

Gibson, E.A. 1982. Effect of the drought of 1976 on the health of cattle, sheep and other farm livestock in England and Wales. Vet. Rec. 11:407.

Gillingham, J.T., M.M. Shirer, J.J. Starnes, N.R. Page, and E.F. McClain. 1969. Relative occurrence of toxic concentrations of cyanide and nitrate in varieties of sudangrass and sorghum-sudangrass hybrids. Agron. J. 61:727.

Gupta, B.K., S.C. Gupta, and I.S. Thind. 1983. Effect of feeding Trianthema monogyna to goats. J. Res. Punjab Agric. Univ. 20:539.

Hageman, R.H., D. Flesher, and A. Gittu. 1961. Diurnal variation and other light effects influencing the activity of nitrate reductase and nitrogen metabolism in corn. Crop Sci. 1:201.

Hale, W.H., F. Hubbert, Jr., and R.E. Taylor. 1961. The effect of concentrate level and nitrate addition on hepatic vitamin A stores and performance of fattening steers. (Abstract). J. Anim. Sci. 20:934.

Harker, K.W., and A.K. Kamau. 1961. A preliminary survey of the nitrate content of various grasses. E. Afr. Agric. and For. J. 27:57.

Harris, D.J., and H.A. Rhodes. 1969. Nitrate and nitrite poisoning in cattle in Victoria. Aust. Vet. J. 45:590.

Helwig, D.M., and B.P. Setchell. 1960. Observations on the diagnosis of nitrate poisoning in sheep. Aust. Vet. J. 36:14.

Hojjati, S.M., T.H. Taylor, and W.C. Templeton, Jr. 1972. Nitrate accumulation in rye, tall fescue, and bermudagrass as affected by nitrogen fertilization. Agron. J. 64:624.

Housholder, G.T., J.W. Dollahite, and R. Hulse. 1966. Diphenylamine for the diagnosis of nitrate intoxication. J. Amer. Vet. Med. Assoc. 148:662.

Houston, W.R., L.D. Sabatka, and D.N. Hyder. 1973. Nitrate-nitrogen accumulation in range plants after massive N fertilization on shortgrass plains. J. Range Manage. 26:54.

Jainudeen, M.R., W. Hansel, and K.L. Davison. 1964. Nitrate toxicity in dairy heifers. 2. Erythropoietic responses to nitrate ingestion during pregnancy. J. Dairy Sci. 47:1382-1387.

Jainudeen, M.R., W. Hansel, and K.L. Davison. 1965. Nitrate toxicity in dairy heifers. 3. Endocrine responses to nitrate ingestion during pregnancy. J. Dairy Sci. 48:217.

Johnson, R.M., and C.A. Baumann. 1947. The effect of thyroid on the conversion of carotene into vitamin A. J. Biol. Chem. 171:513.

Johnson, J.L., N.R. Schneider, C.L. Kelling, and A.R. Doster. 1984. Nitrate exposure in perinatal beef calves. Proc. Ann. Meeting Amer. Assoc. Vet. Lab. Diagn. 26:167.

Kelley, S.T., F.W. Oehme, and S.B. Hoffman. 1974. Effect of chronic dietary nitrates on canine thyroid function. Toxic. Appl. Pharm. 27:200.

Kemp, A., J.H. Geurink, R.T. Haalstra, and A. Malestein. 1977. Nitrate poisoning in cattle. 2. Changes in nitrite in rumen fluid and methemoglobin formation in blood after high nitrate intake. Neth. J. Agric. Sci. 25:51.

Kingsbury, J.M. 1964. Poisonous plants of the United States and Canada. Prentice-Hall, Englewood Cliffs, NJ.

Knight, A.P. 1985. The toxicology of sulfur and nitrate in ruminants. Bovine Pract. 20:121.

Knott, S.G. 1971. Nitrite poisoning in livestock. Queensland Agric. J. 97:485.

Korzeniowski, A., J.H. Geurink, and A. Kemp. 1980. Nitrate poisoning in cattle. 5. The effect of tungsten on nitrite formation by rumen microbes. Neth. J. Agric. Sci. 28:16.

Korzeniowski, A., J.H. Geurink, and A. Kemp. 1981. Nitrate poisoning in cattle. 6. Tungsten as a prophylactic against nitrate-nitrite intoxication in ruminants. Neth. J. Agric. Sci. 29:37.

Kretschmer, A.E., Jr. 1958. Nitrate accumulation in everglades forages. Agron. J. 50:314.

Lawrence, T., H.C. Korven, G.E. Winkleman, and F.G. Warder. 1980. The productivity and chemical composition of altai wild ryegrass (Elymus angustus) as influenced by time of irrigation and time and rate of nitrogen fertilization. Can. J. Plant Sci. 60:1179.

Lawrence, T., F.G. Warder, and R. Ashford. 1968. Nitrate accumulation in intermediate wheatgrass. Can. J. Plant Sci. 48:85.

Lawrence, T., G.E. Winkleman, and F.G. Warder. 1981. Nitrate accumulation in altai wild ryegrass (Elymus angustus), russian wild ryegrass (E. junceus) and crested wheatgrass (Agropyron desertorum). Can. J. Plant Sci. 61:735.

Lee, D.H.K. 1970. Nitrates, nitrites, and methemoglobinemia. Environ. Res. 3:484.

Lewis, D. 1951. The metabolism of nitrate and nitrite in sheep. 2. Hydrogen donators in nitrate reduction by rumen microorganisms in vitro. Biochem. J. 49:149.

McIlwain, P.K., and I.A. Schipper. 1963. Toxicity of nitrate nitrogen to cattle. J. Amer. Vet. Med. Assoc. 142:502.

MacLeod, L.B., and J.A. MacLeod. 1974. Effects of N and K fertilization on the protein, nitrate and non-protein reduced N fractions of timothy and bromegrass. Can. J. Plant Sci. 54:331.

McMahon, T.L., and H.H. Casper. 1984. Forage nitrate analysis: laboratory performance study. J. Assoc. Off. Anal. Chem. 67:1026.

Malestein, A., J.H. Geurink, G. Schuyt, A.J. Schotman, A. Kemp, and A.T. Van't Klooster. 1980. Nitrate poisoning in cattle. 4. The effect of nitrite dosing during parturition on the oxygen capacity of maternal blood and the oxygen supply to the unborn calf. Vet. Q. 2:149.

Malhi, S.S., D.K. McBeath, and V.S. Baron. 1986. Effects of nitrogen application on yield and quality of bromegrass (Bromus inermis) hay in central Alberta, Canada. Can. J. Plant Sci. 66:609.

Marinho, A.A.M. 1986. Nitrate toxicity in ruminants: metabolism of nitrate and nitrite in the rumen. Rev. Port. Cienc. Vet. 81:67.

Muenscher, W. 1951. Poisonous plants of the United States. MacMillan, New York.

Murphy, L.S., and G.E. Smith. 1967. Nitrate accumulation in forage crops. Agron. J. 59:171.

Nielson, F.J.A. 1974. Nitrite and nitrate poisoning with special references to 'Grasslands Tama' ryegrass. N.Z. Vet. J. 22:12.

Nikolic, J.A., and R. Cmiljanic. 1986. The nitrate content of some green forage and its possible effect when used for ruminant nutrition. Acta. Vet. Belgr. 36:107.

Nordkvist, M., C. Rehbinder, S.C. Mukherjee, and K. Erne. 1984. Pathology of acute and subchronic nitrate poisoning in reindeer (Rangifer tarandus). Rangifer 4:9.

O'Donovan, P.B., and A. Conway. 1968. Performance and vitamin A status of sheep grazing high-nitrate pastures. J. Brit. Grassland Soc. 23:228.

O'Hara, P.J., and A.J. Fraser. 1975. Nitrate poisoning in cattle grazing crops. N.Z. Vet. J. 23:45.

Pfander, W.H., G.B. Thompson, G.B. Garner, L.M. Flynn, and A.A. Case. 1964. Chronic nitrate/nitrite toxicity from feed and water. Missouri Agric. Exp. Stat. Spec. Rept. 38:3.

Prasad, B., S.K. Misra, and P.C. Chondhur. 1984a. Therapeutic efficacy of ascorbic acid in acute methemoglobinemia in buffalo calves. Indian J. Vet. Med. 4:90.

Prasad, B., S.K. Misra, and P.P. Gupta. 1984b. Pathological changes in experimental nitrite poisoning in buffalo calves. Indian J. Anim. Sci. 54:1081.

Prasad, J. 1983. Effect of high nitrate diet on thyroid glands in goats. Indian J. Anim. Sci. 53:791.

Ray, I.M., and W.B. Sisson. 1986. Nitrate reductase activity of kleingrass (Panicum coloratum) during drought in the northern Chihuahuan desert. J. Range Manage. 39:531.

Reddy, C.A., and M.J. Allison. 1982. Characterization of nitrite reducing activity in cell extracts of mixed rumen bacteria from sheep adapted to high levels of dietary nitrate, pp. 84. In: Proc. 13th Intl. Cong. Microbiol., Boston, MA.

Setchell, B.P., and A.J. Williams. 1962. Plasma nitrate and nitrite concentration in chronic and acute nitrate poisoning in sheep. Aust. Vet. J. 38:58.

Simon, J., J.M. Sund, F.D. Douglas, M.J. Wright, and T. Kowalczyk. 1959a. The effect of nitrate or nitrite when placed in the rumens of pregnant dairy cattle. J. Amer. Vet. Med. Assoc. 135:311.

Simon, J., J.M. Sund, M.J. Wright, and F.D. Douglas. 1959b. Prevention of noninfectious abortion in cattle by weed control and fertilization practices on lowland pastures. J. Amer. Vet. Med. Assoc. 135:315.

Simon, J., J.M. Sund, M.J. Wright, A. Winter, and F.D. Douglas. 1958. Pathological changes associated with the lowland abortion syndrome in Wisconsin. J. Amer. Vet. Med. Assoc. 132:164.

Sinclair, K.B., and D.I.H. Jones. 1964. The effect of nitrate on blood composition and reproduction in the ewe. Brit. Vet. J. 120:78.

Sinclair, K.B., and D.I.H. Jones. 1967. Nitrite toxicity in sheep. Res. Vet. Sci. 8:65.

Sinha, D.P., and S.D. Sleight. 1971. Pathogenesis of abortion in acute nitrite toxicosis in guinea pigs. Toxicol. Appl. Pharmacol. 18:340.

Smith, G.S., E.E. Hatfield, W.M. Durdle, and A.L. Neumann. 1962. Vitamin A status of cattle and sheep as affected by nitrate added to rations of hay or silage and by supplementation with carotene or preformed vitamin A. (Abstract). J. Anim. Sci. 21:1013.

Smith, G.S., A.L. Neumann, W.G. Huber, H.A. Jordan, and O.B. Ross. 1961. Avitaminosis in cattle fed silage rations supplemented with vitamin A. (Abstract). J. Anim. Sci. 20:952.

Smith, J.E., and E. Beutler. 1966. Methemoglobin formation and reduction in man and various species. Amer. J. Physiol. 210:347.

Smith, R.P., and W.R. Layne. 1969. A comparison of the lethal effects of nitrite and hydroxylamine in the mouse. J. Pharmacol. Exp. Ther. 165:30.

Sokolowski, J.H., U.S. Garrigus, and E.E. Hatfield. 1961. Effects of inorganic sulfur on KNO_3 utilization by lambs. (Abstract). J. Anim. Sci. 20:953.

Sorensen, C. 1971. Influence of various factors on nitrate concentrations in plants in relation to nitrogen metabolism, pp. 229-240. In: R.M. Samish (ed), Recent advances in plant nutrition, Vol. II. Gordon and Breach, New York.

Stahler, L.M., and E.I. Whitehead. 1960. The effect of 2,4-D on potassium nitrate levels in leaves of sugar beets. Science 112:749.

Stritzke, J.F., and W.E. McMurphy. 1982. Shade and nitrogen effects on tall fescue (Festuca arundinacea) production and quality. Agron. J. 74:5.

Sund, J.M., M.J. Wright, and J. Simon. 1960. Nitrates in weeds cause abortion in cattle. In: Proc. 17th North Central Weed Control Conf. No. 31.

Tai-hsiumg, Ch'en, and Ch'uan Yu. 1965. Experimental therapy of methemoglobinemia in nitrite poisoning. Acta Vet. Zootech. Sinica 8:279.

Tillman, A.D., G.M. Sheriha, R.D. Goodrich, E.C. Nelson, and G.S. Smith. 1965a. Effect of injected hydroxylamine upon vitamin A status of sheep. J. Anim. Sci. 24:1136.

Tillman, A.D. G.M. Sheriha, and R.J. Sirny. 1965b. Nitrate reduction studies with sheep. J. Anim. Sci. 24:1140.

Vertregt, N. 1977. The formation of methemoglobin by the action of nitrite on bovine blood. Neth. J. Agric. Sci. 25:243.

Walters, C.L., and R. Walker. 1979. Consequences of accumulation of nitrate in plants, pp. 637-649. In: E.J. Jewott and C.V. Cutting (eds), Nitrogen assimilation of plants. Academic Press, New York.

Wang, Li Chuan, J. Garcia-Rivera, and R.H. Burris. 1961. Metabolism of nitrate by cattle. Biochem. J. 81:237.

Webber, J.J., C.R. Roycroft, and J.D. Callinan. 1985. Cyanide poisoning of goats from sugar gums (Eucalyptus cladocalyx). Aust. Vet. J. 62:28.

Weichenthal, B.A., L.B. Embry, R.J. Emerick, and F.W. Whetzal. 1963. Influence of sodium nitrate, vitamin A, and protein level on feedlot performance and vitamin A status of fattening cattle. J. Anim. Sci. 22:979.

White, L.M., and A.D. Halvorson. 1980. Nitrate levels in vegetative and floral tillers of western wheatgrass and green needlegrass as affected by nitrogen fertilization. Agron. J. 72:143.

Williams, M.C. 1979. Toxicological investigations of Galenia pubescens. Weed Sci. 27:506.

Williams, M.C., and L.F. James. 1983. Effects of herbicides on the concentration of poisonous compounds in plants: a review. Amer. J. Vet. Res. 44:2420.

Winter, A.J. 1962. Studies on nitrate metabolism in cattle. Amer. J. Vet. Res. 23:500.

Winter, A.J., and J.F. Hokanson. 1964. Effects of long-term feeding of nitrate, nitrite, or hydroxylamine on pregnant dairy heifers. Amer. J. Vet. Res. 25:353.

Wolff, I.A., and A.E. Wasserman. 1972. Nitrates, nitrites, and nitrosamines. Science 177:15.

Wright, M.J., and K.L. Davison. 1964. Nitrate accumulation in crops and nitrate poisoning in animals, pp. 197-247. In: A.G. Norman (ed), Advances in agronomy Vol. 16. Academic Press, Inc., New York.

20

Oxalate Poisoning

J.A. Young and L.F. James

INTRODUCTION

On a worldwide basis, oxalate poisoning of grazing animals is usually associated with plants of the wood-sorrel family (Oxalidaceae). This family consists of about ten genera and over 50 species widely distributed in temperate and tropical regions (Munz and Keck 1968). Oxalate poisoning of sheep was reported from Australia in the 1920s (Bull 1929), and it is still considered to be a problem (Dodson 1959).

In North America, oxalate poisoning has been associated with certain members of the chenopod or goosefoot family (Chenopodiaceae). The two species of chenopods that have caused the most problems with oxalate are adapted to similar habitats but have contrasting growth forms and origins. Greasewood (<u>Saracobatus</u> <u>vermiculatus</u>) is a North American endemic shrub that in exceptional cases reaches 2.5 m in height. Halogeton (<u>Halogeton</u> <u>glomeratus</u>) is a low-growing, herbaceous annual that is native to central Asia but has been accidentally introduced to the Intermountain Area between the Sierra Nevada-Cascade and the Rocky Mountains.

GREASEWOOD

Greasewood was suspected of poisoning sheep in Nevada as early as 1916 (Fleming et al. 1928). C.E. Fleming, of the Nevada Agricultural Experiment Station, began feeding trials with greasewood in 1918. Greasewood jumped in prominence as a poisonous plant in May 1920 when a large band of sheep were being trailed out of Nevada into Utah. The sheep had been traveling over an area of range where

261

good feed was scarce. From this degraded salt desert range
of scanty feed they suddenly came upon a large area with a
dense stand of greasewood. The greasewood had just started
growth with the current year's shoots elongated by 5 to 7.5
mm. The abundance of this luxuriant, succulent growth
permitted the sheep to satisfy their hunger quickly. Within
a period of six to ten hours, a large number showed distinct
signs of poisoning in varying degrees of severity.
Eventually most of the band was lost.

Fleming and his associates demonstrated from feeding
trials that greasewood could poison sheep and isolated the
potassium and sodium salts of oxalic acid from the leaves of
the plant (Fleming et al. 1928). The presence of sodium and
potassium salts of oxalic acid in the herbage of greasewood
had previously been demonstrated by J.F. Couch (1922) of the
U.S. Department of Agriculture and confirmed by Marsh
(1923).

Great Basin Habitat

Greasewood is usually found growing on saline/alkaline
soils on lake plains from the margins of playas outward
toward the surrounding mountains until the ground water
table is sufficiently deep that surface wetting does not
reach the ground water level at any season of the year
(Kearney et al. 1914). The lake-plain-playa landscapes are
characteristic of the basin and range topography which
constitutes the Great Basin of western North America. The
Great Basin was named by John Fremont after he determined
there was no drainage from a large portion of the
Intermountain Area south of the Snake River basin in
southern Idaho and north of the Colorado River system in
southern Utah and Nevada (Young and Sparks 1986). Much of
the states of Nevada and Utah along with northeastern
California, southeastern Oregon, and tributary valleys into
Wyoming constitute the Great Basin. The basin is not a
single drainage system surrounded by a mountain rim, but
consists of some 190 mountain ranges that generally are
oriented in a north and south direction with interspersed
valleys.

Pluvial Lakes and Playas

During the Pleistocene, the valleys among the mountain
ranges filled with lakes as evaporation was reduced as the

global climate became cooler (Mifflin and Wheat 1979). The rise of the lakes corresponded with the advance of continental ice sheets and mountain glaciers. Some of the lakes spilled from their basins, and the overflow drained to the eastern and western portions of the Great Basin. The structure of the Great Basin has been likened to a collapsed arch with the lowest elevations on the western and eastern portions and with the higher interior portion consisting of masses of higher mountain ranges.

The lakes that formed during the Pleistocene are termed pluvial, referring to wetter times (Russell 1885). On the eastern edge of the Great Basin, pluvial Lake Bonneville formed and reached over 300 m in depth. During the last rise of Lake Bonneville, its waters crested the divide into the Snake River-Columbia River system and partially drained. In the west, pluvial Lake Lahontan repeatedly formed and dried and never drained to the ocean.

The waters that evaporated from the pluvial lakes were rich in soluble salts that remained in the surface sediments in the bottoms of the former basins. These basins are characterized in their lowest portions by barren flats of very fine textured sediments, the surface of which is often coated with glistening crystals of salt. Occasionally, shallow lakes form on these flat areas which are known in North America by a corruption of the Spanish term, playa.

Occasional flooding and salt accumulations keep the playas nearly free of vegetation. The ground water table is usually very close to the surface of the playa, at least during the winter months. In the various types of dunes that occur near playas, extensive areas of greasewood are found (Young et al. 1986). As soon as seedlings of greasewood become established, fine-textured erosion products from the playa surface (Young and Evans 1986) begin to accumulate. This results in the growth of mounds beneath the plants (Flowers 1934). The soils of the mounds are slightly coarser in texture than the playa sediments, allowing for some leaching of soluble salts as the mounds increase in height. The greasewood plants developed a two story root system with fine roots holding the mound together and apparently providing sites for nutrient exchange while taproots reach down to the ground water table. Greasewood plants have the capability to take up water where the osmotic potential has been greatly lowered by dissolved salts. This is accomplished by allowing soluble salts that would be toxic to other plants to pass through cell membranes with the water. The normally toxic salts are shunted to various parts of the plant, partially as oxalates, or excreted as salts. The deciduous leaves and subtending flower parts that constitute the complex fruits (utricles) of greasewood provide sites for salt deposition within the plant.

Greasewood either dominates or shares dominance with a variety of plants in plant communities reaching out from the playas across the lake plain (Young et al. 1986). The lake plains merge into the outer margins of alluvial fans (Shantz and Piemeisel 1940). The fans often have had their margins reworked by the wave action of Holocene lakes. The rise in elevation that accompanies the encroachment of fill into the basin means the distance between the soil surface and the water table gradually increases. A point is finally reached where the maximum depth of wetting from the soil surface does not reach the water table. This does not permit greasewood roots to reach the water, and some other form of woody chenopod dominates the site, usually a species of Atriplex. Atriplex has undergone explosive speciation as the drying lakes exposed new habitat (Stutz 1978). Greasewood can continue to persist as the non-phreatiphytic dominant of the drier sites as the variety Bailey greasewood (S. vermiculatus var. Baileyi). This form is sometimes elevated to species rank (Billings 1945, 1949).

Greasewood as Forage

Greasewood is, as previously mentioned, a deciduous shrub. The initial current annual growth consists of highly succulent leaves on soft branches. As the season progresses, the branchlets become hardened spines, and the leaves become leathery. Most cases of the poisoning of sheep have occurred in the early spring as the tender growth elongates. The spiescent browse that occurs in the fall is often suspected of causing mechanical damage to cattle.

The bottoms of the pluvial lake valleys evolved as integral parts of the livestock husbandry system of the Great Basin during the 19th century. The arid bottoms of the valleys provide winter ranges when the mountains and foothill sagebrush (Artemisia) ranges are covered with snow (Young and Sparks 1986). The wintering sites which were known as salt desert ranges provide excellent browse from nearly mono-specific patches of semi-woody chenopods such as winterfat (Ceratoides lanata), green molly (Kochia americana), or saltbrush (Atriplex nuttalli). Areas of sand provide Indian ricegrass (Oryzopsis hymenoides). During the early 20th century, both cattle and sheep used this type of vegetation for winter ranges. The animals often were grazed on the ranges in excessive numbers and left on the plant communities late in the spring so the native shrubs had no opportunity to renew reserves and produce seed before the summer drought. Extensive areas of the semi-herbaceous chenopods were severely depleted or left largely bare.

Despite the oxalate content of greasewood herbage, the species makes an important contribution to the forage resources of the salt desert ranges. In the fall, the leaves and fruits of greasewood collect in small windrows under the plants. Cattle and sheep will lick these leaves and fruits from beneath the shrubs. Over time, the soluble salt content of the leaves and fruits is sufficient to change the characteristics of the soil beneath the shrubs (Ricker 1965).

HALOGETON

Halogeton was first collected in North America near Wells, Nevada, in 1934 (Anonymous 1951a). It took some time for the plant to be identified (Standley 1937). Halogeton was already abundant on degraded range sites when it was first collected. Halogeton was considered as somewhat of an oddity because of its fleshy appearance and vivid colors in fruit, but it was just one more of a host of alien annuals to be introduced from Central Asia to the Intermountain Area. Halogeton was even considered to be fair forage, and some thought was given to planting it in firebreaks.

Livestock Losses

In the fall of 1942, Elko County, Nevada, sheepman Nick Goicoa lost 160 head of sheep from a band being grazed near Wells (Burge 1944). There had been a considerable number of sheep losses in Wells during the years prior to 1942 that had been attributed to greasewood poisoning. In the case of the Goicoa sheep, the rumen contents were eventually positively identified as halogeton by C. E. Fleming of the University of Nevada. By March 1943, M.R. Miller of the Agriculture Experiment Station of the University of Nevada reported that dried samples of halogeton herbage contained total oxalates equivalent to 19 percent anhydrous/oxalic acid (Miller 1943).

Research and observations in Nevada soon made it evident that: 1) oxalate content of the plants varied with the phenology of growth; 2) the soils where the halogeton plants were found influenced the oxalate content; and 3) most losses of sheep were directly attributable to poor herd management (Anonymous 1944).

By 1950, halogeton was found in every county in Nevada. The weed had been found in much of northwestern Utah, southern Idaho, southeastern Oregon, northeastern California, and on isolated spots in Wyoming and Colorado (Anonymous 1951b). Halogeton continued its phenomenal rate of spread until 1960, when an estimated 4.6 million hectares were infested in the Intermountain Area (figures compiled by authors). Since the 1960s, outward spread of halogeton has largely ceased. This is probably a result of running out of adapted habitat to invade and the restrictions of day length which limit the latitude extremes of the species.

Federal Research Program

The spread of halogeton from the state of Nevada and other states of the Intermountain Area produced several spectacular poisonings of large numbers of sheep. One of the most well-known occurred in the Raft River Valley of Idaho. This poisoning of sheep resulted in an article in Life Magazine "Sheep Killing Weed" (Anonymous 1951c). This article helped focus national attention on the halogeton problem and contributed to the passage of federal legislation that sponsored research on the halogeton problem. The halogeton act was passed July 14, 1952, and signed by President Truman (Title 7, United States Code, Sections 1651-1656). The act authorized the Agriculture and Interior Secretaries to conduct surveys to detect the presence and effects of halogeton and to plan, organize, direct, and carry out methods to control, suppress, and eradicate the stock-killing weed. States infested with halogeton had to devise cooperative programs before federal monies could be spent to suppress the weed. In actuality, no funds were ever directly appropriated for halogeton control by the federal government as prescribed in the act. Various agencies within each department redirected funds for programs related to halogeton control.

The halogeton research program is historically significant because it heralded a change in policy at the federal level toward research on rangelands. Until the national publicity about halogeton rangeland, research was largely under the administration of the Forest Service, U.S. Department of Agriculture. Exceptions were phases of research, such as poisonous plant research, that dealt with livestock. A major reorganization after the halogeton legislation was enacted resulted in the transfer of several scientists from the Forest Service to the Agriculture Research Service, U.S. Department of Agriculture, and an increasing presence of this agency in range research.

From a plant physiology and ecology standpoint, the halogeton research program had several notable results. Among these were the unraveling of the role of oxalates in the water relations of halogeton (Williams 1960a), the photoperiod control of black and brown seed production by halogeton (Williams 1960b), and the influence of leachate from halogeton litter on the surface soils (Eckert and Kinsinger 1960).

Eradication Attempts

Attempts at large-scale eradication of halogeton were a failure. There was a gradual realization that biological suppression through the seeding of perennial, exotic wheatgrasses (Agropyron spp.) was the most ecologically and economically sound way of reducing losses from halogeton. Halogeton was a symptom of the far more serious problem of range degradation that had been underway for nearly a century. If the range had not been degraded, the spread of halogeton would not have been so rapid or so extensive in area. Biological suppression through seeding was only partially successful because of the lack of adapted species for seeding in degraded salt desert rangelands. For the salt desert areas, the halogeton problems have never gone away. The virtual demise of the range sheep industry coupled with greater awareness of the nature of the problem by some of the remaining sheepman have served to limit the occurrence of spectacular losses of sheep to halogeton. The continued lower level losses of sheep are a major factor in the continued decline of the range sheep industry.

OXALATES

Oxalate poisoning in livestock is of great importance in the Intermountain area. Poisoning occurs primarily from the ingestion of the oxalate-containing plants, halogeton and greasewood.

Oxalic acid is a dicarboxylic acid that readily forms insoluble salts with calcium and magnesium. Its principal soluble salts are sodium, potassium, and ammonium oxalate.

The principal oxalate-containing plants belong to the genus Rumex or genera of the Chenopodiaceae and Oxalidaceae families. Recently, certain grasses have been found to contain troublesome amounts of oxalate.

Plants that produce substantial amounts of oxalate generally fit into one of two groups: (1) those that have a sap pH of about 2 with the oxalate anion present primarily as acid oxalate ($HC_2O_4^-$) (Oxalis spp. and some species of Rumex); and (2) those with a sap pH of about 6 with the oxalate anion present primarily as the oxalate ($C_2O_4^=$) (chiefly Chenopodaceae) (Michael 1959). The oxalate in group 1 exists chiefly as acid potassium oxalate while that in group 2 exists chiefly as soluble sodium oxalate and the insoluble calcium and magnesium oxalate. A few grasses produce oxalate at a much lower level than those genera listed above, and the oxalate probably exists as ammonium oxalate. Setaria sphacelata is an example of an oxalate-producing grass (Seawright et al. 1970; James et al. 1971).

Metabolism

When oxalate is consumed by a ruminant, it may (1) be degraded by rumen bacteria, (2) combine with calcium or magnesium to form insoluble salt, and/or (3) be absorbed from the rumen into the bloodstream where it may combine with calcium to produce hypocalcemia, interfere with other body processes, and/or be excreted (James 1970). The insoluble oxalate salts may accumulate in various tissue, especially the rumen wall and kidneys (Van Kampen and James 1969).

It has been shown by in vivo and in vitro techniques that certain rumen bacteria can degrade oxalate to a nontoxic form. In practical grazing situations, the oxalate degrading bacteria in the rumen will increase (preconditioning) when a ruminant grazes the oxalate-producing plant for about four days to the point where the lethal dose of oxalate is increased substantially in the animal.

Death has been attributed to (1) a hypocalcemia resulting from the formation of calcium oxalate in blood and tissue, (2) uremia resulting from the damage of the kidneys by the oxalate crystals, and/or (3) an interference with energy metabolism due to oxalate interference with succinic dehydrogenase and lactic dehydrogenase, which are involved in energy metabolism (James 1972). At this time, it appears that interference with energy metabolism is the principal cause of death. However, the other two effects are marked and contribute to death.

Oxalate Intoxication

Most oxalate-containing plants are palatable to livestock; therefore, the management of ruminants at risk due to these plants can be critical under certain conditions. Sheep and cattle readily graze both halogeton and greasewood. For example, under conditions where hungry sheep have grazed halogeton intensively, losses of as many as 1,200 have been reported in a single incidence. Many cases where 100 to 800 deaths have occurred have been documented.

Field observations and feeding trials indicate that livestock poisoning by halogeton occurs only when the animal eats too much of the plant too fast. It is assumed that under these conditions the ability of the bacteria to degrade oxalate is exceeded and so it is absorbed. In vitro studies have indicated that under these conditions bacterial activity, in so far as degradation of oxalate is concerned, ceases (James et al. 1967).

Under normal grazing conditions, sheep and cattle can consume large amounts of halogeton or greasewood without apparent harm. Because halogeton is high in soluble salts, its ingestion tends to increase the animal's water intake and urine excretion. It requires 19.8 liters of urine to eliminate 454 gm salt (Cardon et al. 1951). Excessive consumption of halogeton usually occurs when an animal has been allowed to become hungry and then grazes halogeton, or the animal has become thirsty, which can decrease food intake, then is given a drink and allowed to graze in a heavy stand of halogeton. Most cases of halogeton poisoning have occurred when these conditions have been allowed to develop, i.e., animals have become excessively hungry due to withholding feed or water.

Halogeton or greasewood poisoning in sheep is characterized by rapid and labored respiration, depression, weakness, coma, and death (Van Kampen and James 1969). A few animals will have convulsions and some a mild tetany. The gross pathology of these animals includes edema and hemorrhage of the rumen wall; there may be some ascites, and the kidneys may be swollen. On examination of the cut surface of the kidney, oxalate crystals in the renal tubules can be seen with the naked eye. Motility of the gastrointestinal tract slows markedly or ceases shortly after intoxication occurs. Once an animal has become intoxicated on halogeton (oxalates), there is very little that can be done to save it. The only treatment that has been successful is the forcing of water, which is not practical under field conditions (James and Johnson 1970). However, it would seem advisable, if sheep are poisoned on halogeton or greasewood, to ensure they are given an abundant supply of water.

Cattle that have grazed on heavy stands of halogeton and then are driven become stiff, first in the front legs and then the hind ones. If these animals are forced to walk, they may become recumbent. Most live and are able to get to their feet if given feed and water for a period of several days. If driven too fast and hard, they may die (James 1970).

Due to the fact that calcium readily combines with oxalate, early attempts were made at preventing poisoning in sheep by feeding calcium-containing mineral supplements such as dicalcium phosphate (James and Binns 1961). The supplement was mixed with other feeds such as small grains or alfalfa hay and pelleted. It was then fed to sheep that were allowed to graze halogeton. Although this practice gained some popularity, experimental evidence did not support it; therefore, it is not presently recommended (James and Johnson 1970). If supplements are to be fed, they should be to support the general well-being of the animal and not to prevent halogeton or greasewood poisoning.

SUMMARY

Although the livestock industry has, in part, learned to live with halogeton and greasewood, it remains a threat to the individual producer. Successful livestockmen who have grazed their livestock in halogeton-infested areas have suddenly, through a unique set of circumstances or carelessness, sustained catastrophic losses. Although sheep and cattle will graze halogeton and greasewood extensively without apparent harm, livestock operations must be continually aware of the potential of poisoning and be prepared.

The following management practices will minimize death in sheep and cattle due to halogeton poisoning (James and Cronin 1974).

(1) Avoid overgrazing and all other activities that will remove or weaken perennial vegetation. Destroying or weakening this vegetation increases the likelihood of invasion by halogeton.

(2) Develop a grazing plan that will improve the range. Healthy, vigorous, perennial vegetation will resist invasion by halogeton. Halogeton is a poor competitor so will not invade any healthy, vigorous plant community.

(3) Reduce grazing pressure during periods of stress such as drought. Stress on the perennial vegetation during these periods of time weakens them.

(4) Avoid grazing perennials or graze very lightly during the growing season. The growing season for perennials in the cold desert is for a short period of time during the early spring and occasionally into early summer. During most years, growth will be obtained in a period of a few weeks.

(5) Supply ample water and forage of good variety. If it is economical, haul water. Avoid letting animals become hungry.

(6) Have a grazing plan that takes into account halogeton and greasewood.

(7) Adapt sheep to grazing heavier stands of halogeton or greasewood by grazing shadscale on light stands of these plants. This permits time for the rumen bacteria to adapt to the oxalate. These bacteria metabolize the oxalate to nontoxic compounds.

(8) Introduce animals slowly into halogeton infested areas. Even after a short absence from grazing halogeton (3 to 4 days), they should not graze dense stands. Animals are poisoned when they graze too much too fast.

(9) Do not unload animals from trucks into halogeton-infested areas. Let these animals forage until they are full before allowing them to graze halogeton or greasewood for the first time. Then don't allow them to become hungry.

(10) Do not trail thirsty and hungry sheep into watering places surrounded by halogeton or greasewood.

The halogeton problem and oxalate poisoning in general have not gone away. Future users of western rangelands must deal with the possibility of oxalate poisoning in the proper management of rangeland resources.

LITERATURE CITED

Anonymous. 1944. Annual report of the Board of Control for the fiscal year ending June 30, 1944. The University of Nevada Agr. Exp. Sta., Nevada State Printing Office, Carson City, NV.

Anonymous. 1951a. Halogeton discovery. Salt Lake City Tribune Sunday Supplement, June 18. pp. 1, 3.

Anonymous. 1951b. State halogeton control group organized in Reno. Reno Evening Gazette, April 24. p. 11

Anonymous. 1951c. Sheep killing weed. Life Magazine, January 15. pp. 55-56.

Billings, W.D. 1945. The plant associations of the Carson Desert region, western Nevada. Butler University, Botany Studies 7:89-123.

Billings, W.D. 1949. The shadscale vegetation zone of Nevada and eastern California in relation to climate and soils. American Midland Naturalist 42:87-109.

Bull, L.B. 1929. Poisoning of sheep by soursobs (Oxalis cerna): chronic oxalic and poisoning. Australian Vet. J. 5:60.

Burge, L.M. 1944. Report of the Nevada State Dept. of Agr. for the 1942-1944 biennium. State of Nevada, Carson City, NV.

Cardon, C.P., E.B. Stanley, W.J. Pistor, and J.C. Nexbitt. 1951. Ariz. Agr. Exp. Sta. Bull. 239.

Couch, J.F. 1922. The tonic constituent of greasewood (Sarcobatus vermiculatus). Amer. J. Pharm. 94:631-641.

Dodson, M.E. 1959. Oxalate ingestion studies in shop. Australian Vet. J. 35:225.

Eckert, R.E., Jr., and F.E. Kinsinger. 1960. Effects of Halogeton glomeratus leachate on chemical and physical characteristics of soils. Ecology 41:764-772.

Fleming, C.E., M.R. Miller, and L.R. Vauter. 1928. The greasewood (Sarcobatus vermiculatus) a range plant poisonous to sheep. Nevada Agr. Exp. Sta. Bull. 115.

Flowers, S. 1934. Vegetation of the Great Salt Lake region. Botanical Gazette 95:353-418.

James, L.F. 1970. Locomotor disturbance in cattle grazing Halogeton glomeratus. J. Amer. Vet. Med. Assoc. 156:1310-1312.

James, L.F. 1972. Oxalate toxicosis. Clinical Tox. 2:231-243.

James, L.F., and W. Binns. 1961. The use of miner supplements to prevent halogeton poisoning in sheep. Proc. West Sect. Amer. Soc. Animal Sci. 12 LXVIII.

James, L.F., and E.H. Cronin. 1974. Management practices to minimize death losses of sheep grazing halogeton infested ranges. J. Range Manage. 27:424-426.

James, L.F., and A.E. Johnson. 1970. Prevention of fatal Halogeton glomeratus poisoning in sheep. J. Amer. Vet. Med. Assoc. 157:437-442.

James, L.F., J.C. Street, and J.E. Butcher. 1967. In vitro degradation of oxalate and of cellulose by rumen ingesta from sheep fed Halogeton glomeratus. J. Amer. Sci. 26:1438-1444.

James, M.P., A.A. Seawright, and D.P. Steele. 1971. Experimental acute ammonium oxalate poisoning of sheep. Aust. Vet. J. 47:9-17.

Kearney, T.H., L.J. Briggs, H.L. Shantz, J.W. McLaine, and R.L. Piemeisel. 1914. Indicator significance of vegetation in Tooele Valley. Utah. Agr. Res. 1:365-417.

Marsh, C.D. 1923. Greasewood as a poisonous plant. Circ. 279. USDA Circ. 279.

Michael, P.W. 1959. Oxalate ingestion studies in the sheep. Aust. Vet. J. 35:431-432.

Mifflin, M.D. and M.M. Wheat. 1979. Pluvial lakes and estimated pluvial climates of Nevada. Bull. 99. Nevada Bureau of Mines and Geology. Mackay School of Mines and Geology, University of Nevada, Reno.

Miller, M.R. 1943. Halogeton glomeratus, poisonous to sheep. Science 97:2516.

Munz, P.A., and D.D. Keck. 1968. A California flora with supplement, University of California Press, Berkeley, CA.

Ricker, W.H. 1965. The influence of greasewood on soil moisture and soil chemistry. Northwest Sci. 39:36-42.

Russell, I.C. 1885. Geological history of Lake Lahontan, a Quaternary lake of northwestern Nevada. U.S. Geo. Survey Monogr. 11.

Seawright, A.A., S. Groenendyh, and K.I.N.G. Silva. 1970. Aust. Vet. J. 46:293-296.

Shantz, H.L., and R.L. Piemeisel. 1940. Types of vegetation in Escalante Valley, Utah, as indicators of soil conditions. USDA Tech. Bull. 713.

Standley, P.C. 1937. Studies of American plants. Field Mus. Nat. History (Botany Publ.) 17:225-248.

Stutz, H.C. 1978. Explosive evolution of perennial Atriplex in western America. Great Basin Nat. Mem. 2:161-168.

Van Kampen, K.R., and L.F. James. 1969. Acute halotegon poisoning of sheep: pathogenesis of lesions. Amer. J. Vet. Res. 30:1779-1783.

Williams, M.C. 1960a. Effects of sodium and potassium on growth and oxalate content of halogeton. Plant Physiology 35:500-505.

Williams, M.C. 1960b. Biochemical analyses, germination, and production of black and brown seed of Halogeton glomeratus. Weeds 8:452-461.

Young, J.A., and R.A. Evans. 1986. Erosion and deposition of fine textured particles from playas. J. Arid Environ. 10:103-118.

Young, J.A., R.A. Evans, B.A. Roundy, and J.A. Brown. 1986. Dynamic landforms and plant communities in a pluvial lake basin. Great Basin Naturalist 46:1-21.

Young, J.A., and B.A. Sparks. 1986. Cattle in the Cold Desert. Utah State University Press, Logan, UT.

21

Ecological Aspects of
Selenosis on Rangelands

H.F. Mayland and L.F. James

INTRODUCTION

Plants containing high concentrations of selenium (Se) have long been recognized for their toxic effects on animals. Marco Polo, traveling in China in 1295, was probably describing signs of Se poisoning when he wrote that the hooves of his livestock became swollen and dropped off when they grazed plants growing in certain areas (Rosenfeld and Beath 1964). Loss of hair and nails in humans, presumably suffering from chronic Se ingestion was described in Colombia by Father Simon Pedro in 1560 [National Research Council (NRC) 1976, 1983]. Guang-Qi (1987) has also described and illustrated chronic selenium toxicosis in some Chinese people.

Selenium poisoning was probably first described in the United States in 1856 by Dr. T. C. Madison, a physician with the U.S. Calvary stationed near the Missouri River in the Nebraska Territory. Dr. Madison described loss of mane and tail hair and sloughing of hooves in calvary horses grazing in areas later shown to be high in Se (Rosenfeld and Beath 1964). Reports of similar intoxications over the next 75 years led to investigation of Se by the South Dakota and Wyoming Agricultural Experiment Stations and the U.S. Department of Agriculture. These reports demonstrated that these maladies were due to the consumption of seleniferous plants by foraging livestock (Moxon 1937; Anderson et al. 1961; Beath 1982). Major areas of selenosis in the United States are identified in Figure 21.1.

Low levels of Se may also be a problem. Selenium was shown to be an essential nutrient in 1957 (NRC 1983). Since then, many areas of the United States and the world have been identified where plants or feeds do not contain

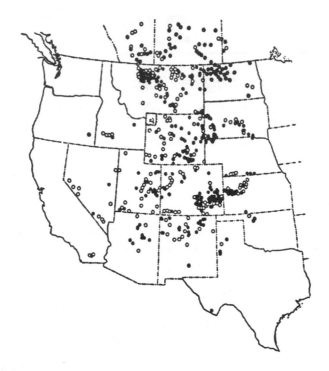

FIGURE 21.1
Distribution of seleniferous vegetation in the western United States and Canada (adapted from Rosenfeld and Beath 1964). Each open dot represents the place of collection of a plant specimen containing 50 to 500 mg Se/kg; each solid dot, specimens containing more than 500 mg Se/kg.

sufficient Se to meet animal requirements. When this occurs, animal performance suffers, e.g., reduced weight gain and poor reproduction. Young growing animals may develop potentially fatal white muscle disease, while poultry show depressed growth and exudative diathesis. In these situations, Se intake may be supplemented by fertilizing pastures or crops, supplementing feed, or providing an injectable form of Se directly to the animal. Readers interested in the nutritional requirements are directed to NRC (1983), Gissel-Nielsen et al. (1984), and Combs and Combs (1986). This chapter describes the ecological aspects of excess Se on rangelands.

SELENIUM IN SOIL

The concentration of total Se in most soils lies within the range of 0.1 to 2 mg Se/kg. However, high concentrations of total Se (1,200 mg Se/kg), of which 38 mg/kg occurred as water-soluble selenate, have been reported in seleniferous areas of the world (Lakin and Byers 1948, Swaine 1955). Soils developed from the Cretaceous shales of South Dakota, Montana, Wyoming, Nebraska, Kansas, Utah, Colorado, and New Mexico tend to have high Se values ranging from 2 to 10 mg Se/kg.

The total Se taken up by plants is closely correlated to the amount of ammonium-DTPA extractable soil Se (Soltanpour and Workman 1980). This soluble portion may not relate to the total soil Se (Lakin 1972). For example: the Se content of some Hawaiian surface soils varies from 1 to 20 mg/kg, but this Se is unavailable for vegetation because of its complexation with the iron and aluminum minerals (Anderson et al. 1961; Rosenfeld and Beath 1964).

The chemical forms of Se present in soils and sediments are closely related to the oxidation-reduction potential and pH of the soil (van Dorst and Peterson 1984). Inorganic Se exists in the selenate or selenite form in aerated, alkaline soils that predominate on western rangelands. In poorly aerated acidic soils, inorganic Se exists as selenide or elemental Se. The bioavailability of inorganic Se is selenate > selenite >> elemental Se = selenide.

SELENIUM IN WATER

Selenium generally occurs as a minor constituent in drinking water in a concentration range of 0.1 to 100 µg/liter (NRC 1983). Samples rarely exceed the 10 µg Se/liter upper limit established by the 1977 Safe Drinking Water Act of the Environmental Protection Agency (EPA). Rivers draining some of the seleniferous regions may contain as much as 10 µg total Se/liter (R.A. Smith, personal communication). Water from wells drilled into the Cretaceous Colorado formations of central Montana may contain as much as 1,000 µg Se/liter (Donovan et al. 1981). Waters that are high in Se are also likely to be high in total salts (Donovan et al. 1981).

Rumble (1985) reported 2.2, 10.6 and 1.1 µg Se/liter of water in coal and bentonite mine impoundments, and livestock ponds, respectively, in the Northern High Plains. Mayland (unpublished research) found 0.4 to 0.7 µg Se/liter in most of the sampled livestock waters in the seleniferous areas of Montana and Wyoming; however, several reservoir samples contained as high as 270 µg Se/liter. That high value is many times greater than the EPA (1977) allowable value of 10 µg Se/liter in drinking water, but it is only three times the 100 µg/liter level of Se in blood considered adequate for animal requirements.

SELENIUM IN PLANTS

Accumulators vs. Nonaccumulators

Rosenfeld and Beath (1964) and Shrift (1973) divided plants into three groups on the basis of their ability to accumulate Se when grown on high-Se soils. These groupings are somewhat arbitrary as some of the plants listed may appear in more than one of the groups.

Group I plants include the so-called Se accumulator or indicator plants. These plants grow well on soil containing high levels of available Se, and their presence is generally indicative of seleniferous soils. Plants in this group include 26 species of Astragalus (Rosenfeld and Beath 1964) plus many species in the genera of Machaeranthera, Haplopappus, and Stanleya. These plants absorb relatively high concentrations of Se that may be in the hundreds and occasionally even thousands of milligrams per kilogram dry weight. Lesser amounts of Se are often found in these plants (Figure 21.2).

Plants in Group II include species of the genera Aster, Atriplex, Castilleja, Grindelia, Gutierrezia, Machaeranthera, Mentzelia, and some species of Astragalus. These plants rarely accumulate more than 50 to 100 mg Se/kg.

Plants in Group III include grains, grasses, and many forbs that may accumulate toxic amounts of Se from 5 to 12 mg/kg, but seldom more than 50 mg Se/kg, even when grown on seleniferous soil (Figure 21.2). These plants are perhaps more important than the Se accumulating plants in overall livestock management, because they are more palatable and more likely to be grazed.

Some plants that grow on seleniferous soils accumulate only low levels of Se. Examples of these plants include white clover (Trifolium repens), buffalo grass (Buchloe dactyloides), and grama grasses (Bouteloua spp.). High sulfur (S) containing plants like the Brassica species (mustard, cabbage, broccoli, and cauliflower) and other Cruciferae are relatively strong concentrators of Se.

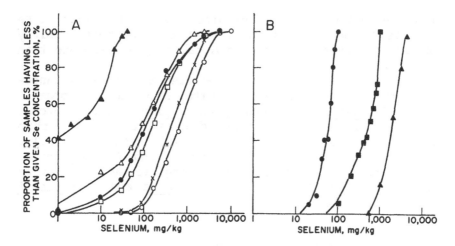

FIGURE 21.2
Proportion of samples having less than given Se
concentration (%).

A. ▲—▲ data for western wheatgrass (Pascopyrum
smithii) and Sandberg bluegrass (Poa secunda)
sampled from seleniferous areas of Montana and
Wyoming by Mayland (unpublished data); △—△
Stanleya spp.; ●—●Xylorrhiza section of
Machaeranthera; □—□Astragalus bisulcatus; X—X
Astragalus pectinatus; and O—O Oonopsis section
of Haplopappus: latter data adapted from
Rosenfeld and Beath (1964).

B. ●—●data for vegetative wheat (Triticum
aestivum), ■—■Astragalus bisulcatus, and ▲—▲
Astragalus pectinatus reported in plants from
North Dakota by Lakin and Byers (1948).

Absorption of Se and S by plants may be correlated because
of their similar chemistries (Shrift 1973).

Plant Response to Selenium Species

Gissel-Nielsen (1973) reported that more selenate than
selenite was taken up by plants. There was a rapid decline
in the bioavailability of the selenate form, presumably
because it was converted to less available forms. Adding
sulfate to the soil greatly decreased the uptake of selenate

but had a lesser effect on selenite uptake. The ratio of Se in roots to Se in tops shows that Se was more readily translocated from the roots when taken up as selenate than as selenite.

Hamilton and Beath (1963) measured Se uptake by rangeland forbs and grasses from selenite, selenate, and organic (ground Astragalus bisulcatus plus A. preussii) Se sources. Selenate was generally absorbed more efficiently than selenite, and organic Se was absorbed least efficiently. However, plants identified in Groups I and II (see above) readily absorbed large amounts of organic Se provided as finely ground mixtures of A. bisulcatus and A. preussii added to the soil.

Plant Response to High Bioavailable Selenium

Selenium is known to be required for animal health but is not yet considered an essential element for plant growth. Several studies suggest that Se is required at least for the accumulator plant species. Trelease and Trelease (1938, 1939) reported a pronounced stimulating effect in the growth of Astragalus racemosus and A. pattersonii (grown in nutrient culture) when up to 9,000 µg Se/liter, as selenite, was provided. Both species are identified as Se accumulator plants. However, A. crassicarpus (identified as A. succulentus by Rosenfeld and Beath 1964), which is a nonaccumulator, was poisoned by Se at rates as low as 300 µg/liter.

These findings led Trelease and Trelease (1938, 1939) and others (Rosenfeld and Beath 1964, Shrift 1969) to suggest that Se might be an essential element for the Se indicator plants. This apparent Se requirement by the accumulator species may be confounded by a Se-phosphorus interaction, whereby the increased growth in the presence of Se could have been due to a depression of phosphorus toxicity by selenite (Broyer et al. 1972).

If either A. bisulcatus (accumulator) or A. crotolariae (nonaccumulator) has a requirement for Se, the critical level probably would be less than 80 µg Se/kg dry plant tops (Broyer et al. 1972). This value is of the same order reported for alfalfa and/or subterranean clover (Trifolium subterraneum) (Broyer et al. 1966) and much less than the value suggested by Trelease and Trelease (1939).

Selenium Compounds in Plants

Shrift (1973) has summarized the findings on many chemical compounds of Se isolated from plants. Much of the Se in nonaccumulating species is found in the form of protein-bound selenomethionine. In contrast, the Se in accumulator plants is mostly water soluble and is not associated with protein. Formation of Se-methylselenocysteine has been found to occur only in accumulators and has been suggested as a chemotaxonomic basis for distinguishing accumulators from nonaccumulators. Another distinction between these plant groups is the existence of a large amount of selenocystathionine in accumulator species and only trace amounts in nonaccumulator plants (Shrift 1969, 1973).

There is little incorporation of the selenoamino acids into the proteins of the accumulator plants tested (Peterson and Butler 1962). However, extensive protein incorporation of the selenoamino acids occurred in ryegrass (Lolium perenne), wheat (Triticum aestivum), and clover. Lewis (1976) suggested that Se accumulator species have evolved a detoxification mechanism, whereby Se is excluded from protein incorporation, while nonaccumulator species do not have this mechanism. Selenium incorporation into proteins could result in alteration of the protein structure, inactivation of the protein, and eventual poisoning of the plant (Peterson and Butler 1962).

The Se metabolites in plants are analogs of S compounds. Nevertheless, Se metabolism in plants cannot be identified from known mechanisms involving S metabolism (Shrift 1973). It is now known that many isolated enzyme systems can utilize S and Se analogs interchangeably. What was traditionally thought to be the mechanism of Se toxicity, namely, a general interference with the enzymes involving S assimilation, has proved to be more complex.

SELENIUM ABSORPTION IN ANIMALS

Selenium toxicosis in aquatic and terrestrial animals has been well described in several reviews (NRC 1983; Sorensen 1986; Ohlendorf et al. 1986). Generalizations concerning the acute toxicity of Se to animals are exceedingly difficult to make because of the large array of chemical compounds, aqueous solubility, methods of administration, animal species, and variability within species (Combs and Combs 1986). However, selenite, selenate, and several seleno-amino acid sources have the greatest bioavailability, while the reduced forms of Se have the least bioavailability. Feedstuffs vary with respect to Se bioavailability, with most plant sources being moderately to highly available, while the Se in soybean, meat, and fish is only slightly available. The various sources produce similar toxicological effects (Moxon and Rhian 1943).

The toxicity of Se-containing plants cannot be evaluated solely on total Se content. There are several Se-accumulating plants such as the seleniferous Astragalus and Happlopappus that contain additional toxins that may complicate the evaluation of Se toxicity. In general, Se as selenomethionine is more readily absorbed than when ingested as selenite, selenate, or selenocystine (NRC 1983). Nearly one-half of the Se in wheat grain is present as selenomethionine (Olson et al. 1970). The Se in wheat is more readily available to animals than is the Se in selenocystine, Astragalus, or fish meal (Combs and Combs 1986, NRC 1983). Selenium from plant forms is generally more available to animals than that from animal forms (Combs and Combs 1986).

As in plants, Se in animals interacts with other trace elements. Frank et al. (1986) found 20% lower liver-copper concentrations in cattle supplemented with 0.58 mg Se/kg diet compared with 0.18 mg Se/kg in the control diet over a seven to eight month period. This has practical applications because copper deficiency has been reported in areas where cattle and sheep are often supplemented with Se. A synergistic relationship was reported by Mayland et al. (1986) when high, but not toxic, levels of Se increased the absorption of lead and its subsequent toxicity in mature sheep.

Increasing dietary S intake has been observed to increase the clinical incidence of white muscle disease. The addition of 3.3g S/kg (as Na_2SO_4) to the diet reduced the availability of 0.17 mg Se/kg (as Na_2SeO_3) to lambs (Hintz and Hogue 1964). Halverson et al. (1962) reported that the addition of sodium sulfate to diets containing 10 mg Se/kg reduced the toxicity to rats (Rattus rattus) when Se was added as selenate but not when it was added as selenite or as seleniferous wheat. However, the efficacy of sulfate-S supplementation of larger animals ingesting seleniferous feeds has not been tested.

Selenium toxicity (i.e., that of SeO_2 or Na_2SeO_3) has been reduced in rats by feeding rather high levels of arsenic, copper, iron, silver, or some organic feedstuffs like Torula yeast and linseed oil meal (Combs and Combs

1986). The active factor in linseed oil meal appears to be a cyanogenic glycoside (Palmer et al. 1980). Arsenic appears to increase the biliary elimination of Se (Combs and Combs 1986). There are many interacting factors associated with increasing the tolerance of animals to various forms and sources of Se. In large animals, arsanilic acid supplementation has been somewhat effective in protecting pigs against Se toxicosis during reproduction and growth (Wahlstrom and Olson 1959a, 1959b). Even though some livestock grazing on seleniferous forage are supplemented with arsenic (H.F. Mayland, unpublished research), there is no experimental evidence that arsenic or any of the other factors discussed above have any beneficial effects on reducing Se toxicosis in grazing animals.

FECAL AND URINARY SELENIUM

Urine is the primary route of Se excretion by monogastric animals, regardless of whether the Se is given orally or injected. The main route of Se excretion in ruminants, though, is a function of the method of administration and the age of the animal (NRC 1983). When Se is ingested by ruminants, most of it is excreted in feces. In contrast, Se that is injected either intravenously or subcutaneously into ruminants is excreted mostly in urine. Lambs (Ovis aries) and, presumably, calves (Bos taurus) that have not developed rumen function can excrete 66 to 75% of the orally ingested Se in the urine. It is likely that exogenous fecal Se is predominantly Se which has been reduced to an unavailable form such as elemental Se (Langlands et al. 1986). Rumen organisms undoubtedly contribute to this age effect.

Nearly all of the Se excreted in the feces of ruminants is in an insoluble form, and very little is available for uptake by plants. Peterson and Spedding (1963) showed that during a 75-day period, less than 0.3% of the Se taken up by three pasture species originated from the Se contained in sheep manure.

Trimethylselenonium ion (TMSe$^+$) is the major urinary Se metabolite (NRC 1983). When added to nutrient solutions, TMSe$^+$ was absorbed and translocated to leaves and stems but not to the grain of wheat (Olson et al. 1976a). Large differences were observed in Se uptake by barley, wheat, and alfalfa when TMSe+ was applied in a soil-pot study in the greenhouse. Olson et al. (1976a) noted that very little of the Se from TMSe$^+$ was absorbed by plants. In addition, some of the TMSe[+] was lost to the atmosphere through volatilization from the plant. It is likely that the TMSe[+] excreted in animal urine contributes little biologically active Se to plants because it is not metabolized.

A portion of the TMSe+ added to soil was biologically volatilized, and this loss was increased by liming of the soil at rates to increase the pH from 5.45 initially to values approaching pH 7.05 (Olson et al. 1976b). In addition, 30 to 50% of the TMSe[+] added to several different soils was sorbed to the soil particles during a 21-day period. The biologically inactive TMSe[+] in urine plus the stimulation of plant growth by the added nitrogen and sulfate in urine may explain the lowered Se content in grass growing on urine patches (Joblin and Pritchard 1983).

VOLATILE SELENIUM

Volatile Se compounds are naturally released into the atmosphere as a result of biological activity in aquatic (Chau et al. 1976) and terrestrial ecosystems (Abu-Erreish et al. 1968; Shrift 1973; Doran and Alexander 1977). The pathways for Se volatilization in higher plants have been reviewed by Lewis (1976). The volatile compounds include dimethyl selenide (DMSe), which produces the garlicky odor associated with Se accumulator plants. Also included are dimethyl diselenide, dimethylselenone, methane selenol, and hydrogen selenide (Zieve and Peterson 1981, 1984c). Dimethyl selenide is volatilized from soils by microbial activity (Zieve and Peterson 1984b) and sorbed on the surfaces of soil colloids or lost to the atmosphere (Zieve and Peterson 1985). Dimethyl selenide is also absorbed by plants through the leaves (Zieve and Peterson 1984a), but this source accounts for only a small portion of the Se in most plants (Zieve and Peterson 1986).

SELENIUM INTOXICATION

All livestock and humans are known to be susceptible to Se poisoning. However, poisoning is most likely to occur in animals grazing seleniferous forage. Poisoning may occur in poultry and swine as a result of including seleniferous grain in their diet. The Food and Nutrition Board of the National Academy of Sciences (1980) has suggested 5 mg Se/kg diet as the critical level between toxic and nontoxic feeds.

Selenium intoxication in grazing animals has been classified by Rosenfeld and Beath (1964) as follows: (1) acute intoxication; and (2) chronic intoxication, the latter including two syndromes identified as alkali disease and blind staggers.

Acute Intoxication

Acute Se poisoning usually results from the ingestion of excessive amounts of primary or indicator plants containing high amounts of Se. Because plants containing high levels of Se are relatively unpalatable, acute poisoning is uncommon. Poisoning is characterized by abnormal posture and movement, watery diarrhea, labored respiration, abdominal pain, prostration, and death. The signs observed are related to the acuteness of the intoxication.

Chronic Intoxication

Chronic Se intoxication is divided into two syndromes, alkali disease and blind staggers. One of the distinctions that has been made between alkali disease and blind staggers is in the kind of seleniferous forage consumed. Alkali disease in cattle, horses, hogs, and poultry is said to be associated with the consumption of seleniferous forages such as grasses and crops in which the Se is bound to protein and is relatively insoluble in water. Blind staggers is said to occur in cattle and possibly sheep and is thought to be associated with the consumption of Se indicator plants wherein the Se is in a water soluble form (Rosenfeld and Beath 1964).

Alkali Disease. The condition described by Dr. T.C. Madison in 1856 (Rosenfeld and Beath 1964) was later known as alkali disease and is now known to be chronic Se poisoning. This form of Se poisoning is still referred to as alkali disease. The term alkali disease was coined by the early settlers of the semi-arid Great Plains of the United States (Moxon 1937). The early pioneers associated the disease with waters from alkali or saline seeps and with alkali spots in the soil. In fact, water from some of the saline seeps in the northern Great Plains may contain an average of 300 µg Se/liter (Miller et al. 1981).

It was not until the early 1930s that alkali disease was shown to be caused by grazing seleniferous forages (Moxon 1937). Alkali disease, according to Rosenfeld and Beath (1964), results from the ingestion by livestock of plants such as grasses and small grains containing 5 mg Se/kg over an extended period of time. According to these investigators, the Se producing this disease condition is bound to plant protein and is therefore relatively insoluble in water. However, alkali disease has been produced by feeding soluble sodium selenate (Moxon 1937, Hartley et al. 1984). Alkali disease is characterized by dullness, lack of vitality, emaciation, rough coat, loss of hair (especially the long hair), hoof changes, and lameness. Cattle, horses, and swine will all develop alkali disease when fed seleniferous feeds. Sheep do not respond in the same manner. They show neither loss of body cover nor do they develop hoof lesions (Rosenfeld and Beath 1964).

According to Olson (1978), reduced reproductive performance is the most significant effect of alkali disease in livestock. The effect on reproduction may be quite marked, yet the animal involved may not show typical signs of alkali disease. Reproduction has been adversely affected in cattle (Dinkel et al. 1963), swine (Wahlstrom and Olson 1959a, 1959b), and mice and rats (Schroeder and Mitchener 1971, Halverson 1974). The forages responsible for the reproduction losses in cattle probably contained between 5 and 10 mg Se/kg.

As in many other situations, sheep respond differently to Se than cattle. Glenn et al. (1964) and Maag and Glenn (1967) reported that feeding as much as 0.55 mg Se/kg body weight as selenate did not lower the conception rate nor noticeably affect the developing fetuses of two-year-old ewes. This rate (0.55 mg/kg) corresponds to a dietary intake of 15 to 20 mg Se/kg forage. The Se was administered with a linseed oil meal carrier which has since been shown to contain several compounds that reduce the toxicosis of Se (Palmer et al. 1980). Thus, compounds in the linseed oil may have reduced any toxicological effect of the selenate on reproduction in sheep.

Ranch managers have learned through experience that some areas or pastures are "hotter" or more seleniferous than others and that grazing time on these areas should be limited to a few weeks to minimize the effects of the high Se levels in the forage. Many of these ranges are now utilized by stocker cattle rather than cow-calf units (H.F. Mayland, unpublished research). This shift away from cows may be a response to reduced reproductive performance and/or other economic factors (Wahlstrom and Olson 1959a, 1959b). More individual attention is often given to horses than to cattle, and managers are very careful which pastures these animals are allowed to graze. Even so, the effective life of these animals may be shortened because of chronic selenosis.

Blind Staggers. This disorder, according to Rosenfeld and Beath (1964), results from livestock grazing moderate amounts of indicator plants over extended time periods. Blind staggers has been described in cattle and sheep but not in horses, hogs, or poultry. Many of the selenium compounds occurring in these plants are readily extractable with water.

Blind staggers is characterized by three stages. During stage 1, the animal frequently wanders in circles, disregards objects in its path, becomes anorexic, and shows evidence of impaired vision. In stage 2, stage 1 signs intensify, and the front legs become increasingly weak. During stage 3, the tongue becomes partially or totally paralyzed, there is inability to swallow, varying degrees of blindness, labored respiration, abdominal pain, grinding of the teeth, salivation, emaciation, and death (Rosenfeld and Beath 1964). The three stages of intoxication are not clearly defined in sheep, as sheep exhibit few clinical signs and may die suddenly when intoxicated (Moxon and Olson 1974).

Blind staggers has been reported in Wyoming but not in other western states having problems of Se poisoning in livestock (Moxon and Rhian 1943). Jensen et al. (1956), in discussing polioencephalomalacia in cattle and sheep, point out that a condition in Colorado known as forage poisoning was clinically identical to blind staggers, but they were unable to associate Se with the forage poisoning condition. The comparison made was as follows:

"Polioencephalomalacia, a noninfectious disease of pasture and feedlot cattle and sheep, is charac-terized by multiple foci of necrosis in the cerebral cortex. In Colorado, the disease is known as forage poisoning. In Wyoming, where the disease has been studied extensively, it is known as blind staggers from Se poisoning. The clinical syndromes of the disease in Colorado and Wyoming are identical. The cause of the disease in cattle and sheep of Colorado has not been studied adequately, while the neuropathology of the disease in cattle and sheep of Wyoming has not been reported. Although it is assumed that blind staggers reported from Wyoming, and forage poisoning reported from Colorado, are a single entity, the appellation polioencephalomalacia is appropriate until the etiological and pathological factors are clearly established." (p. 321)

CONCLUSIONS

Large areas of the semiarid western United States are underlain with seleniferous geological materials of Cretaceous origin. Soil, water, and plants associated with these materials are also likely to be seleniferous. Plants growing on these areas may vary significantly in their ability to accumulate Se. Those plants accumulating high (hundreds of mg Se/kg) or moderate (50 to approximately 100 mg Se/kg) amounts are termed primary and secondary accumulators, respectively. Most of the Se in these plants occurs in water-soluble forms like selenate and several seleno-amino acids. Ingestion of these plants at a rate providing about 10 to 15 mg Se/kg body weight can be lethal. Fortunately, these plants are generally not very palatable. The third group is composed of the grasses, small grains, and many of the palatable forbs. These plants may contain 5 to 10 mg Se/kg, much of which is incorporated into protein.

Contrary to earlier conclusions that Se was an essential element for plant growth, it has been shown that the presence of Se in the soil solution ameliorates excess phosphorus uptake and subsequent phosphorous poisoning in the accumulator plants.

Animal intoxication varies between and within animal species and Se sources. Both chronic and acute forms are recognized. Chronic forms are identified as alkali disease and blind staggers. Alkali disease (characterized by loss of hair, inappetence, and elongated hooves) occurs when animals, over long periods of time, graze Group III forages containing organic Se in moderate amounts. Consumption of plants in Groups I and II may produce blind staggers in cattle and sheep but not horses and swine. This neurological disorder has been attributed to the high levels of inorganic Se present in these plants. However, some of these plants also contain other toxic principles that may produce blind staggers. Research is needed to determine the cause of this neurological disorder.

LITERATURE CITED

Abu-Erreish, G.M., E.I. Whitehead, and O.E. Olson. 1968. Evolution of volatile selenium from soils. Soil Sci. 106:415-420.

Anderson, M.S., H.W. Lakin, K.C. Beeson, F.F. Smith, and E. Thacker. 1961. Selenium in agriculture. Agriculture Handbook No. 200.

Beath, O.A. 1982. The story of selenium in Wyoming. Wyo. Agr. Exp. Bull. No. 774.

Broyer, T.C., C.M. Johnson, and R.P. Huston. 1972. Selenium and nutrition of Astragalus. I. Effects of selenite or selenate supply on growth and selenium content. Plant and Soil 36:635-649.

Broyer, T.C., D.C. Lee, and C.J. Asher. 1966. Selenium nutrition of green plants. Effect of selenite supply on growth and selenium content of alfalfa and subterranean clover. Plant Physiol. 41:1425-1428.

Chau, Y.K., P.T.S. Wong, B.A. Silverberg, P.L. Luxon, and G.A. Bengert. 1976. Methylation of selenium in the aquatic environment. Science 192:1130-1131.

Combs, G.F., Jr., and S.B. Combs. 1986. The role of selenium in nutrition. Academic Press, Inc., Orlando, FL.

Dinkel, C.A., J.A. Minyard, and D.E. Ray. 1963. Effects of season of breeding on reproductive and weaning performance of beef cattle grazing seleniferous range. J. Animal Sci. 22:1043-1045.

Donovan, J. J., J.L. Sonderegger, and M.R. Miller. 1981. Investigations of soluble salt loads, controlling mineralogy, and factors affecting the rates and amounts of leached salts. Montana Water Resources Research Center Report No. 120. Montana State University, Bozeman, MT.

Doran, J.W., and M. Alexander. 1977. Microbial formation of volatile selenium compounds in soil. Soil Sci. Soc. Amer. J. 41:70-73.

Environmental Protection Agency (EPA). 1977. National interim primary drinking water regulations. Environmental Protection Agency, Office of Water Supply, EPA-57019-76-003.

Food and Nutrition Board. 1980. Recommended dietary allowances, pp. 162-164. National Academy of Sciences, Washington, DC.

Frank, A., B. Pehrson, and L.R. Petersson. 1986. Concentration of some important elements in the liver of young cattle supplemented with selenite enriched feed. J. Vet. Med. Assoc. 33:422-425.

Gissel-Nielsen, G. 1973. Uptake and distribution of added selenite and selenate by barley and red clover as influenced by sulfur. J. Sci. Fd. Agric. 24:649-655.

Gissel-Nielsen, G., U.C. Gupta, M. Lemand, and T. Westermarck. 1984. Selenium in soils and plants and its importance in livestock and human nutrition. Adv. Agron. 37:397-461.

Glenn, M.W., R. Jensen, and L.A. Griner. 1964. Sodium selenate toxicosis: The effects of extended oral administration of sodium selenate on mortality, clinical signs, fertility, and early embryonic development in sheep. Amer. J. Vet. Res. 25:1479-1499.

Guang-Qi, Y. 1987. Historical notes on a discovery of selenium related endemic diseases, pp. 9-32. In: G.F. Combs, Jr., J.E. Spallholz, O.A. Levander, and J.E. Oldfield (eds), Selenium in biology and medicine, part A. Van Nostrand Rheinhold Publ., New York.

Halverson, A.W. 1974. Growth and reproduction with rats fed selenite-Se. Proc. S.D. Acad. Sci. 53:167-177.

Halverson, A.W., P.L. Guss, and O.E. Olson. 1962. Effect of sulfur salts on selenium poisoning in the rat. J. Nutr. 77:459-464.

Hamilton, J.W., and O.A. Beath. 1963. Uptake of available selenium by certain range plants. J. Range Manage. 16:261-264.

Hartley, W.J., L.F. James, H. Broquist, and K.E. Panter. 1984. Pathology of experimental locoweed and selenium poisoning in pigs, pp. 141-149. In: A.A. Seawright, M.P. Hegarty, L.F. James, and R.F. Keeler (eds), Plant toxicology. Proc. Aust.-U.S.A. Poisonous Plants Symp., Brisbane. Queensland Poisonous Plants Comm., Ani. Res. Inst., Yeerongpilly.

Hintz, H.F., and D.E. Hogue. 1964. Effect of selenium, sulfur and sulfur amino acids on nutritional muscular dystrophy in the lamb. J. Nutr. 82:495-498.

Jensen, R., L.A. Griner, and O.R. Adams. 1956. Polioencephalomalacia of cattle and sheep. J. Amer. Vet. Med. Assoc. 129:311-321.

Joblin, K.N., and M.W. Pritchard. 1983. Urinary effect on variations in the selenium and sulphur contents of ryegrass from pasture. Plant and Soil 70:69-76.

Lakin, H.W. 1972. Selenium accumulation in soils and its absorption by plants and animals. Geol. Soc. Amer. Bull. 83:181-189.

Lakin, H.W., and H.G. Byers. 1948. Selenium occurrence in certain soils in the United States, with a discussion of related topics: Seventh Report. USDA Tech. Bull. 950.

Langlands, J.P., J.E. Bowles, G.E. Donald, and A.J. Smith. 1986. Selenium excretion in sheep. Aust. J. Agric. Res. 37:201-209.

Lewis, B.G. 1976. Selenium in biological systems, and pathways for its volatilization in higher plants, pp. 389-409. In: J.O. Nriagu (ed), Environmental biogeochemistry. Ann Arbor Science, Ann Arbor, MI.

Maag, D.D., and M.W. Glenn. 1967. Toxicity of selenium: farm animals, pp. 127-140. In: O.H. Muth, J.E. Oldfield, and P.H. Weswig (eds), Selenium in biomedicine. AVI Publ. Co., Westport, CT.

Mayland, H.F., J.J. Doyle, and R.P. Sharma. 1986. Effects of excess dietary selenite on lead toxicity in sheep. Biological Trace Element Res. 10:65-75.

Miller, M.R., P.L. Brown, J.J. Donovan, R.N. Bergatino, J.L. Sonderegger, and F.A. Schmidt. 1981. Saline seep development and control in the North American Great Plains--hydrogeological aspects. Agricultural Water Management 4:115-141.

Moxon, A.L. 1937. Alkali disease or selenium poisoning. S. Dak. Agr. Exp. Sta. Bull. 311:1-84.

Moxon, A.L., and O.E. Olson. 1974. Selenium in agriculture. In: R.A. Zingaro and W.C. Cooper (eds), Selenium. Van Nostrand Reinhold Co., New York.

Moxon, A.L., and M. Rhian. 1943. Selenium poisoning. Physiol. Rev. 23:305-337.

National Research Council. 1976. Selenium--medical and biologic effects of environmental pollutants. National Academy of Science, Washington, DC.

National Research Council. 1983. Selenium in nutrition (revised). Board on Agriculture, National Academy of Sciences, Washington, DC.

Ohlendorf, H.M., R.L. Hothem, C.M. Bunck, T.W. Aldrich, and J.F. Moore. 1986. Relationships between selenium concentrations and avian reproduction. Trans. N.A. Wildl. and Nat. Res. Conf. 51:330-342.

Olson, O.E. 1978. Selenium in plants as a cause of livestock poisoning, pp. 121-134. In: R.F. Keeler, K.R. Van Kampen, and L.F. James (eds), Effects of poisonous plants on livestock. Academic Press, New York.

Olson, O.E., E.E. Cary, and W.H. Allaway. 1976a. Absorption of trimethylselenonium by plants. Agron. J. 68:805-809.

Olson, O.E., E.E. Cary, and W.H. Allaway. 1976b. Fixation and volatilization by soils of selenium from trimethylselenonium. Agron. J. 68:839-843.

Olson, O.E., E.J. Novacek, E.I. Whitehead, and I.S. Palmer. 1970. Investigations on selenium in wheat. Phytochemistry 9:1181-1188.

Palmer, I.S., O.E. Olson, A.W. Halverson, R. Miller, and C. Smith. 1980. Isolation of factors in linseed oil meal protective against chronic selenosis in rats. J. Nutr. 110:145-150.

Peterson, P.J., and G. Butler. 1962. The uptake and assimilation of selenite by higher plants. Aust. J. Biol. Sci. 15:126-146.

Peterson, P.J., and D.J. Spedding. 1963. The excretion by sheep of Se-75 incorporated into red clover: the chemical nature of the excreted selenium and its uptake by three plant species. NZ J. Agric. Res. 6:13-22.

Rosenfeld, I., and O.A. Beath. 1964. Selenium. Geobotany, biochemistry, toxicity and nutrition. Academic Press, New York.

Rumble, M.A. 1985. Quality of water for livestock in manmade impoundments in the Northern High Plains. J. Range Manage. 38:74-77.

Schroeder, H.A., and M. Mitchener. 1971. Toxic effects of trace elements on the reproduction of mice and rats. Arch. Environ. Health 23:102-106.

Shrift, A. 1969. Aspects of selenium metabolism in higher plants. Ann. Rev. Plant Physiol. 20:475-494.

Shrift, A. 1973. Selenium compounds in nature and medicine. E. Metabolism of selenium by plants and microorganisms, pp. 763-814. In: D.L. Klayman and W.H.H. Gunther (eds), Organic selenium compounds: their chemistry and biology. Wiley-Interscience, New York, NY.

Soltanpour, P.N., and S.M. Workman. 1980. Use of NH_4HCO_3-DTPA soil test to assess availability and toxicity of selenium to alfalfa plants. Commun. Soil Sci. Plant Anal. 11:1147-1156.

Sorensen, E.M.B. 1986. The effects of selenium on freshwater teleosts, pp. 59-116. In: E. Hodgson (ed), Reviews in environmental toxicology 2. Elsevier, New York.

Swaine, D.J. 1955. The trace-element content of soils. Technical Communication No. 48. Commonwealth Bureau of Soil Sci. Harpenden, England.

Trelease, S.F., and H.M. Trelease. 1938. Selenium as a stimulating and possibly essential element for indicator plants. Amer. J. Botany 25:372-380.

Trelease, S.F., and H.M. Trelease. 1939. Physiological differentiation in Astragalus with reference to selenium. Amer. J. Botany 26:530-535.

van Dorst, S.H., and P.J. Peterson. 1984. Selenium speciation in the soil solution and its relevance to plant uptake. J. Sci. Food Agric. 35:601-605.

Wahlstrom, R.C., and O.E. Olson. 1959a. The effect of selenium on reproduction in swine. J. Animal Sci. 18:141-145.

Wahlstrom, R.C., and O.E. Olson. 1959b. The relation of pre-natal and pre-weaning treatment to the effect of arsanilic acid on selenium poisoning in weanling pigs. J. Anim. Sci. 18:578-582.

Zieve, R., and P.J. Peterson. 1981. Factors influencing the volatilization of selenium from soil. Sci. Total Environ. 19:277-284.

Zieve, R., and P.J. Peterson. 1984a. The accumulation and assimilation of dimethylselenide by four plant species. Planta (Berlin) 160:180-184.

Zieve, R., and P.J. Peterson. 1984b. Volatilization of selenium from plants and soils. Sci. Total Environ. 32:197-202.

Zieve, R., and P.J. Peterson. 1984c. Dimethylselenide--an important component of the biogeochemical cycling of selenium. Trace Substances in Environmental Health 18:262-267.

Zieve, R., and P.J. Peterson. 1985. Sorption of dimethylselenide by soils. Soil Biol. Biochem. 17:105-107.

Zieve, R., and P.J. Peterson. 1986. An assessment of the atmosphere as a source of plant selenium: Application of stable and radioactive selenium isotopes. Toxicol. Environ. Chem. 11:313-318.

22

Some Other Major Poisonous Plants of the Western United States

Joseph L. Schuster and Lynn F. James

INTRODUCTION

Poisonous plants constitute one of the most significant management problems on western and southwestern ranges. Kingsbury (1964), in an extensive review of poisonous plants in the U.S. and Canada, reported that over 1,000 species are toxic to livestock in some way. In the Southwest, Sperry et al. (1964) described 80 poisonous plant species in Texas; Gay and Dwyer (1967) listed 100 poisonous species for New Mexico; and Schmutz et al. (1968) identified 300 species as being poisonous in Arizona. Of the Arizona plants, 30 were believed to cause major livestock losses, and 36 to be of secondary importance.

Thirty-three of the more common poisonous species found on western U.S. ranges, along with their distribution, toxic principles, and livestock affected, are listed in Table 22.1. Of these species, Astragalus, Cicuta, Conium maculatum, Delphinium, Gutierrezia sarothrae, Hymenoxys odorata, Quercus, and Senecio longilobus are treated in detail elsewhere in this symposium. We will discuss briefly the ecology and significance of the remaining species in the following paragraphs.

These species, by no means, represent all the important plants poisonous to livestock. However, they are widespread, and are important animal poisoning species in the western U.S. Many are toxic at all times, whereas others are toxic only under certain conditions. The presence of these species, even in small amounts, can create serious management problems for the livestock producer. To prevent livestock losses, forage that could otherwise be grazed may have to be left ungrazed; grazing systems may have to be altered; and additional improvements such as fencing, herding, or plant control measures may be required.

295

TABLE 22.1
Common toxic plants of southwestern and western United States.

Name	Distribution	Toxic Principle	Livestock Affected
Apocynum cannabinum Indian hemp	Throughout U.S.	Cardiac glycosides	All, especially sheep
Asclepias spp. Milkvetch	Throughout U.S.	Cardenolides	All
Astragalus spp. Oxytropis spp. Locoweeds	Western U.S.	Miserotoxin or Swainsonine	All
Baileya multiradiata Desert Baileya	Southwestern U.S. and Mexico	Unknown	Sheep and goats
Cicuta spp. Waterhemlock	Wet habitats	Cicutoxin	All and humans
Conium maculatum Poison-hemlock	Introduced; throughout U.S.	Piperidine alkaloids	All and humans
Cymopterus watsonii Spring parsley	Intermountain States	Furocoumarins	Sheep and cattle
Delphinium spp. Larkspur	Throughout U.S.	Diterpenoid alkaloids	All, but mostly cattle
Drymaria spp. Inkweed	Southwestern U.S. and Mexico	Alkaloids	All, especially cattle
Gutierrezia sorathrae Snakeweed	Western U.S.	Saponins	Sheep and cattle
Halogeton glomeratus Halogeton	Introduced; Intermountain States	Oxalates	All, but mostly sheep
Helenium hoopesii Orange sneezeweed	Western U.S.	Sesquiterpene lactones	Mostly sheep, cattle less
Hymenoxys odorata Bitterweed	Southwestern U.S.	Hymenoxon	Sheep
Hymenoxys richardsonii Pinque	Western U.S. and Canada	Sesquiterpenes	All, especially sheep

Plant	Distribution	Toxic Principle	Animals Affected
Hypericum perforatum St. Johnswort	Introduced; throughout U.S.	Hypericin	All, but goats least
Isocoma wrightii Jimmyweed	Southwestern U.S. and Mexico	Tremetol	All
Lupinus spp. Lupine	Throughout U.S. and Canada	Quinolizidine alkaloids	All, also deer
Oxytenia acerosa Copperweed	Southwestern U.S.	Alkaloids	All, especially cattle
Prunus virginiana Chokecherry	Western U.S.	Cyanogenic glycosides	All, especially sheep
Psilostrophe spp. Paperflowers	Southwestern U.S. and Mexico	Unknown	Sheep primarily, cattle suspected
Pteridium aquilinum Bracken-fern	Throughout U.S.	Thiaminase-horses Unknown-cattle	All, especially cattle and horses
Quercus havardii Quercus gambelii Shinnery and scrub oak	Southwestern and central western U.S.	Tannins	All, especially cattle
Sarcobatus vermiculatus Greasewood	Western U.S. and Mexico	Oxalates	All, but mostly sheep
Senecio longilobus Threadlead groundsel	Southwestern U.S. and Mexico	Pyrrolizidine alkaloids	All, cattle and horses most
Solanum spp. Nightshade	Throughout U.S.	Steroidal alkaloids	All
Tetradymia canescens Tetradymia glabrata Horsebrush	Western U.S.	Furanoeremophilanes	Sheep
Triglochin maritima Triglochin glabrata Arrowgrass	Western U.S., Canada, and Mexico	Hydrocyanic acid	All
Veratrum californicum Veratrum	Pacific Coast, Northern Rocky Mountain States	Steroidal alkaloids	All
Zigadenus spp. Deathcamas	Western U.S.	Steroidal alkaloids	All, especially sheep

INDIAN HEMP

Indian hemp (Apocynum cannalbinum), also called dogbane, hemp dogbane, and American hemp, is widely distributed throughout the United States and Canada. This perennial herb affects primarily sheep, but all other classes of livestock are susceptible. Its leaves are poisonous at all times, even when dry. Indian hemp, an erect plant, grows up to 1.5 m tall and is commonly found on coarse soil and along streams at elevations up to 2,130 m (Kingsbury 1964; James et al. 1980). Hemp dogbane is unpalatable and generally grazed when animals are hungry and/or other forage is scarce. Poisoning often occurs when animals are trailed through heavy stands of it. Signs of poisoning may include increased pulse rate, vomiting, cyanosis, body weakness, convulsions, coma, and death.

MILKWEED

Several species of milkweed (Asclepias) are poisonous to range animals. The principal poisonous species are A. labriformis, A. subverticillata, A. eriocarpa, and A. fascicularis. Poisoning occurs more frequently in sheep and cattle than in horses. Since milkweeds are distasteful to livestock, most losses occur as a result of animals being forced to eat the plant through mismanagement or from hay containing milkweed. Signs of poisoning include respiratory difficulty, rapid pulse, muscular weakness, spasms, bloat, and death. Losses occur at any time, but milkweed is most dangerous during the active growing season. Milkweed is found in sandy soils of plains and foothills throughout the southwestern and western U.S. (James et al. 1980).

DESERT BAILEYA

Desert Baileya (Baileya multiradiata), a small, showy, weak perennial of the sunflower family, is found throughout the southwestern U.S. and Mexico. It is locally common on sandy and gravelly soils up to 1,520 m feet elevation. Its poisoning principle is unknown, but toxicity occurs in both green and dry states. Signs of poisoning include depression, anorexia, slobbering of a green material, extreme weakness, rapid heartbeat, trembling of the limbs, arching of the back, and recumbency. Poisoning under range conditions is generally limited to sheep, although goats are known to be poisoned. Losses occur when there is a scarcity of forage. Poisoning typically results from ingestion of large amounts of green plant material. Animals taken off this plant refuse to eat for a few days, but most animals regain their appetite. Mortality as high as 25% has been reported in bands of sheep in Texas (Dollahite 1960).

SPRING PARSLEY

Spring parsley (Cymopterus watsonii) is confined to the Intermountain Region, primarily in Nevada, southeastern Oregon, southwestern Idaho, and southwestern Utah. It causes photosensitization in sheep and, infrequently, in cattle. Few animals actually die from eating spring parsley, but losses occur when affected ewes or cows with blistered, sore udders refuse to nurse their young. Poisoning occurs from early spring until plants mature and dry in early summer. Spring parsley grows on well-drained soils on rolling foothills and with sagebrush, pinyon pines, and junipers at elevations between 1,220 and 2,440 m. Treatment requires removal of the animals from infested ranges and providing shade until recovery (James et al. 1980).

INKWEED

Inkweed or thickleaf drymary (Drymaria pachyphylla) is a succulent, grayish-green glabrous summer annual found on alkaline clay soil in western Texas, southern New Mexico and Arizona, and Mexico. The seeds germinate following summer rainfalls when the soil is moist. Overgrazing has led to the establishment of heavy populations in some areas. It affects primarily cattle. Because of its unpalatability, inkweed is generally eaten only under harsh grazing conditions (Little 1937). Its poisonous principles are unknown, but symptoms of poisoning are a loss of appetite, diarrhea, restlessness, arched back (tucked up abdomen), depression, coma, and death. Animals poisoned generally die before symptoms are noticed (Kingsbury 1964). Serious and dramatic cattle losses occur from the plant in the U.S. and Mexico.

ORANGE SNEEZEWEED

Orange sneezeweed (<u>Helenium hoopesii</u>), an erect herbaceous perennial, has increased rapidly over large areas of range in recent years. All classes of livestock are affected, but sheep are the most frequently poisoned species. It is a major problem for sheep producers on the summer ranges of the central Rocky Mountains (Kingsbury 1964). Sneezeweed is found throughout the Mountain West at elevations of 1,520 to 3,650 m on moist slopes and well-drained meadows. It begins growth in early spring and matures in the late summer and early fall. Being an aggressive invader, once established it will inhibit growth of other plants.

The poisoning is cumulative, affecting animals eating small amounts over long periods of time (James et al. 1980). Signs of poisoning include emaciation, lips stained green from vomiting, stiffening limbs, depression, coughing, wasting, and eventual death. Sickness and death result from sheep eating small amounts of sneezeweed over a long period of time (usually over ten days). Recovery and prevention can be obtained if the sheep are removed from the infested range at the very earliest appearance of signs of poisoning for a period of about two weeks.

PINQUE

Pinque or western rubberweed (<u>Hymenoxys richardsonii</u>), a thick-rooted perennial herb, inhabits dry soils at elevations of 1,830 to 2,440 m in the mountains and foothills in the Rocky Mountain Region. It has caused extensive loss of sheep in Colorado, New Mexico, and Arizona (Kingsbury 1964). Losses are heaviest when hungry animals are placed on overgrazed ranges with heavy infestations of pinque. Toxicity builds up gradually, but its effects are cumulative. Most poisoning occurs in the spring or fall after animals have had one to two weeks access to pinque. Signs of poisoning include depression, weakness, vomiting, bloating, frothing at the mouth, green discharge from the nose, and coma (James et al. 1980). Pinque contains the same toxin as sneezeweed, and management practices to prevent poisoning are the same.

JIMMYWEED

Jimmyweed or rayless goldenrod (Isocoma wrightii) is confined to western Texas and southern Arizona and New Mexico. It is commonly found in riparian zones along river valleys and drainages. Jimmyweed is an erect, bushy, unbranched perennial shrub growing from 0.6 to 1.2 m tall, and is toxic to horses, cattle, sheep, and goats. Its toxic substance, tremetol, is excreted in the milk of lactating animals so that the young as well as humans may be poisoned by the consumption of contaminated milk. It produces the condition known as "trembles" after an animal consumes 1 to 1.5% of its body weight in jimmyweed for one to three weeks. Rayless goldenrod is not readily grazed. Death usually results if the animals are not removed from the affected areas (James et al. 1980).

The signs of poisoning include depression, standing in a humped-up stance, stiff gait, progressive weakness, inactive recumbency, coma, and finally death. Trembling, constipation, labored breathing, and dabbling urine may also occur. White snakeroot (Euporatorum rugosum) of the Midwest contains the same toxin as does rayless goldenrod and produces the same intoxication.

LUPINE

The lupines (Lupinus spp.) are widely distributed throughout the western U.S. All classes of livestock are affected, but losses are greatest with sheep. Literally hundreds of thousands of sheep have died from lupine poisoning. Lupines are dangerous throughout their growth cycle, but younger plants are more so than older ones. The toxicity of the various species of lupines varies markedly. The seeds are especially dangerous because of the high alkaloid content.

Signs of poisoning include dry coat, nervousness, depression, reluctance to move, difficult respiration, muscular twitching, loss of muscle control, frothing, convulsion, coma, and often death. Sheep poisoned on lupine should not be moved until signs of recovery become apparent. Moving them may cause death which otherwise might not occur. Lupine ingestion is also responsible for marked congenital deformities known as "crooked calf disease" (Shupe et al. 1967). Lupines grow in foothills and in sagebrush and aspen areas of the Mountain States and where abundant, present a major problem to livestock production.

COPPERWEED

Copperweed (Oxytenia acerosa), an erect perennial herb, is toxic to all species of livestock, especially cattle. It is common on alkaline soils or streambeds and desert ranges and foothills of southern California, southwestern Colorado, southeastern Utah, and northern Arizona. The plant is unpalatable and grazed only when other feed is scarce. Cattle are more likely to eat the plant in the fall when they are being trailed from summer ranges. Sheep are occasionally poisoned in the fall and winter by eating fallen leaves (Kingsbury 1964).

CHOKECHERRY

Chokecherry (Prunus virginiana), distributed widely throughout the western U.S., continues to be an important poisonous plant to cattle and sheep. Other toxic species of Prunus can be found throughout the U.S. Animals become poisoned upon eating considerable quantities of leaves and twigs in a short time. Acute poisoning can occur when animals eat large amounts of chokecherry leaves prior to watering. Chokecherry occurs in thickets on hillsides and canyon slopes or as a shrub or small tree with other riparian vegetation. The toxic substance, hydrocyanic acid, is found principally in the leaves. Toxicity decreases with maturity, but bruising, wilting, withering, or frost damage of leaves increases its toxicity (James et al. 1980).

Signs of poisoning include distress, cyanosis, rapid breathing, muscular twitching and convulsions, coma, and death. Death may occur quickly after a lethal dose is consumed in a short period of time. Chokecherry is one of the more economically important poisonous plants of the United States. Although it is grazed extensively at times, significant losses can occur when the plant is eaten in large quantities under stressed conditions.

BRACKEN-FERN

Bracken-fern or western bracken (Pteridium aquilinum) is distributed throughout the United States. It is poisonous to all classes of livestock, especially cattle and horses. Cattle poisoning often occurs during late summer when other feed is scarce or when animals are fed hay containing bracken-fern. Acute symptoms of poisoning are manifested in cattle after they have consumed large amounts of the plants, but horses exhibit a more chronic reaction. Bracken-fern contains a thiaminase which causes a thiamine deficiency in horses and can be treated with thiamine. Livestock losses from bracken-fern can be high, especially in the Pacific Coast States, and in the northeastern, southeastern, and midwest U.S. Bracken-fern grows on burned-over areas in woodlands and other shaded places and on hillsides. Because the poisoning is cumulative, live-stock are affected only after they have eaten considerable amounts of bracken-fern for two to four weeks (James et al. 1980). Bracken-fern is unpalatable and must be grazed over a long period of time before poisoning occurs. Livestock rarely graze bracken-fern if good forage is available.

GREASEWOOD

Greasewood (<u>Sarcobatus</u> <u>vermiculatus</u>) is a deciduous shrub adapted to heavy saline soils of the western United States. It causes problems only when large amounts are consumed over short periods of time. The toxic principle is an oxalate, and toxicity increases as the growing season advances. Losses in sheep can be high in the fall if sheep eat large quantities of leaves that have fallen to the ground. Poisoning occurs four to six hours after an animal eats a large amount of greasewood. A sheep may die if it eats as little as 900 g of green leaves in a short period of time (James et al. 1980). The signs of poisoning are depression, weakness, reluctance to move, increased respiration, drooling, recumbency, and death.

Cattle grazing heavy stands of greasewood may die from what is known locally as black greasewood poisoning. It is not known if this is due to oxalate poisoning, malnutrition, or possibly to a condition related to the consumption of the branches or stems of the plant which are very hard. This could possibly result in a condition similar to hardware disease in cattle.

NIGHTSHADE

Several species of nightshade (<u>Solanum</u>) are found throughout the U.S. All should be considered poisonous. The plants contain the glyco-alkaloid solanine which is most highly concentrated in the unripe berries. It is extremely toxic, even in small amounts, so that when animals are forced to eat the berries, poisoning can occur with all species. Purple nightshade (<u>Solanum elaeagnifolium</u>) has been found toxic to cattle if as little as 0.1% of the animal's weight is eaten (Kingsbury 1964).

HORSEBRUSH

Two species of horsebrush, littleleaf (<u>Tetradymia glabrata</u>) and gray or spineless (<u>Tetradymia canescens</u>), affect sheep in the western U.S. Photosensitization results when sheep feed on horsebrush and are exposed to bright sunlight. Sheep must graze black sage (<u>Artemisia nova</u>) prior to or in conjunction with consumption of the <u>Tetradymia</u> in order for photosensitization to occur. The grazing of <u>Tetradymia</u> alone may cause severe liver damage and even death, but the photosensitization requires the action of both plants. The resulting disease is called "bighead" because of the characteristic swelling of the head of the exposed animals.

Most losses from horsebrush occur when sheep are trailed through heavily infested areas. Although it has not been experimentally demonstrated, field observations suggest that poisoning occurs principally during periods of approaching storms or during them. The plant is especially dangerous during bud stage because of its palatability to sheep at that time. All plant parts of both species are poisonous, but sheep eat only buds, leaves, and fine stems. Littleleaf horsebrush is most abundant on benchlands, well-drained slopes, and low elevations on sheep winter ranges in the Intermountain Region. Spineless horsebrush is most abundant in sagebrush areas and foothills in New Mexico, Arizona, Wyoming, Montana, and Washington (James et al. 1980).

Signs of poisoning include depression, weakness, recumbency, and in cases of severe poisoning, death may occur. In less severe cases, photosensitization develops and is manifested by swelling of the head. Usually only the head swells because the skin on the face is exposed to the sun, and the rest of the body is protected by the wool. The exposed parts of the body may itch, and the sheep may develop a star-gazing stance. The ears, lips, and face swell; the skin may crack; and large pieces of skin may peel off. The animal may show extreme discomfort and seek shade or protection from the sun. Abortions may occur.

ARROWGRASS

The species of arrowgrass which most commonly poison livestock are Triglochin maritima and T. palustris. They are perennial forbs widely distributed in marshy areas throughout the United States. The plants are not highly toxic until growth is stunted during dry periods. All classes of livestock are susceptible, but cattle and sheep are the most affected. Poisoning only occurs if animals eat large amounts in a short period of time. Arrowgrass cut for hay can be toxic. Arrowgrass grows best on soils covered with water or in moist soils near springs (James et al. 1980). Hydrocyanic acid is the toxic principle, thus the signs of poisoning are the same as for chokecherry.

VERATRUM

Veratrum or false hellebore (Veratrum californicum) is found primarily in the Pacific West and the Rocky Mountain States. It is grazed readily by sheep and goats. Veratrum is poisonous from the time it initiates growth in the spring until senescence in the fall, but toxicity decreases as plants mature. It grows on moist open meadows and hillsides at elevations of 1,830 to 3,350 m. Intoxication of pregnant ewes results in congenital deformities of the head of offspring commonly called "monkey-faced lambs." Binns et al. (1963) and Binns et al. (1965) reported that veratrum causes cyclopian and related malformations in lambs born to ewes that ingested the plant on the 14th day of gestation. Signs of poisoning include salivation, body weakness, irregular gait, vomiting, irregular heartbeat, coma, and death.

DEATHCAMAS

Deathcamas species (Zigadenus) are perennial herbs and occur throughout the western U.S. They affect all classes of livestock, especially sheep. Because growth begins early, deathcamas may be heavily grazed and can cause severe losses in sheep even under good range conditions. Deathcamas contains toxic alkaloids throughout the plant and thus is dangerous at all times. Poisoning in sheep from species of this plant have occasionally been reported in which losses ranged between 500 and 2,000 animals. Less devastating losses are frequent (Kingsbury 1964). The signs of poisoning include salivation, vomiting, muscular weakness, increased heart rate, convulsion, coma, and death.

COCKLEBUR

Cocklebur (<u>Xanthium</u> <u>strumarium</u>) is a widely distributed annual plant that grows around reservoirs, river bottoms, lake beds, flooded areas, in fields, and other such areas that are favored by good moisture and sandy soil. Poisoning occurs primarily in cattle, sheep, horses, and swine. The plant is toxic only during the cotyledonary stage of growth. Poisoning is quite common. Signs of poisoning include rapid pulse, difficult respiration, nausea, vomiting, and some spasmodic contraction of muscles.

SUMMARY

Experience of the authors and contact with range managers, practicing veterinarians, and diagnostic laboratories indicate that these plants continue to be a hazard to ranchers grazing livestock on ranges infested with them. We know that many livestock die each year from intoxication by these plants, yet neither we nor anyone else, including the ranchers, know of the full impact of these plants on range livestock production. However, we do know that the economic impact is significant.

LITERATURE CITED

Binns, W., L.F. James, J.L. Shupe, and G. Everett. 1963. A congenital type malformation in lambs induced by maternal ingestion of a range plant, <u>Veratrum californicum</u>. Amer. J. Vet. Res. 24:1164-1175.

Binns, W., J.L. Shupe, R.F. Keeler, and L.F. James. 1965. Chronologic evaluation of teratogenicity in sheep fed <u>Veratrum californicum</u>. J. Amer. Vet. Med. Assoc. 147:839-842.

Dollahite, J.W. 1960. Desert Baileya poisoning in sheep, goats and rabbits. Tex. Agr. Exp. Sta. Prog. Rep. 2149.

Gay, C.W., and D.D. Dwyer. 1967. Poisonous range plants. Coop. Ext. Serv. Cir. 391, New Mexico State Univ., Las Cruces.

James, L.F., R.F. Keeler, A.E. Johnson, M.C. Williams, E.H. Cronin, and J.D. Olsen. 1980. Plants poisonous to livestock in the western states. USDA Agri. Info. Bull. 415.

Kingsbury, J.M.   1964.   Poisonous plants of the U.S. and Canada.   Prentice-Hall, Englewood Cliffs, NJ.

Little, E.L.   1937.   A study of poison drymaria on southern New Mexico ranges.   Ecology 18:416.

Schmutz, E.M., B.N. Freeman, and R.E. Reed.   1968. Livestock poisoning plants of Arizona.   Univ. Arizona Press.

Shupe, J.L., W. Binns, L.F. James, and R.F. Keeler.   1967. Lupine, a cause of crooked calf disease.   J. Amer. Vet. Med. Assoc. 151:198-203.

Sperry, O.E., J.W. Dollahite, G.O. Hoffman, and B.J. Camp. 1964.   Texas plants poisonous to livestock.   Tex. Agr. Exp. Sta. Bull. B-1028.

23

Noxious Weeds That Are Poisonous

Russell J. Lorenz and Steven A. Dewey

INTRODUCTION

Noxious weed terminology has not been standardized, therefore definitions are many and varied. "Noxious" is a legal designation, rather than a biological or botanical term. A noxious weed is one that has been legislated against in some way because it has an undesirable property or trait. A noxious weed may be an annual, biennial, or perennial species, a grass or a broadleaf, woody or herbaceous, thorny or smooth, unsightly or beautiful, nutritious or poisonous. There is no single trait that automatically qualifies a plant to be declared noxious by legislative bodies or governmental agencies. The same species may be designated as noxious in one state or county, and not considered noxious in another. A plant species may be noxious in a particular state this year, while a year earlier or a year later the same species may not be noxious in the same state.

Even the economic impact caused by a particular weed or plant species does not always seem to correlate with its presence or absence on noxious weed lists. For example, many serious poisonous range plants are totally omitted from all noxious weed laws. Other common poisonous plants, such as halogeton, larkspur, locoweeds, and lupine, have only been declared noxious in one or two states, yet they cause very significant losses throughout much of the U.S.

All of these factors lead to confusion when biological and legal terminology are combined. The definition of "noxious species" provided by the Soil Conservation Society of America is: "A plant that is undesirable because it conflicts, restricts, or otherwise causes problems under the management objectives. Not to be confused with species declared noxious by laws." This definition is basically the same as that provided by the Society for Range Management.

The following definitions for noxious weed species are examples taken from state weed laws:

"A plant considered to be extremely destructive or harmful to agriculture and so designated by law." Nebraska Weeds. Revised 1975, Reprinted 1977. Nebraska Department of Agriculture.

"Any plant the Commissioner determines to be especially injurious to public health, crops, livestock, land, or other property." Utah Noxious Weed Act, Title 4, Chapter 17, Utah Code annotated 1953, amended 1985.

"Any plant which is determined by the Director to be injurious to public health, crops, livestock land or other property." Noxious Weed Law of Idaho, Title 22, Chapter 24, Idaho Code.

"Any plant propagated by either seed or vegetative parts which is determined by the commissioner (Agriculture) after consulting with the state cooperative extension service, or a county weed board after consulting with the county extension agent, to be injurious to public health, crops, livestock, land, or other property." North Dakota Noxious Weeds Law with Regulations, Chapter 63-01.1, ND Century Code with Regulations.

"Weeds, seeds or other plant parts that are considered detrimental, destructive, injurious, or poisonous, either by virtue of their direct effect or as carriers of diseases or parasites that exist within this state, and are on the designated list. 'Designated list' means the list of weeds and pests from time to time designated by joint resolution of the board and the Wyoming Weed and Pest Council." Wyoming Weed and Pest Control Act of 1973, Title II, Chapter 5.

"Any plant which is determined by the Director to be injurious to public health, crops, livestock, land or other property. Noxious weeds for the purpose of this policy shall be rated 'A', 'B',

'C', or 'Q'. (1) 'A' Pest--a weed of known economic
importance known to occur in the state in small enough
infestations to make eradication practicable; or, not
known to occur, but its status in surrounding states
makes future occurrence seem imminent. 'B' Pest--a
weed of known economic importance and of limited
distribution in the state and is subject to intensive
control or eradication where feasible at the county
level. 'C' Pest--a weed of known economic importance
and of general distribution that is subject to
intensive control or eradication as local conditions
warrant. 'Q' Pest--a newly detected weed species of
which economic significance to Oregon agriculture has
yet to be determined." Plants; Infestation,
Quarantine, Pest and Weed Control Law, Policy Statement
of 1983, and Weed Control Policy and Classification
System. ORS Chapter 570, 505, 1985.

The phrase "not to be confused with species declared
noxious by laws" in the definition provided by the Soil
Conservation Society of America is understandable after
reading the weed and seed laws of the 50 states. The
complexity of these laws, the disparities in both the letter
and intent of the laws, and the designated mechanism for
declaring a plant to be noxious, complicates and frustrates
definition and eventually the control process. Weeds do not
respect political boundaries; therefore, if control programs
are to be effective, county lines, state lines, and even
international boundaries must provide no more hindrance to
control than they do to the spread of weeds.
Generally, a plant has attained the status of being a
serious problem before it is legislated as being noxious.
Often, little or no control is initiated until the law
designates the weed as being noxious. For highly aggressive
weed species, prevention is the best and often the only
economical control, so using the legal designation as the
trigger for a control problem is ineffective and costly.
All 50 states have seed laws which designate certain
weed seeds as noxious, and restrict or prohibit the
transport or sale of commercial seed contaminated with
noxious weed seeds. But less than half of the states have
weed laws that designate the actual plants of certain weedy
species as noxious, and require control of these plants in
the field. Weed laws require landowners to take whatever
means necessary to prevent proliferation and spread of
noxious weeds growing on their land; they also regulate the
transport or sale of crops, soils, or equipment contaminated
with noxious weeds or weed parts.

More than 500 plant species appear on noxious weed lists supported by state or provincial laws. Of these, more than 125 have been documented as being poisonous to livestock (Appendix 23.1). Many of the remaining noxious species may be poisonous but have not yet been tested for potential livestock toxicity. Some of the weed species listed in Appendix 23.1 are found primarily in crop-land, and therefore might be omitted from lists of plants considered detrimental to ruminant animals. But, past experience has shown that many of the cropland weeds introduced to North America soon became serious problems in grasslands. Other "cropland" weeds can become problems to livestock when they are part of crop vegetation harvested for hay or other stored forage; and especially so when drought induced feed shortages lead to harvesting feed from unusual areas for use as emergency feed.

Almost all weeds designated as noxious somewhere in North America are not native to the area in which they are a problem. In most cases, these weeds were introduced by accident as contaminants in seed grain, in ship's ballast, or in various other items transported to North America. In some cases, weeds were introduced intentionally for their ornamental or medicinal value. An important aspect contributing to the biological success of introduced weed species is the fact that the natural diseases and predators that keep them reasonably controlled in their native countries were, in most cases, not introduced with the plants.

The following are but a few of the species designated as noxious and capable of causing livestock illness or death. In addition to causing the economic losses associated with livestock poisoning, these weeds are highly competitive against desirable forage species, and can significantly reduce the carrying capacity on grazing lands.

SPECIFIC EXAMPLES OF NOXIOUS WEED PROBLEMS

Leafy Spurge

Leafy spurge (_Euphorbia esula_) is a long-lived herbaceous perennial weed introduced to North America from Europe and Asia. It grows 0.6 to 1.2 m tall from a woody crown and has an extensive, deep, and extremely persistent root system. Roots are commonly found to a depth of 4.6 m, and they contain stored nutrient reserves capable of supporting growth for many years, making the plant highly resistant to mechanical or chemical defoliation. It reproduces from seed and from numerous vegetative buds on the crown and on creeping, horizontal roots (Barkley 1986). Leafy spurge begins growth early in the spring, making it a vigorous competitor with forage and range plants. Upright, branched stems bear numerous linear-shaped smooth leaves. The plant has a bluish-green to yellow-green color during the growing season, and it generally turns to a rusty reddish-orange color in late summer. The flat-topped clusters of yellow bracts appear in late May and the true leafy spurge flowers develop above the bracts about 7 to 10 days later. Seed pods contain three smooth, oval seeds, ranging in color from gray to brown to speckled. When mature, the seed pods forcibly open, throwing seed up to 4.6 m from the plant. Each stem produces about 120 seeds. Seeds remain viable in the soil for about eight years (Best et al. 1980). All plant parts contain a milky latex at all stages of growth.

Leafy spurge contains toxic substances that cause skin inflammation in cattle; prolonged exposure causes swelling of the tongue and lips, reducing the ability of the animal to graze. When ingested, it acts as a digestive tract irritant and purgative. The resulting scours and weakness can lead to death of cattle. It has produced inflammation of the feet of horses after they walked in freshly mowed leafy spurge (Watson 1985). Sheep often eat leafy spurge with little effect, but death of sheep has occurred following consumption (Landgraf et al. 1984).

The highly competitive nature of leafy spurge rapidly reduces production of desirable species, and it becomes established and spreads rapidly on even the best managed grassland. It grows on all sites and in all situations, from wetlands to arid ridgetops, from level plains to steep river banks, from heavy clay soils to sandhills and sandy deltas, and from open prairie to shaded woodlands (Dunn 1979).

Control of leafy spurge is extremely costly and difficult. The nature of the plant makes chemical control relatively ineffective. Present chemical control technology will not eliminate stands that have been established for more than three years. At best, it has slowed the rate of spread (Lyn and Messersmith 1987). Further complicating control is the fact that leafy spurge in North America came from a wide variety of sources in Europe and Asia; consequently, it is a complex of species, sub-species, and taxonomic variations. Some types are more susceptible to herbicides than others, and some potential biological control agents are specific to only certain types of leafy spurge. There is need for research on the chemistry, physiology, morphology, and anatomy of the leafy spurge complex in order to facilitate development of more cost-effective control measures.

Leafy spurge seeds provide feed for many birds. A small percentage of the seed passes through these birds as germinable. Consequently, spread of the weed has been unusually rapid. Plants starting from seed in rangeland are often difficult to find until they have developed into patches covering several square decimeters. By that time, they have developed a deep root system and have produced at least two crops of seed. Left uncontrolled, a few such patches in a square kilometer of rangeland have the potential for dominating the vegetation in that square kilometer in a few years, and for providing the seed to establish new patches for several kilometers in all directions from the original infestation.

One of the means of introduction of leafy spurge to North America was as an impurity in seed grain brought by early settlers. Consequently, the problem was originally on cropland, but tillage operations and annual use of herbicides on cropland reduced the problem after WWII. For several years, the problem appeared to be insignificant on grasslands, but without tillage, with limited use of herbicides, and with resistance of leafy spurge to grazing, the problem has grown to tremendous proportions on grasslands of the northern United States and southern Canada. Farmers, ranchers, and land managers are spending millions of dollars each year in an attempt to control leafy spurge. At best, these efforts are slowing the spread of the weed, with very few, if any, examples of complete control.

Cattle avoid grazing palatable forage species growing among leafy spurge plants, and the competitive nature of leafy spurge rapidly reduces forage production. A research study in North Dakota showed that usable forage production was reduced by 75% on leafy spurge infested rangeland. Forage utilization was 20% by midsummer on areas without leafy spurge, but utilization was reduced to 2% on low leafy spurge densities (42 stems/m^2). Zero forage utilization occurred at the moderate (112 stems/m^2) and high (170 stems/m^2) densities. Some use was made of the forage in the leafy spurge stands late in the season when active growth of the leafy spurge decreased, but at that time, quality of the forage was significantly lower than it was earlier in the season, resulting in further economic loss (Lyn and Kirby, 1987).

Estimates of direct losses in forage and livestock production are difficult to establish, but figures from Montana and North Dakota were $2.5 and $19 million annually, respectively (Reilly and Kaufman 1979, Messersmith and Lyn 1983). Leafy spurge is a serious problem in at least 12 states and five Canadian provinces, and it is rapidly expanding to new areas (Dunn 1979). It has become a national and an international problem.

Leafy spurge respects no boundary, physical or man made. National parks, state parks, campgrounds, and other public areas are being invaded by leafy spurge. The aesthetic and recreational loss, and the loss of natural vegetation within infested areas, will have social and economic impacts on urban and rural people alike (Reilly and Kaufman 1979). The impact of large acreages of leafy spurge on wildlife is not known, but it is highly probable that at least the ruminant wildlife species will be impacted. Urban areas from small towns to large cities are also being invaded by leafy spurge. The expense of control on highway rights-of-way, city streets, in public parks, and in other public lands has to be borne by everyone through taxes. Urban homeowners' lawns and gardens are easily invaded by leafy spurge, causing added expense and limiting the use of lawns as play areas for children and as general family recreation areas.

Vigilance is especially important in controlling an aggressive, persistent weed like leafy spurge. Early detection and prompt treatment is the only hope for avoiding a costly problem. No patch of leafy spurge is too small to spray.

At this time, properly applied herbicide is the best control for leafy spurge (Alley et al. 1984; Eberlein et al. 1982). Recommendations on chemical to use and rate and time of application change as research produces new data upon which to base recommendations. Check with the Extension service or county weed officers on the latest recommendations for your area.

A biological control program is being developed for leafy spurge, using host-specific insects and diseases from Europe and Asia (Watson 1985). However, the process of collecting, screening, and testing to be certain that the predator does not attack other plant species in North America is costly and time consuming. Increasing the populations to large enough numbers for release takes many years of research and development. In the meantime, landowners cannot afford not to spray.

Diffuse and Spotted Knapweed

Both diffuse and spotted knapweed (Centaurea diffusa and C. maculosa) were introduced to North America from Europe and Asia, primarily as a contaminate in alfalfa seed. Early reports of infestations in the United States were from the Pacific Northwest in about 1900 (Hickman and Lyn 1985).

Spotted knapweed has spread faster and has caused more problems than diffuse knapweed. Both reproduce primarily by seed, but spotted knapweed has a biennial or short-lived perennial habit while diffuse knapweed has an annual or biennial growth habit. Spotted knapweed has a stouter taproot than does diffuse knapweed and can reproduce vegetatively through lateral shoots from the main stem just below the soil surface. Both species have upright branched stems with flower heads at the tips of the branches, and with no spines on the stems or leaves. Flower color of spotted knapweed is generally pink, purple, or occasionally white; diffuse knapweed flowers are often white or sometimes pink to lavender. The major distinguishing feature of spotted knapweed is black-tipped flower bracts (USDA 1971).

The most common method of spreading these two species is by transport of the seed on vehicles or machinery. Consequently they are usually found first on disturbed sites along road ditches, utility lines, railroads, buried cables, or pipelines. The mature plants become brittle and stiff, facilitating lodging of plant parts on vehicle undercarriages and on machinery. The seed is held loosely in the dried flower head and is scattered as the plant material is transported. Both species commonly grow in hayfields, especially in alfalfa. Transport of hay containing these weeds is another common method of spread.

Diffuse and spotted knapweed have not been proven toxic to livestock; but at least two other knapweed species within the Centaurea genus are known to be toxic to horses (Kingsbury 1964; Young et al. 1970), giving some reason to suspect that diffuse and spotted knapweed may contain similar toxic properties. Perhaps the lack of reported poisonings is due to the fact that both diffuse and spotted knapweed are very unpalatable, with few leaves and with stiff, branched stems. Sharp stiff bracts on mature diffuse knapweed make the plants very spiny, and can deter even the hungriest livestock animals. The dense stands formed by both knapweed species limit access by animals to desirable forage species. The knapweeds are also allelopathic, which eventually eliminates all other plant species from an infested area.

A primary concern is the ability of these species to spread rapidly. In Montana, the increase in acreage between 1920 and 1980 has been at an average rate of 27.4% per year (Lacey and Fay 1984). North Dakota and Nebraska have both experienced rapid increases in counties reporting the problem, and in rate and spread from the original infestations. The largest infestations in North America are in northwestern United States and southwestern Canada; northern Minnesota, Wisconsin, Michigan, and New York are also infested. As people learn to recognize the plants, it will be reported in many other states.

The knapweeds rapidly reduce the usable forage on infested sites. Research reports indicate 50 to 75% reduction in usable forage, with eventual losses of 90 to 100% in heavy infestations (Lacey and Fay 1984). Cost of chemical control is sizeable because, even though the plants can be killed with herbicide, each plant can produce 1,000 or more seeds. Seed longevity in the soil is five years or more, so the potential for reinfestation from seed exists for several years. The allelopathic nature of the plant will eliminate desirable species, which will necessitate range re-seeding as part of a control program.

In addition to loss of forage to grazing livestock, hay fields containing even a few knapweed plants will reduce the value of the hay crop by limiting sales outside the area of production. Knowledgeable buyers will avoid hay containing troublesome weeds.

Control of the knapweeds as soon as small patches are found is the only way to prevent a larger problem. Education of the public is necessary, so everyone can recognize these weeds and know what the consequences are if control measures are not initiated as soon as the plants are found.

Chemical control with 2,4-D, dicamba, or picloram has been effective, when properly used, but control of large acreages in diverse habitats is often not economical or practical using herbicides. Biological control is being developed, using insects collected in Europe and Asia (Hickman and Lyn 1985). Some degree of success has been attained from releases in Montana, Oregon, and Idaho, but large scale control has not been attained.

Yellow Starthistle

Yellow starthistle (Centaurea solstitialis) is an annual or winter annual weed, probably introduced from the

Middle East. The plant is usually 0.3 to 0.6 m tall, loosely branched at the base, with reduced leaves and winged stems. Cotton-like hairs cover leaf and stem surfaces, giving the plant a grayish green color. Long spine-tipped bracts surround the many yellow flower heads on each plant (Whitson 1987). Yellow starthistle has been declared noxious according to weed or seed laws in most, if not all, areas of the U.S. where infestations occur.

Archeological findings indicate the presence of yellow starthistle in California as early as 1824 (Maddox and Mayfield 1985). Since its introduction into the U.S., this aggressive plant has spread rapidly over millions of hectares of rangeland, pasture, and wasteland in the western states. Surveys conducted in California in 1958 and again in 1985 showed a 640% increase in infested area, from 0.5 million to 3.2 million hectares in just 27 years (Maddox and Mayfield 1985). Though a relatively new invader in Idaho, a field survey conducted in 1981 revealed over 75,000 hectares of rangeland heavily infested with yellow starthistle (Callihan et al. 1982).

Yellow starthistle is a highly competitive weed species, and is reported to produce allelopathic compounds (Maddox et al. 1985). Grazing often further increases the incidence of yellow starthistle by selectively removing the competitive influence of other desirable species. Yellow starthistle causes a nervous disorder called "chewing disease" or nigropallidal encephalomalacia in horses (Kingsbury 1964). Symptoms include muscle stiffness, swelling and eventual paralysis of throat and mouth, and muscle twitching. Death usually occurs through starvation or dehydration.

Yellow starthistle can be effectively controlled by a combination of pasture management and herbicides (Hill and Hackett 1987). As with the weeds previously mentioned, prevention, early detection, and eradication of new infestations are the most cost-effective control measures, and should be given the highest priority.

Dyers Woad

Dyers woad (Isatis tinctoria) is described as a deep tap-rooted winter annual, biennial, or short-lived perennial from 0.6 to 1.2 m tall (King 1966). Yellow flowers and blue-green foliage make it easy to recognize when it blooms in May and June. Dyers woad is a prolific seed producer. Seeds are contained in one-celled winged pods which turn dark brown or black at maturity. Crown buds may survive from year to year, allowing individual plants to persist for three or more seasons (Callihan et al. 1984). Part of the competitive and aggressive nature of this weed is attributed to production of allelopathic compounds (Young and Evans 1971).

Dyers woad is an example of a rapid invader capable of dominating sites and forming dense infestations that compete vigorously with desirable plant species. It is adapted to a wide variety of environmental niches, from irrigated cropland and pastures to semi-arid rangeland and rugged mountains. It seems especially suited to rocky or gravelly disturbed sites. Since its introduction into the U.S. in the 19th century, dyers woad has spread rapidly in many western states (King 1967). The first report in Idaho was of a single plant collected in 1938. An extensive field survey in 1983 found over 9500 hectares infested (Callihan et al. 1984). A 1981 report from Utah indicated the number of infested hectares had doubled in the last ten years (Evans and Chase 1981). Discovered in Montana in the 1950s, this weed has now spread into at least six counties. Just one infestation near Dillon, Montana, increased from 0.8 to more than 40 hectares in two years (Aspevig et al. 1985). It has been declared noxious according to weed laws in several western states.

It is estimated that in just three counties in northern Utah, dyers woad is costing farmers and ranchers in excess of $2 million annually in the form of reduced crop yields and range production (Evans and Chase 1981). Though not reported as toxic to livestock, it is a candidate for toxicological investigations; and is closely related to other known toxic weeds such as the wild mustards, tansy mustard, field pennycress, and whitetop (Kingsbury 1964).

Several herbicides will effectively control dyers woad if applied at the proper growth stage of the weed. Digging or hand rogueing small or light infestations can be effective if care is taken to remove the deep taproot (Evans and Chase 1981).

Kochia

Kochia (Kochia scoparia) is an early emerging summer annual weed reproducing and spreading exclusively by seed. Flowers are small and inconspicuous. Plants may range in height from 0.3 to 1.8 m, producing an average of over 14,000 seeds per plant. When mature, the multibranched plants often assume a spherical shape and are driven by the wind like tumbleweeds. Leaves often turn a reddish or purplish color in the autumn. It is not to be confused with the perennial kochia (Kochia prostrata), which is considered a desirable forage and ground cover in some arid regions of the U.S. (Smith et al. 1983).

Kochia is a rapidly growing, drought tolerant species introduced into the U.S. as an ornamental. It escaped cultivation and soon became a serious weed problem in crops and on rangeland in many western states. Time-series distribution maps based on herbaria collections in five northwestern states show one report of kochia in a single Wyoming county in 1890 (Forcella 1985b). By 1940, kochia specimens had been submitted from 44 counties in all five states. By 1980, kochia had been recorded in over 70 counties, having essentially overrun the states of Montana and Wyoming, and spreading into many additional counties in the other three states. In another study, kochia was reported to have the highest rate of spread of 40 species of alien or exotic weeds examined (Forcella 1985a).

One kochia plant in 3 m of row has been shown to reduce sugarbeet yields by as much as 26% (Smith et al. 1983). In addition to the aggressive and competitive nature of Kochia scoparia, it is reported toxic to livestock. Cattle consuming kochia may develop polioencephalomalacia and photosensitization (Dickie and James 1983). Saponins, oxalates, alkaloids, and nitrates are some of the toxic agents that have been identified in kochia. In spite of its proven toxic properties, kochia is sometimes promoted commercially as a desirable alternative forage. Some advertisements actually suggest that it be used to replace alfalfa.

Efforts to prevent the intentional or accidental introduction of kochia into previously uninfested areas should be given a high priority. Control of existing infestations can be achieved with a number of herbicides approved for cropland, pastures, or rangeland; treatment in an early stage of growth is essential for effective control.

Black Henbane

Black henbane (Hyoscyamus niger) may behave as an annual or biennial. Plants range in height from 0.3 to 1.2 m. Leaves on rosettes and mature plants are covered with soft long hairs, and have a "foul" odor. Flowers are tan to cream-colored, with a dark purple throat and purple netted petal lobes. Seeds are contained in distinctive "pineapple-shaped" fruits, and fruits are arranged in rows along each flower stalk (Whitson 1987). Branches of dried henbane fruits are often used as part of decorative flower arrangements. Henbane is a native of Europe. It was introduced into North America and cultivated as an ornamental and medicinal plant. It is now spreading throughout much of the U.S. and is becoming a common weed in pastures, fencerows, and along roadsides. Henbane tends to be opportunistic in nature, typically invading disturbed or heavily grazed sites. Though not reported to be a highly competitive aggressor, it is capable of invading pastures and rangeland.

Low doses of the tropane alkaloids found in black henbane have been used throughout history for various medicinal purposes (Kingsbury 1964). At higher or uncontrolled rates, these alkaloids are toxic to man and animals. Livestock, fowl, and human poisonings from consuming the foliage or seeds have been reported. Because animals generally find the plants distasteful or unpalatable, livestock poisonings occur less often than might otherwise be expected. Alkaloids found in black henbane include hyoscyamine, scopolamine, and atropine.

Control of henbane with herbicides is best achieved in the seedling or rosette growth stage. Cutting or digging isolated plants before seed production occurs can also be effective.

LITERATURE CITED

Alley, H.P., N.E. Humburg, J.K. Forstrom, and M. Ferrell. 1984. Leafy spurge repetitive herbicide treatments. Res. in Weed Sci., Wyo. Agr. Exp. Sta. Res. J. 192:90-93.

Aspevig, K., P. Fay, and J. Lacey. 1985. Dyers woad: A threat to rangeland in Montana. Montana State University Extension Bulletin, MT-8523.

Barkley, T.M., (ed). 1986. Flora of the Great Plains. Great Plains Flora Association. University Press of Kansas.

Best, K.F., G.G. Bowes, A.G. Thomas, and M.G. Maw. 1980. The biology of Canadian weeds. 39. Euphorbia esula L. Can. J. Plant Sci. 60:651-663

Callihan, R.H., S.A. Dewey, J.E. Patton, and D.C. Thill. 1984. Distribution, biology, and habitat of dyers's woad (Isatis tinctoria) in Idaho. J. Idaho Acad. Sci., Vol. 20, No. 1-2, June-December, pp. 18-32.

Callihan, R.H., R.L. Sheley, and D.C. Thill. 1982. Yellow starthistle identification and control. University of Idaho Current Information Series No. 634.

Cheeke, P.R., and L.R. Shull. 1985. Natural toxicants in feeds and poisonous plants. AVI Publishing Co. Inc., Westport, CN., p.492.

Dickie, C.W., and L.F. James. 1983. Kochia scoparia poisoning in cattle. J. Amer. Vet. Med. Assoc. 183(7): 765-768.

Dunn, P.H. 1979. The distribution of leafy spurge (Euphorbia esula) and other weedy Euphorbia spp. in United States. Weed Sci. 27:509-516.

Eberlein, C.V., R.G. Lyn, and C.G. Messersmith. 1982. Leafy spurge identification and control. North Dakota Coop. Ext. Service W-765.

Evans, J.O., and R.L. Chase. 1981. Dyers woad control. Utah State Univ. Ext. Bull. EL-199.

Forcella, F. 1985a. Final distribution is related to rate of spread in alien weeds. Weed Research 25:181-191.

Forcella, F. 1985b. Spread of kochia in the northwestern United States. Weeds Today 16(4):4-6.

Halls, M.R., and T.D. Crane. 1983. Plants poisonous to animals - A bibliography for the world literature 1960-1979. Commonwealth Bureau of Animal Health, New Haw, UK, p.160.

Hickman, M.V., and R.G. Lyn. 1985. Spotted Knapweed: A threat to northwestern rangelands. Weeds Today 16(4):1-2

Hill, K., and I. Hackett. 1987. Spotted knapweed, diffuse knapweed and yellow starthistle control in Nevada. University of Nevada-Reno Ext. Bull. 87-5.

King, L.J. 1966. Weeds of the world: Biology and control. Interscience Publications Inc., NY.

King, L.J. 1967. The distribution of woad (Isatis tinctoria L., Cruciferae) in North America. Proc. N.E. Weed Control Conf. (NY) 21:589-593.

Kingsbury, J.M. 1964. Poisonous plants of the United States and Canada. Prentice-Hall, Inc., Englewood Cliffs, NJ.

Lacey, J.R., and P.K. Fay (eds). 1984. Proceedings of the knapweed symposium. Great Falls, MT., April 3-4, Plant and Soil Science Dept. and Coop. Ext. Ser., Montana State Univ. Bull. 1315.

Landgraf, B.K., P.K. Fay, and K.M. Havstad. 1984. Utilization of leafy spurge (Euphorbia esula) by sheep. Weed Sci. 32:348-352.

Lyn, R.G., and D.R. Kirby. 1987. Foraging behavior of cattle in leafy spurge (Euphorbia esula) infested rangeland. Weed Technology (In Press).

Lyn, R.G., and C.G. Messersmith. 1987. Leafy spurge control with resulting forage production from several herbicide treatments. Res. Prog. Rep. West. Soc. Weed Sci., pp. 19-20.

Maddox, D.M., and A. Mayfield. 1985. Yellow starthistle infestations are on the increase, California Agriculture, Nov-Dec, pp. 10-12.

Maddox, D.M., A. Mayfield, and N.H. Poritz. 1985. Distribution of yellow starthistle (Centaurea solstitialis) and Russian knapweed (Centaurea repens). Weed Sci. 33:315-327.

Messersmith, C.G. 1983. The leafy spurge plant. North Dakota Farm Res. 40(5):3-7.

Messersmith, C.G., and E.F. Lyn. 1983. Distribution and economic impacts of leafy spurge in North Dakota. North Dakota Farm Res. 40(5):8-13.

Reilly, W., and K.R. Kaufman. 1979. The social and economic impact of leafy spurge in Montana, pp. 21-24. In: Proc. leafy spurge symp., Bismarck, ND. June 26-27.

Smith, L.J., S.A. Dewey, D.C. Thill, and R. Callihan. 1983. Kochia scoparia, a serious crop pest and potential threat to Idaho's livestock. Univ. Idaho Ext. Bull. 722.

U.S. Department of Agriculture. 1971. Common weeds of the United States. Dover Publications, Inc., NY.

Watson, A.K. (ed). 1985. Leafy spurge. Monograph series, Weed Sci. Soc. of America, No. 3. 309 West Clark St., Champaign, IL 61820.

Whitson, T. (ed). 1987. Weeds and poisonous plants of Wyoming and Utah. Univ. Wyoming and Utah State Univ.

Young, S., W.W. Brown, and B. Klinger. 1970. Nigropallidal encephalomalacia in horses fed Russian knapweed (Centaurea repens L.). Amer. J. Vet. Res. 31:1393-1404.

Young, J.A., and R.A. Evans. 1971. Germination of dyer's woad. Weed Sci. 19:76-78.

324

APPENDIX 23.1
Weeds designated as noxious[1] according to weed and/or seed
laws in at least one U.S. state or Canadian province.
Scientific and common names appear as printed in the various
laws and regulation documents. Names of plants reported to
be poisonous[2] are preceded by an asterix. Many plants on
this list have not been tested for potential toxicity to
livestock.

Prohibited Noxious Weeds and Weed Seeds
United States and Canada - 1987

Scientific Name	Common Name(s)
Abutilon theophrasti	velvetleaf
Acacia armata	kangaroothorn
Acaena anserinifolia	biddy biddy
Acaena novae-zelandiae	biddy biddy
Acaena pallida	biddy biddy
Acaena sanguisorbae	piripiri
Acanthosperum hispidum	bristly starbur
Aegilops cylindrica	jointed goatgrass
Aegilops ovata	ovate goatgrass
Aegilops spp.	goatgrass
Aegilops triuncialis	barb goatgrass
Aeschynomeme indica	Indian jointvetch
Aeschynomene virginica	curly indigo
Aeschynomene spp.	curly indigo
* Agropyron repens	quackgrass
* Agrostemma githago	corn cockle
Alfombrilla	lightning weed
Alhagi camelorum	camelthorn
Alhagi pseudalhagi	camelthorn
* Allium canadense	wild onion
Allium spp.	onion, wild or garlic
Allium paniculatum	panicled onion
Allium vineale	wild garlic
Alopecurus myosuroides	slender foxtail
Alternanthera philoxeroides	alligatorweed
Amaranthaceae sp.	amaranth
Amaranthus albus	tumble pigweed
Amaranthus blitoides	prostrate pigweed
* Amaranthus retroflexus	redroot pigweed
Ambrosia artemisiifolia	common ragweed
Ambrosia grayi	wollyleaf bursage
Ambrosia psilostachya	western ragweed (perennial)

APPENDIX 23.1 (cont.)
Prohibited Noxious Weeds and Weed Seeds, U.S. and Canada

Scientific Name	Common Name(s)
Ambrosia spp.	ragweed
Ambrosia tenuifolia	false ragweed
Ambrosia tomentosa	skeletonleaf bursage
Ambrosia trifida	giant ragweed
Anaphalis margaritacea	pearly everlasting
Anchusa arvensis	small bugloss
Andropogon bicornis	West-Indian foxtail grass
Anoda cristata	spurred anoda
Antennaria spp.	pussytoes
Anthemis arvensis	corn chamomile
Apocynum sibiricum	prairie dogbane
* Apocynum androsaemifolium	spreading dogbane
* Apocynum cannabinum	hemp dogbane
Araujia sericofera	bladderflower
Arctium minus	common burdock
Arctium spp.	burdock (giant)
Arctotheca calendula	capeweed
Artemisia absinthium	absinth wormwood (common)
Artemisia biennis	biennial wormwood
Artemisia cana	silver sagebrush
Aruneus sp.	goatsbeard
* Asclepias galioides	whorled milkweed
* Asclepias incarnata	swamp milkweed
* Asclepias speciosa	showy milkweed
* Asclepias spp.	milkweed
* Asclepias subverticillata	western whorled milkweed
* Asclepias syriaca	common milkweed
* Asclepias verticillata	eastern whorled milkweed
Astragalus crassicarpus	Indian-pea
* Astragalus sp.	locoweed
* Avena barbata	slender oats
* Avena fatua	wild oats
Avena sterilis	sterile oats
Axyris amaranthoides	Russian pigweed
Azolla pinnata	pinate mosquitofern
* Barbarea vulgarsi	yellow rocket
* Barbarea spp.	yellow rocket
Bassia hyssopifolia	fivehook bassia
Berberis vulgaris	European barberry (common)
Berberis spp.	barberry

APPENDIX 23.1 (cont.)
Prohibited Noxious Weeds and Weed Seeds, U.S. and Canada

Scientific Name	Common Name(s)
Berteroa incana	hoary alyssum
* Bidens spp.	beggarticks
Borreria alata	broadleaf buttonweed
Boussingaultia gracilis	Madeira vine
* Brassica arvensis	wild mustard
* Brassica campestris	bird rape
* Brassica hirta	white mustard
* Brassica juncea	India mustard
* Brassica kaber	wild charlock mustard
* Brassica nigra	wild mustard
* Brassica rapa	birdsrape mustard
* Brassica spp.	mustards
* Bromus commutatus	hairy chess
Bromus secalinus	cheat
Bromus tectorum	downy brome
Calonyction muricatum	giant morningglory
Calystegia sepium	hedge bindweed
Camelina spp.	falseflax
Campanula rapunculoides	creeping bellflower
* Cannabis sativa	marijuana
Caperonia castaneaefolia	Mexican weed
Capsella bursa-pastoris	shepherdspurse
Cardaria chalepensis	lens-podded whitetop
* Cardaria draba	hoary cress
Cardaria pubescens	hairy whitetop
Cardaria spp.	hoary cress
Cardiospermum halicacabum	balloonvine
Carduus acanthoides	plumeless thistle
Carduus nutans	musk thistle (nodding)
Carduus pycnocephalus	Italian thistle
Carduus spp.	nodding thistle
Carduus tenuiflorus	slenderflower thistle
Carthamus baeticus	smooth distaff thistle
Carthamus lanatus	woolly distaff thistle
Carthamus leucocaulos	whitestem distaff thistle
Carthamus oxyacantha	carthamus
* Cassia obtusifolia	sicklepod
Cenchrus echinatus	southern sandbur
Cenchrus incertus	field sandbur (coast)
Cenchrus longispinus	longspine sandbur (mat)
Cenchrus pauciflorus	sanbur
Cenchrus spp.	sandbur

APPENDIX 23.1 (cont.)
Prohibited Noxious Weeds and Weed Seeds, U.S. and Canada

Scientific Name	Common Name(s)
Centaurea calcitrapa	purple starthistle
Centaurea cyanus	ragged robin
Centaurea diffusa	diffuse knapweed
Centaurea iberica	Iberian starthistle
Centaurea maculosa	spotted knapweed
Centaurea melitensis	malta starthistle
Centaurea picris	Russian knapweed
* Centaurea repens	Russian knapweed
* Centaurea solstitialis	yellow starthistle
Centaurea sulphurea	Sicilian starthistle
Centaurea triumfetti	squarrose knapweed
Cerastium arvense	field chickweed
Cerastium vulgatum	mouseeared chickweed
Cereus peruvianus	spiny tree cactus
* Chenopodium album	common lambsquarters
Chenopodium berlandieri	netseed lambsquarters
Chenopodium gigantospermum	mapleleaf goosefoot
* Chenopodium glaucum	oakleaf goosefoot
Chenopodium sp.	small-seeded lambsquarters
Chondrilla juncea	rush skeletonweed
Chorispora tenella	blue mustard
Chrysanthemum leucanthemum	oxeye daisy
Chrysopogon aciculatus	pilipiliula
Chrysopsis villosa	hairy goldenaster
Cichorium intybus	chicory
* Cicuta bulbifera	bulbous waterhemlock
* Cicuta douglasii	Western waterhemlock
* Cicuta maculata	spotted waterhemlock
Cirsium altissimum	tall thistle
* Cirsium arvense	Canada thistle
Cirsium flodmanii	Flodman thistle
Cirsium ochrocentrum	yellowspine thistle
Cirsium undulatum	wavyleaf thistle
Cirsium vulgare	bull thistle
Clidemia hirta	Koster's Curse
Cnicus benedictus	blessed thistle
Commelina benghalensis	tropical spiderwort
* Conium maculatum	poison-hemlock
Conringia orientalis	haresear mustard
* Convolvulus arvensis	field bindweed
Convolvulus sepium	hedge bindweed
* Convolvulus spp.	bindweed

APPENDIX 23.1 (cont.)
Prohibited Noxious Weeds and Weed Seeds, U.S. and Canada

Scientific Name	Common Name(s)
* Conyza canadensis	horseweed (marestail)
Corydalis aurea	golden corydalis
Crepis tectorum	narrowleaf hawksbeard
Crotalaria longisrostrata	long-beaked rattle pod
* Crotalaria sagittalis	rattlebox
* Crotalaria spectabilis	showy crotalaria
* Crotalaria spp.	crotalaria
Crupina vulgaris	common crupina
Cucumis melo	dudaim melon
Cucumis myriocarpus	paddy melon
Cuscuta epilinum	flax dodder
Cuscuta epithymum	clover dodder
Cuscuta indecora	large seeded dodder
Cuscuta pentagona	field dodder
Cuscuta planiflora	small seeded dodder
Cuscuta racemosa	Chilean dodder
Cuscuta reflexa	giant dodder
Cuscuta spp.	dodder
Cycloloma atriplicifolium	winged pigweed
Cymbopogon refractus	barbwire grass
Cynara cardunculus	artichoke thistle
* Cynodon dactylon	bermudagrass
Cynodon spp.	bermudagrass
* Cynoglossum officinale	houndstongue
Cyperus esculentus	yellow nutsedge
Cyperus rotundus	purple nutsedge
Cyperus spp.	nutgrass tubers
Cytisus monspessulanus	French broom
* Cytisus scoparius	Scotch broom
* Datura stramonium	jimsonweed
* Daucus carota	wild carrot
* Delphinium barbeyi	tall larkspur
* Delphinium nuttalianum	low larkspur
* Delphinium spp.	larkspur
* Descurainia pinnata	pinnate tansymustard
* Descurainia richardsonii	Richardson tansymustard
* Descurainia sophia	flixweed
Dichrostachys nutans	marabu
Digitaria insularis	sourgrass
Digitaria ischaemum	smooth crabgrass
Digitaria sanguinalis	large crabgrass
Digitaria scalarum	blue couch
Digitaria velutina	velvet fingergrass

APPENDIX 23.1 (cont.)
Prohibited Noxious Weeds and Weed Seeds, U.S. and Canada

Scientific Name	Common Name(s)
Dipsacus spp.	teasel
Draba nemorosa	wood whitlowgrass
Dracocephalium parviflorum	American dragonhead
Dracocephalium thymiflorum	thyme-leaved dragonhead
* Drymaria arenarioides	alfombrilla
* Echinochloa crusgalli	barnyardgrass
Echinocystis lobata	wild cucumber
Echium vulgare	blueweed
Eichhornia azurea	anchored waterhyacinth
Elephantopus mollis	elephantopus
Emex spinosa	spiny emex
* Equisetum arvense	field horsetail (rush)
* Equisetum palustre	marsh horsetail
Erigeron philadelphicus	Philadelphia fleabane
Erodium cicutarium	Redstem filaree
Erodium sp.	cranesbill
Erucastrum gallicum	dog mustard
* Erysimum cheiranthoides	wallflower mustard
* Eupatorium adenophorum	croftonweed (Maui pamakani)
Eupatorium maculatum	spotted joepyeweed
Eupatorium riparium	Hamakua pamakani
* Euphorbia cyparissias	cypress spurge
* Euphorbia esula	leafy spurge
* Euphorbia helioscopia	sun spurge
Euphorbia lucida	leafy spurge
* Euphorbia maculata	spotted spurge
Euphorbia oblongata	oblong spurge
Euphorbia prunifolia	painted euphorbia
Euphorbia serrata	serrate spurge
Euphorbia serpyllifolia	thymeleaf spurge
Fagopyrum tataricum	tartary buckwheat
Franseria discolor	creeping ragweed
Franseria sp.	bur ragweed
Franseria tomentosa	woolly-leaf povertyweed
Fumaria officinalis	fumitory
* Galega officinalis	goatsrue
Galeopsis tetrahit	common hempnettle
Galinsoga ciliata	hairy galinsoga
Galinsoga parviflora	smallflower galinsoga
Galium aparine	catchweed bedstraw
Galium boreale	northern bedstraw
Galium mollugo	smooth bedstraw

APPENDIX 23.1 (cont.)
Prohibited Noxious Weeds and Weed Seeds, U.S. and Canada

Scientific Name	Common Name(s)
Galium spp.	bedstraw
Galium tricorne	bedstraw
Galium verum	yellow bedstraw
Gaura coccinea	scarlet gaura
Gaura odorata	scented gaura
Gaura sinuata	wavyleaf gaura
* Glechoma hederacea	ground ivy
Glycyrrhiza lepidota	wild licorice
Grevillea banksii	Kahili flower
* Grindelia squarrosa	curlycup gumweed
Gypsophila paniculata	babysbreath
Hackelia sp.	stickseed
Halimodendron halodendron	Russian salttree
* Halogeton glomeratus	halogeten
Harrisia martinii	red-fruited harrisia
Helianthus ciliaris	Texas blueweed
Helianthus petiolaris	prairie sunflower
Heliotropium curassavicum	seaside heliotrope var.
Heracleum mantegazzianum	------
Heracleum sp.	cowparsnip
Heteropogon contortus	tanglehead
Hieracium aurantiacum	orange hawkweed
Hieracium florentinum	kingdevil hawkweed
Hieracium floribundum	yellowdevil hawkweed
Hieracium pilosella	mouseear hawkweed
Hieracium pratense	yellow hawkweed
Hieracium vulgatum	common hawkweed
Hoffmannseggia densiflora	pignut
* Hordeum jubatum	foxtail barley
Hydrilla verticillata	hydrilla
Hygrophila polysperma	Indian hygrophila
Hymencphysa pubescens	ballcress
* Hyoscyamus niger	black henbane
* Hypericum perforatum	common St. Johnswort
* Hypericum sp.	St. Johnswort
Hyptis suaveolens	wild spikenard
Hyssopus officinalis	hyssop
Imperata brasiliensis	Brazilian satintail
Imperata brevifolia	satintai
Imperata cylindrica	cogongrass (Kogon)
Ipomoea aquatica	swamp morningglory
Ipomoea hederacea	ivyleaf morningglory
Ipomoea hirsutula	morningglory

APPENDIX 23.1 (cont.)
Prohibited Noxious Weeds and Weed Seeds, U.S. and Canada

Scientific Name	Common Name(s)
Ipomoea purpurea	wild morningglory
* Ipomoea spp.	morningglory
Ipomoea triloba	threelobe morningglory
Ipomoea turbinata	purple morningglory
Iris douglasiana	Douglas iris
* Iris missouriensis	western blue flag
Isatis tinctoria	dyers woad
Ischaemum rugosum	saramollagrass
Iva axillaris	poverty sumpweed
* Kalmia angustifolia	sheep-laurel
* Kochia scoparia	kochia
Lactuca canadensis	tall lettuce
Lactuca pulchella	blue lettuce
Lactuca serriola	prickly lettuce
Lagascea mollis	acuate
Lappula echinata	European sticktight
Lathyrus tuberosa	tuberous vetchling
Lepidium campestre	field peppercress
Lepidium densiflorum	greenflower pepperweed
* Lepidium draba	perennial peppergrass
Lepidium latifolium	perennial pepperweed
Lepidium perfoliatum	clasping pepperweed
Leptochloa chinensis	Chinese sprangletop
Limnophila sessiliflora	limnophila
Linaria dalmatica	Dalmatian toadflax
Linaria vulgaris	yellow toadflax
Lithospermum arvense	corn gromwell
Lithospermum ruderale	western gromwell
* Lobelia inflata	indiantobacco
Lolium persicum	Persian darnel
* Lolium temulentum	poison ryegrass (darnel)
* Lupinus argenteus	silver lupine
* Lupinus spp.	lupine
Lychnis alba	white campion
* Lygodesmia juncea	skeletonweed
Lysimachia ciliata	fringed loosestrife
Lythrum salicaria	purple lythrum
Lythrum sp.	yellow loosestrife
Malachra alceifolia	malachra
Malva neglecta	common mallow
Matricaria maritima	false chamomile (scentless)
Medicago lupulina	black medic

APPENDIX 23.1 (cont.)
Prohibited Noxious Weeds and Weed Seeds, U.S. and Canada

Scientific Name	Common Name(s)
Melastoma spp.	Melastoma
* Melilotus alba	white sweetclover
* Melilotus officinalis	yellow sweetclover
Mikania cordata	African mile-a-minute
Mikania micanthra	mile-a-minute
Mimosa invisa	giant sensitiveplant
Miscanthus japonicus	cane grass
Monochoria hastata	arrowleaved monochoria
Monochoria vaginalis	monochoria (nuttall
Monolepis nuttalliana	Nuttall povertyweed
Montanoa hibiscifolia	tree daisy
Muhlenbergia schreberi	nimblewill
Neslia paniculata	ball mustard
Nothoscordum inodorum	false garlic
Nymphaea mexicana	Mexican waterlily (banana)
Odontites verna	red bartsia
Onopordum acanthium	Scotch thistle
Onopordum spp.	onopordum thistles
Orobanche ludoviciana	Cooper's broomrape
Orobanche ramosa	hemp broomrape (branched)
Orobanche spp.	broomrape
Oryza longistaminata	--------
Oryza punctata	--------
Oryza rufipogon	-------
Oryza sativa	red rice
Panicum antidotale	blue panicgrass
* Panicum capillare	witchgrass
Panicum dichotomiflorum	fall panicum
Panicum repens	torpedograss
Passiflora pulchella	winged-leaf passionflower
* Pastinaca sativa	wild parsnip
* Peganum harmala	harmel
* Pennisetum clandestinum	kikuyugrass
* Pennisetum pedicellatum	kyasumagrass
Pennisetum polystachiyon	missiongrass
Pennisetum setaceum	crimson fountaingrass
* Phalaris canariensis	canarygrass
Physalis heterophylla	clammy groundcherrry
Physalis lobata	purple-flower groundcherry

APPENDIX 23.1 (cont.)
Prohibited Noxious Weeds and Weed Seeds, U.S. and Canada

Scientific Name	Common Name(s)
Physalis subglabrata	smooth groundcherry
Physalis viscosa	grape groundcherry
Piper aduncum	--------
Pittosporum undulatum	pittosporum
Plantago aristata	bracted plantain
Plantago lanceolata	buckhorn plantain
Plantago major	broadleaf plantain (common)
Plantago spp.	plantain
Pluchea camphorata	stinkweed
Poa annua	annual bluegrass
Polygonum aviculare	prostrate knotweed
Polygonum cilinode	blackfringe knotweed
Polygonum coccineum	swamp smartweed
Polygonum convolvulus	wild bickwheat
Polygonum cuspidatum	Japanese knotweed
Polygonum erectum	erect knotweed
* Polygonum hydropiper	marshpepper smartweed
Polygonum lapathifolium	pale smartweed
* Polygonum pennsylvanicum	Pennsylvania smartweed
Polygonum perfoliatum	mile-a-minute weed
Polygonum persicaria	ladysthumb
Polygonum polystachyum	Himalayan knotweed
Polygonum sachalinense	Sakhalin knotweed (giant)
Polygonum sagittatum	arrowleaf tearthumb
Polygonum scabrum	green smartweed
Polygonum spp.	smartweed
* Portulaca oleracea	common purslane
Potentilla recta	sulfur cinquefoil
* Prosopis spp.	mesquite
Prosopis strombulifera	creeping mesquite
* Pteridium aquilinum	brackenfern
Pueraria lobata	kudzu
* Ranunculus acris	tall buttercup
* Ranunculus repens	creeping buttercup
* Raphanus raphanistrum	wild radish
Rapistrum rugosum	common giant mustard
* Rhamnus cathartica	European buckthorn
* Rhamnus frangula	alder buckthorn
* Rhamnus sp.	buckthorn
Rhodomyrtus tomentosus	downy rosemyrtle
* Rhus radicans	poison-ivy

APPENDIX 23.1 (cont.)
Prohibited Noxious Weeds and Weed Seeds, U.S. and Canada

Scientific Name	Common Name(s)
Rhynchaspora spp.	spearhead
Rorippa austriaca	Austrian fieldcress
Rosa multiflora	multiflora rose
Rottboellia exaltata	itchgrass
Rubus penetrans	blackberry
Rubus spp.	evergreen blackberry
* Rudbeckia hirta	blackeyed-susan
* Rumex acetosella	red sorrel (sheep)
Rumex alitissimus	smooth dock
* Rumex crispus	curly dock
Rumex maritimus	golden dock
Rumex obtusifolius	broadleaf dock
* Rumex spp.	dock and/or sorrel
Saccharum spontaneum	wild sugarcane
Sagittaria sagittifolia	arrowhead
Salsola australis	common Russian thistle
Salsola collina	spineless Russian thistle
* Salsola iberica	Russian thistle
* Salsola kali	Russian thistle
Salsola pestifer	Russian thistle
Salsola vermiculata	Mediterranean saltwort
Salvia aethiopis	Mediterranean sage
Salvia pratensis	meadow sage
* Salvia sp.	pasture sage
Salvinia auriculata	giant salvinia
Salvinia biloba	--------
Salvinia herzogii	--------
Salvinia molesta	karibaweed
Scirpus validus	softstem bulrush (tussocck)
Scleranthus annuus	knawel
Scolymus hispanicus	golden thistle
* Senecio jacobaea	tansy ragwort
Senecio squalidus	Oxford ragwort
* Senecio vulgaris	common groundsel
Sesbania exaltata	hemp sesbania
Setaria faberi	giant foxtail
Setaria glauca	yellow foxtail
Setaria viridis	green foxtail
Sicyos angulatus	burcucumber
Sida hederacea	alkali sida (mallow)
Sida rhombifolia	teaweed

APPENDIX 23.1 (cont.)
Prohibited Noxious Weeds and Weed Seeds, U.S. and Canada

Scientific Name	Common Name(s)
Silene alba	white campion (cockle)
Silene csereii	biennial campion
Silene cucubatus	bladder campion
Silene noctiflora	nightflowering catchfly
Silene vulgaris	bladder campion
Silybum marianum	blessed milk thistle
Sinapis arvensis	wild mustard
Sisymbrium altissimum	tumble mustard
Sisymbrium loeselii	tall hedge mustard
* Sium suave	waterparsnip
Solanum cardiophyllum	heartleaf nightshade
* Solanum carolinense	horsenettle (Carolina)
* Solanum dimidiatum	robust horsenettle
* Solanum dulcamara	bitter nightshade
* Solanum eleagnifolium	silverleaf nightshade
Solanum lanceolatum	lanceleaf nightshade
Solanum marginatum	white-margined nightshade
Solanum ptycanthum	Eastern black horsenettle
* Solanum rostratum	buffalobur
Solanum torvum	turkeyberry (terongan)
Solanum xantii	purple nightshade
Sonchus arvensis	perennial sowthistle
Sonchus oleraceus	annual sowthistle
Sonchus spp.	sowthistle
* Sorghum almum	sorghum-almum
Sorghum bicolor	shattercane
* Sorghum halepense	johnsongrass
* Sorghum spp.	sorghum
Sparganium erectum	branched burreed
Spartium junceum	Spanish broom
Spergula arvensis	corn spurry
Sphaerophysa salsula	swainsonpea
Stachys palustris	woundwort
* Stellaria media	common chickweed
Stipa brachychaeta	puna grass
Stratiotes aloides	crabsclaw
Striga asiatica	witchweed
Striga spp.	witchweed
Symphytum asperum	prickly comfrey (rough)
Symphytum officinale	common comfrey
Taeniatherum caput-medusae	Medusahead

APPENDIX 23.1 (cont.)
Prohibited Noxious Weeds and Weed Seeds, U.S. and Canada

Scientific Name	Common Name(s)
Tagetes minuta	wild marigold
* Tanacetum vulgare	common tansy
Taraxacum laevigatum	redseeded dandelion
Taraxacum officinale	dandelion
* Thlaspi arvense	field pennycress
Tragopogon spp.	goat's-beard
* Tribulus terrestris	puncturevine
Tridax procumbens	coat buttons
* Triglochin palustris	marsh arrowgrass
Triumfetta bartramia	bur bush
Triumfetta semitriloba	Sacramento bur
Ulex europaeus	gorse
Urena lobata	cadillo (Aramina)
Urtica dioica	stinging nettle
* Urtica spp.	nettles
Utricularia vulgaris	common bladderwort
Vaccaria pyramidata	cowcockle
Verbascum thapsus	common mullein
Vicia cracca	tufted vetch
* Viscum album	European mistletoe
Xanthium commune	cocklebur
Xanthium pennsylvanicum	cocklebur
* Xanthium spinosum	spiny cocklebur
* Xanthium spp.	cocklebur
* Xanthium strumarium	common cocklebur
* Zigadenus sp.	camas
Zygophyllum fabago	Syrian beancaper

1/ Information taken from weed and/or seed laws aquired from each U.S. state and Canadian province.

2/ According to Cheeke and Shull (1985), Hails and Crane (1983), and Kingsbury (1964).

24

The Importance of Diagnosing Poisoning from Plants

E. Murl Bailey, Jr.

The diagnosis of animal diseases is extremely important from a treatment and a management standpoint. There are reasons why animals get sick and die. Before treating and/or managing animal diseases, diagnostic procedures must be instituted, regardless of whether the etiologic agent in the disease process is toxic or biologic in nature. The diagnostician must ascertain whether the disease process is an individual animal problem or a herd problem. Once the cause of a problem is known, adequate treatment and/or preventive measures may be instituted. Newly instituted management procedures are often the only measures available to contain the problem when dealing with poisoning from plants.

TYPES OF ANIMAL DISEASES

Animal diseases are initially categorized according to the duration of time from onset until completion, i.e., death or recovery. These categories are designated as acute, subacute (or subchronic), or chronic.

Acute diseases are those conditions in which affected animals rapidly become ill and possibly die within 24 to 48 hours. In acute diseases, the affected animals develop clinical signs which have a sudden onset and are obvious to the animal owner. These signs may include anorexia (going off feed), diarrhea, dehydration, or depression. Other clinical signs may develop within a few hours. An animal with an acute disease, because of its sudden onset, must be treated rapidly and with vigor, or a large percentage of those affected may succumb. Many animals which survive an acute disease become "poor doers" and, consequently, poor producers and economic liabilities. Some disease conditions are designated peracute because the animals die so rapidly. The animal owner only finds dead animals and no ill animals. In these cases, the diagnoses are made from the necropsy or postmortem examination of the carcasses.

Subacute animal diseases have a somewhat slower onset with the animals gradually developing clinical signs. While the affected animals may become as sick as animals with acute diseases, these diseases are generally not as severe but will cause severe economic losses until the animal recovers (sometimes these animals also die). Animals with subacute diseases may gradually become anorexic and have a rapid weight loss over a few days. In other animals affected with subacute diseases, there may be no weight loss--only the development of clinical signs with increasing severity, i.e., neurological signs.

Chronic animal diseases are those which may take weeks or months to develop in affected animals. In chronically affected animals, there are gradual losses of weight and/or production, and marked clinical signs may develop. In breeding animals, the first sign may be prolonged anestrus or reduced breeding efficiency.

ORGAN SYSTEMS AFFECTED BY POISONOUS PLANTS

Toxins from plants may interfere with most organ systems in an animal's body. These organ systems include the cardiopulmonary, nervous, gastrointestinal, renal, hepatic, musculoskeletal, and reproductive systems. In addition, plants may accumulate toxic metals or be infested with mycotoxin producing fungi (Appendix 24.1). Therefore, the disease syndromes associated with plants can affect the same organ systems which might be affected by microorganisms; they can also mimic the diseases produced by the microorganisms.

Plants which affect the blood or blood-forming organs (bone marrow) may release hemoglobin from blood cells, causing hemoglobinuria, while others may cause the release of blood into the urine. Both of these conditions are known as "red water disease" and can be caused by Clostridium hemolyticum, leptospirosis, or anaplasmosis. Sudden death in animals can be caused by several plants which affect the heart and must be differentiated from lightning strikes or acute bacterial diseases such as anthrax. Some plants can cause disease syndromes affecting the respiratory tract, which must be differentiated from acute shipping fever or diseases caused by respiratory viruses. There are large numbers of plants which cause nervous syndromes. These must be differentiated from the bacterial and viral conditions which affect the brain. There are a large number of bacterial and viral disease conditions which affect the gastrointestinal (GI) system. Likewise, there are multitudes of plants which can also affect the GI tract. Fortunately, there are few infectious conditions which cause renal disease(s). There is, however, a large number of plants which can affect the kidneys.

Photosensitization (an abnormal sensitivity to sunlight) can be caused by several plants. One form of photosensitization, primary photosensitization, must be differentiated from vesicular diseases, such as vesicular exanthema, foot and mouth disease, blue tongue, soremouth, or infectious bovine rhinotracheitis. The more common hepatogenous photosensitization is normally associated with jaundice and is easily differentiated from infectious diseases.

There are large numbers of plants which have an effect on the reproductive system and cause poor breeding efficiency, abortion, or teratogenesis. These disease conditions are also often caused by viral or bacterial organisms.

The multitude of disease conditions in animals can be caused by many types of etiologies, such as bacterial, viral, or toxic conditions. It is important that a diagnosis be made rapidly in order to establish proper management or treatment strategies.

MAKING A DIAGNOSIS OF PLANT INTOXICATIONS

Disease syndromes in animals may be produced by viruses, bacteria, fungi (or mycotoxins), chemicals, or plants. Therefore, the determination of the disease etiology is often a task unto itself. The diagnosis of plant poisoning in livestock is a difficult task and requires a detailed investigation. The mere presence of a toxic plant in a pasture or a history of plant intoxications in other years is not prima facie evidence of a plant poisoning.

Many factors must be considered when attempting to ascertain whether a plant-related disease is occurring in animals. If a suspect plant is present, a determination must be made if the animal species concerned has grazed the suspect plant. Other factors which must be considered include: plane of nutrition and presence of an adequate water supply; climatic conditions (i.e., rainfall, frost, etc.); and whether the animals of interest are familiar to the area or newly introduced.   Animals may consume large quantities of plant material just prior to or subsequent to a storm.   Very thirsty animals will attempt to ingest large quantities of any plant species after drinking water. Starved animals will ingest many materials including toxic plants.

The following information is requisite for determining the presence or absence of a plant intoxication:

1. adequate history including management, number of animals involved, etc.;
2. presence of syndrome and/or clinical signs;
3. suspected plant material has been grazed by the species concerned;
4. identification of seeds or plant material in rumen, abomasal contents, or feces on autopsy;
5. postmortem lesions;
6. presence of plant chemical in tissues, if known.

PREVENTION OF ANIMAL DISEASES

Producers and veterinarians alike make strong attempts to prevent the occurrence of any disease syndromes in domestic animals (i.e., "an ounce of prevention...!!!"). Many, if not most, bacterial and viral disease conditions can be prevented by the strategic utilization of prophylactic measures such as vaccines while a few must be handled by management schemes.   Intoxications, whether plant or chemical induced, can only be prevented by the application of strong management measures.   The treatment regimens for most intoxications are often impossible to apply on a herd basis, or the therapeutic measures are unknown. Management strategies for poisonous plants include rotation grazing, chemical and/or mechanical control, denying access to infested premises, or grazing less susceptible species.   The economic impact of the various strategies for controlling disease syndromes produced by toxic plants will vary according to management, climatic, and geographic conditions.

CONCLUSIONS

The diagnosis of animal diseases, regardless of the etiology, is extremely important in order for a producer to overcome the economic reversal caused by the disease occurrence. Animal disease conditions due to poisonous plants are commonly diagnosed when animals die or become ill under range conditions. This type of diagnosis is often tendered because of lack of previous history of infectious diseases or because of the known presence of poisonous plants in a pasture. When animal deaths occur, it is extremely important that the causes of deaths be determined rapidly because postmortem changes may prevent the proper diagnosis. Although plant intoxications occur in animals under range conditions, there are other disease conditions which might occur, such as chemical poisoning and, especially, infectious diseases. The utilization of proper animal health and range management techniques will prevent the occurrence of animal disease and deaths under range conditions and, equally important, allow the animal agricultural producer to curb economic losses and produce more salable goods from rangelands.

LITERATURE CITED

Bailey, E. Murl, Jr. 1978. Physiologic responses of livestock to toxic plants. J. of Range Manage. 31:(5)343-347.

ADDITIONAL READING

Cheeke, P.R., and L.R. Shull. 1985. Natural toxicants in feeds and poisonous plants. AVI Publishing Co., Inc., Westport, CN.

Keeler, R.F., K.R. Van Kampen, and L.F. James (eds). 1978. Effects of poisonous plants on livestock. Academic Press, New York.

Kingsbury, J.M. 1964. Poisonous plants of the United States and Canada. Prentice-Hall, Inc., Englewood Cliffs, NJ.

Seawright, A.A., M.P. Hegarty, L.F. James, and R.F. Keeler (eds). 1985. Plant toxicology. Proc. Australia-U.S.A. Poisonous plants symp., Queensland Poisonous Plants Comm., Anim. Res. Inst., Yeerongpilly.

APPENDIX 24.1
Organs, systems, and body functions affected by poisonous plants

Organ/System/ Syndrome	Scientific Name	Common Name

ABORTION

	Astragalus spp.	Locoweed
	Gutierrezia spp.	Perennial broomweed
	(Xanthocephalum spp.)	
	Iva angustifolia	Sumpweed
	Pinus ponderosa	Ponderosa pine

BIRTH DEFECTS

	Astragalus spp.	Locoweed
	Conium maculatum	Poison hemlock
	Lathyrus odoratus	Sweet pea
	Lupinus sericeus	Lupine, Bluebonnet
	Nicotiana spp.	Tobacco
	Veratrum spp.	False hellebore, Skunk cabbage
	Vicia spp.	Vetch

BLOOD

Anti-clotting

| | Melilotus spp. | Sweet clover |

Bone Marrow Dysfunction

| | Pteridium aquilinum | Bracken-fern |

Hemolysis

| | Allium spp. | Onion |

GASTROINTESTINAL TRACT

	Centaurium spp.	Mountain pink
	Euphorbia spp.	Spurges
	Flourensia cernua	Blackbrush, Tarbush
	Jatropha spp.	Nettlespurge

(Continued)

APPENDIX 24.1 (Cont.)

Organ/System/ Syndrome	Scientific Name	Common Name

GASTROINTESTINAL TRACT (Cont.)

	Melia azedarach	Chinaberry
	Phytolacca americana	Pokeweed
	Quercus spp.	Oaks
	Ricinus spp.	Castorbean
	Sesbania spp.	Sesbane, Rattlebox
	Solanum spp.	Nightshade
	Xanthium spp.	Cocklebur

HEART

	Asclepias spp.	Milkweed
	Baileya multiradiata	Desert baileya
	Datura spp.	Jimson weed
	Drymaria spp.	Drymary, Alfombrilla
	Kalanchoe spp.	Kalanchoe
	Nerium oleander	Oleander
	Taxus cuspidata	Yew
	Veratrum spp.	False hellebore, Skunk cabbage

KIDNEYS

	*Amaranthus spp.	Pigweed, Carelessweed
	*Beta Vulgaris	Beets
	*Halogeton glomeratus	Halogeton
	*Kochia scoparia	Summer cypress
	Lantana camara	Lantana
	Quercus spp.	Oaks
	*Sarcobatus vermiculatus	Greasewood

(Continued)

APPENDIX 24.1 (Cont.)

Organ/System/ Syndrome	Scientific Name	Common Name
LIVER		
	Agave lecheguilla	Lecheguilla
	**Amsinckia intermedia	Fiddleneck
	**Crotalaria spp.	Crotalaria
	Cycas spp.	Cycads
	**Eichium plantagineum	Viper's bugloss
	Eupatorium rugosum	White snakeroot
	**Heliotropium europaeum	Heliotrope
	Lantana camara	Lantana
	Phyllanthus abnormis	Abnormal leafflower
	Sartwellia flaveriae	Sartwellia
	**Senecio spp.	Groundsel
	Xanthium spp.	Cocklebur
LUNGS		
	Astragalus emoryanus	Red-stemmed peavine
	Astragalus miser	Milkvetch
	Brassica napus	Rape
	Panicum antidotale	Blue panic grass
	Perilla frutescens	Beef steak plant
	Salvia spp.	Sage
MALNUTRITION SYNDROME		
	Allium spp.	Onions
	Alloysia lycoides	Whitebrush
	Prosopis spp.	Mesquite
MUSCULOSKELETAL		
	Cassia spp.	Senna
	Cestrum diurnum	Night blooming jessamine
	Lathyrus odorata	Sweet pea
	Solanum malacoxylon	

(Continued)

APPENDIX 24.1 (Cont.)

Organ/System/ Syndrome	Scientific Name	Common Name

NERVOUS

	Scientific Name	Common Name
	Acacia berlandieri	Guajillo
	Aconitum spp.	Monkshood
	Aesculus spp.	Buckeye
	Astragalus spp.	Locoweed
	Asclepias spp.	Milkweeds
	Centaurea spp.	Yellow star thistle, Russian knapweed
	Cicuta spp.	Waterhemlock
	Claviceps spp.	Ergot
	Conium spp.	Poison-hemlock
	Cycas spp.	Cycads
	Cynodon dactylon	Bermudagrass
	Datura spp.	Jimsonweed
	Delphinium spp.	Larkspur
	Eupatorium rugosum	White snakeroot
	Helenium spp.	Sneezeweeds
	Hymenoxys spp.	Bitterweed
	Isocoma wrightii	Rayless goldenrod
	Kallstroemia spp.	Caltrops
	Lathyrus spp.	Singletary pea
	Lobelia spp.	Indian Tobacco
	Lupinus spp.	Lupines, Bluebonnet
	Nicotiana spp.	Tobacco
	Nothalaena sinuata var. cochisensis	Jimmy fern
	Oxytropis spp.	Locoweed
	Peganum harmala	African rue
	Phalaris spp.	Harding grass
	Psilostrophe spp.	Paperflower
	Pteridium aquilinum	Brackenfern
	Solanum dimidiatum	Threadsalve, Potatoweed
	Sophora spp.	Mescalbean, Mountain laurel
	Strychnos nuxvomica	Strychnine
	Trifolium repens	White clover
	Vicia villosa	Hairy vetch
	Zigadenus spp.	Deathcamas

(Continued)

APPENDIX 24.1 (Cont.)

Organ/System/ Syndrome	Scientific Name	Common Name

PHOTOSENSITIZATION

Hepatogenous

	Scientific Name	Common Name
	Agave lecheguilla	Lecheguilla
	Avena sativa	Oats
	Eupatorium rugosum	White snakeroot
	Cynodon dactylon	Bermudagrass
	Lantana spp.	Lantana
	Medicago spp.	Alfalfa
	Nolina spp.	Sacahuista
	Panicum coloratum	Kleingrass
	Tetradymia spp.	Horsebrush
	Tribulus terrestris	Goat head, Puncture vine

Primary

	Scientific Name	Common Name
	Ammi majus	Bishop's weed
	Cooperia pedunculata	Rain lily
	Cymopterus watsonii	Desert parsley
	Hypericum perforatum	St. Johnswort
	Polygonium fagopyrum	Buckwheat
	Thamnosma texana	Dutchman's breeches

SELENOSIS

	Scientific Name	Common Name
	Astragalus spp.	Locoweeds
	Atriplex spp.	Saltbrush
	Conopsis spp.	Goldenweed
	Stanleya pinnata	Prince's plume
	Xylorrhiza spp.	Woody aster

* Oxalate-producing plants.
** Plants contain pyrrolizidine alkaloids.
*** Photosensitization may or may not occur.

25

Use of Plant Toxin Information in Management Decisions

Richard F. Keeler and William A. Laycock

INTRODUCTION

Managers and users of public or private grazing lands realize that grazing hazards may be present because of poisonous plants. Information about the toxins those plants contain and their effects on livestock can facilitate intelligent management decisions to reduce losses. The suggestions contained in this paper are designed to show what kind of information is needed and what approach to take both before and after the fact so that toxin information can help rather than hinder the solution to field problems.

HOW TO MAKE INTELLIGENT POISONOUS PLANT MANAGEMENT
DECISIONS BEFORE TROUBLE STARTS

If there are poisonous plants on a range where animals graze, problems can be expected. What can be done before trouble starts, and what facts can be gathered that will help develop a preventive management strategy?

Poisonous plants present on the range in use or being managed should be identified. These plants can be cataloged by carefully examining the range several times during the year, by interviewing previous users of the range to identify plants of intermittent abundance, and by examining neighboring ranges. The potential hazards are then known before trouble starts.

A determination should be made as to whether these plants are poisonous and palatable to the class of livestock to be grazed. Most poisonous plants are poisonous to all classes of livestock, but differences exist because of palatability, excretion, or detoxification. For example,

cattle are much more susceptible than sheep to the toxins in
Delphinium spp. Sheep are more often poisoned by _Lupinus_
spp. toxins than are cattle. Pregnant cows grazing certain
Lupinus spp. may deliver deformed calves, but lambs are not
similarly affected when ewes graze the plant; susceptibility
to poisoning by _Conium maculatum_ varies greatly among
livestock species. Because of susceptibility and preference
differences, it may be possible to utilize mixed grazing or
sequential grazing to minimize hazards.

TABLE 25.1
Number of toxins in a given genus

Poisonous Genus	Toxin or Toxin Class	Approximate No. of Different Toxins
Halogeton	oxalic acid	1
Astragalus	selenium compounds indolizidine alkaloid nitropropanol glycoside	3
Conium	piperidine alkaloids	7
Veratrum	steroidal alkaloids	40
Lupinus	quinolizidine alkaloids	70

Virtually all poisonous plants contain several, not
just one, toxins (Table 25.1). Toxic effects may be quite
different among the several toxins. Once the poisonous
plants on the range are cataloged, then the known toxins
should be cataloged. What will be the toxic signs in
animals poisoned by these plants and their toxins,
remembering that because some of the plants may contain
several toxins, the toxic signs could vary because of toxin
variation?

Next, when will the identified poisonous plants be
expected to thrive? Will any of them start growing before
good forage does; will they survive water deprivation; will
they be less affected by frost than will desirable plants?
In other words, under what special weather conditions will
these plants be particularly hazardous, and will good forage
be in short supply at that time? Usually toxin concentra-

tion in most poisonous plants either rises or declines gradually. For instance, in Veratrum californicum, both total and individual alkaloid concentrations decline in above-ground parts of the plant as the season progresses (Keeler and Binns 1971). A similar situation exists for Lupinus spp. (Keeler et al. 1976). In both cases, however, seeds are very high in alkaloid. Some plants, such as Lupinus and Conium, are known to exhibit diurnal fluctuations in concentration of alkaloids that can be of considerable magnitude (Fairbairn and Suwal 1961; Wink and Hartmann 1982) (Table 25.2). Other parameters that can cause sudden or drastic change in toxin level include herbicide treatment (Williams and Cronin 1963) or frost (Keeler and Binns 1971).

TABLE 25.2
Diurnal fluctuation in alkaloid concentration in 4 plants

	2-4 AM	7-8 AM	12 NOON	3-4 PM
Conium maculatum (µg/fruit)[1]				
γ-coniceine	1	6	15	1
coniine	90	2	4	120
Baptisia australis (µg/g)[2]				
total	360	990	1350	2700
Lupinus polyphyllus (µg/g)[2]				
total	1365	840	2625	2940
Lupinus hartwegii (µg/g)[2]				
total	100	200	260	160

Data adapted from:
[1] Fairbairn & Suwal 1961
[2] Wink & Hartmann 1982

Consumption of any of the identified poisonous plants can be affected by management practices, stress, fill of water or feed, etc. Hungry or thirsty sheep unloaded in the middle of an abundant stand of a succulent poisonous plant like Halogeton glomeratus will eat their fill and certainly die. It is safer to trail hungry sheep than cows through Delphinium spp.; trailing hungry sheep through podded

Lupinus invites trouble, and so on. Poisonous plant abundance should be carefully checked, particularly in bed grounds, trailing areas, around waterholes, etc.

WHAT SHOULD BE DONE WHEN RANGE ANIMALS DIE OR SHOW ILLNESS AND A POISONOUS PLANT IS SUSPECTED?

With advance planning, much trouble from poisonous plants can be avoided. But despite all care, preliminary planning, study, and management strategy, there will still be some animals affected by the toxins of poisonous plants. What should be done under those circumstances? A decision to keep grazing or to remove animals can have serious economic implications, especially if the decision is incorrect or is based on inadequate information.

First a determination must be made of what poisonous plants were eaten and how much was eaten. A pretty good guess of which one is responsible can be made because toxic signs among survivors may compare favorably with toxic signs reported for one of the poisonous plants found on that range. But consideration must be given also to the possibility that several different toxins of variable concentration may give rise to different signs in any given plant. In other words one must be skeptical about conclusions. A careful determination must be made of which of the poisonous plants have actually been eaten, whether there seems to have been enough eaten to account for the toxic signs and deaths, and whether those eaten can actually induce the signs seen. Did the dead or sick animal eat those plants? One way to be certain is to determine if there are fragments of the suspected plants in the upper gastrointestinal tract of the dead animals. If the rumen contents or fecal samples can be examined shortly after death, there may well be enough fragment evidence still present.

Next, what were the circumstances surrounding the deaths? Were there any obvious mistakes made? For instance, were hungry or thirsty animals unloaded into, trailed through, or allowed to water in abundant stands of the suspected plants? Were the plants at a particularly hazardous stage or animals in a particularly hazardous state? A careful review of the circumstances may be as important as any other single factor in ascribing cause.

Finally, who can help and when is it needed? Consultation with veterinarians or poisonous plant experts can help rule out other possibilities and help rule in the suspected plant cause. These colleagues can help decide whether additional evidence is needed, such as a thorough autopsy or chemical analyses on blood or tissues from dead or dying animals or on suspected plants.

The bottom line in any before-the-fact and after-the-fact efforts and investigations centers on the effects of the plant toxins on animals.

HOW POISONOUS PLANT TOXIN INFORMATION CAN BE MISLEADING

Plant toxin information can either help or hinder in solving poisonous plant problems on the range. Problems can arise from placing heavy reliance on toxin information unless some of the constraints attending that use are considered.

Conclusions may not be firm even when toxic signs in dying animals are consistent with those reported for the plant suspected as causing the problem. Remember, some plants may contain several different toxins that may induce different toxic effects. For instance, Astragalus spp. may contain selenium, nitro compounds, or the locoweed alkaloid, swainsonine, and signs of poisoning from each are different. By contrast, several different plants can induce similar toxic signs from similar or even dissimilar toxins. For instance Lupinus, Thermopsis, Sophora, Genista, and certain other genera all contain quinolizidine alkaloid toxins (Pelletier 1970) and may induce similar toxic signs. Veratrum and Zygadenus genera contain similar steroidal alkaloid toxins. Delphinium and Aconitum genera contain similar diterpenoid alkaloid toxins, etc. A good example of dissimilar toxins that induce the same condition relates to toxins from certain members of three genera that give rise to clinically similar congenital limb and spinal deformities in cattle. The toxins are specific quinolizidine alkaloids from Lupinus, piperidine alkaloids from Conium, and pyridino-piperidines from Nicotiana (Keeler 1978; Keeler and Crowe 1985).

Caution is appropriate when applying results of a "total" analysis of any particular chemical class. Toxins in a given class, and even in a given subclass, may vary in toxic potency by several orders of magnitude. The LD_{50}s of the various Lupinus alkaloid toxins so far tested differ by only about ten times (Table 25.3), and so a total alkaloid analysis might be of some limited value. But the LD_{50}s of Veratrum alkaloid toxins (as approximated from hypotensive

352

activity) differ among those alkaloids by over 10,000 times
(Table 25.4), and so a total alkaloid analysis to measure
toxicity is useless. A minor change in a potent alkaloid,
enough to cause death, would not even be reflected in a
total alkaloid analysis.

TABLE 25.3
Relative approximate LD_{50}s in guinea pigs of various lupine
alkaloids by intraperitoneal injection 1/

Lupine Alkaloid	Approximate LD_{50}
Sparteine	25 mg/kg
Lupanine	30 mg/kg
13-hydroxylupanine	230 mg/kg

1/ Data adapted from Couch 1926

TABLE 25.4
Relative hypotensive potency of various Veratrum alkaloids
in dogs 1/

Alkaloid	Hypotensive Potency
germitrine	11
protoveratrine	4.7
germine	0.004
jervine	0.004
isorubijervine	0.0005

1/ Data adapted from Kerwin et al. 1962

When seeking analytical help, be realistic in the
request for toxin analyses of plant or animal tissues or
fluids. Do not ask "are there any toxins in it" without
providing further information. Rather, ask for a specific
kind of analysis. There are several thousand different
kinds of plant toxins. An analysis for each might be
realistically valued at $50/toxin/sample. If an open
invitation is given to a pay-for-services laboratory to "see

if there are any toxins in it," the bill could be staggering. Likewise, making such a request for free analytical help may invite the smell, the eyeball, or the sink test. Furthermore, there is little use in measuring the wrong toxin; be specific.

If analytical help is sought, what assay accuracy and precision will meet the needs? In some cases there may not be a need for as much accuracy or precision as in other cases. For example, supposing several cows are suspected to have died from eating a particular plant containing alkaloid A as its only toxin. If one asks for an analysis of plant material to determine concentration of alkaloid A, there is no need to ask for accuracy to four significant figures if there is no information on whether the cows ate 0.1 or 10 lbs each. On the other hand, if alkaloid A is one of 50 related alkaloids in the plant and represents only 0.1% of the total alkaloid but is the most toxic by four orders of magnitude, then that is an analytical problem requiring a great deal of accuracy.

Crisis analysis for toxins is being done all the time on human samples like drug abuse cases. Why won't the same thing work for poisonous plant toxins with livestock? Let's consider how and why the situation with humans works and why a similar system is less likely to work with grazing livestock. An example of a well-run program for human drug abuse cases is in Chicago. When an unconscious person is admitted to a Chicago hospital and is suspected to be a drug abuse case, a blood sample is quickly drawn, stabilized, and flown to the University of Illinois drug testing laboratory at Urbana. A gas chromatographic/mass spectrometric analysis is performed and the drug is quickly identified by fragmentation analysis from among the few dozen known drugs of abuse. Results are phoned to the physician who can institute well-established therapy within a couple of hours after the blood is drawn.

Why can't this be done for livestock ingesting poisonous plants on the range? In the case of humans, the blood is drawn in a prescribed fashion, stabilized to prevent sample degradation, and analyzed within a couple of hours. With animals on the range, a lab would be lucky to get a sample within a day or two--a sample that had been taken by who knows what means, unstabilized, and left lying about in a jeep or saddlebag long enough to cause serious sample degradation. When the lab finally gets that deteriorated sample, they get along with it the usual request, "Are there any toxins in it?". The lab is now faced with an analytical chore several orders of magnitude more complex than in the drug abuse case--specifically, what toxin of several thousand plant toxins to look for in a highly degraded sample. Even if the lab is able to identify the toxin before all the animals have finally died, what therapy would be appropriate? The sad part is that there are very few therapeutic measures available for animals sick from poisonous plant toxins.

Finally, with regard to plant samples, if there is some analytical information of reasonable accuracy, one should make certain that plant toxin variability has not spoiled the conclusions. Could concentration in the plant have been different at time of sampling and time of ingestion? As indicated, diurnal or other abrupt changes could be misleading.

HOW POISONOUS PLANT TOXIN INFORMATION CAN HELP SOLVE FIELD PROBLEMS

The maximum usefulness of plant toxin information is based on using existing information derived from solid research to avoid problems rather than relying on results of a crisis analysis to solve a problem after trouble starts.

We will now cite some examples of how existing toxin information can help prevent poisonous plant problems. These examples serve to illustrate several principles: (1) plant toxins vary in concentration in the plant; (2) toxins vary in potency; (3) animal species vary in susceptibility to toxins; (4) a given animal species may vary in susceptibility with time or state of adaptation; and (5) similar toxicoses can be induced by widely different plants and toxins. The examples to be cited illustrate the wisdom of developing a preventive strategy based on known toxin information rather than waiting for trouble to start.

The first example deals with Halogeton glomeratus which grows in great abundance on certain desert ranges, particularly where soil has been disturbed (James et al. 1980). The toxin of halogeton is oxalic acid (or oxalate salts). Excessive ingestion by livestock causes death from deposition of oxalate crystals in the kidney and from hypocalcemia (James 1972). However, rumen microorganisms of known identity (Allison 1985) can degrade oxalate before a ruminant absorbs it. The practical key is to get enough oxalate toxin degraded before absorption. Fortunately the oxlate-degrading rumen organisms proliferate when the ingesta includes small amounts of oxalate. Consequently, with time, the animal adapts to higher and higher levels of oxalate. Early studies suggested adapted sheep could handle nearly double the amount of halogeton compared to levels before adaptation (James 1972). More recent studies suggested adapted sheep could handle even greater levels (Allison 1985). Let sheep adapt slowly to halogeton and avoid trouble.

The second example deals with certain members of the Lupinus genus (lupine). Lupine is a poisonous plant palatable to both sheep and cattle (James et al. 1980). Cattle seldom ingest a lethal dose, but sheep will. Toxic signs in poisoned livestock include depression, labored breathing, tremors, irregular heart rate and irregular gait. Sheep like lupine seeds and pods, the very parts of the plant highest in the quinolizidine alkaloid toxins. If hungry sheep are unloaded in or trailed through an abundant stand of podded lupine, the risk of poisoning is high. Lupine in the flower or post seed stage is not much of a hazard to sheep because the alkaloid content is low. Alkaloids in lupine vary in their death-inducing potency as indicated earlier, but only by about ten times--nowhere near the variation in potency found in Veratrum alkaloid toxins. For lupine, a total alkaloid assay may provide some useful information if sheep are being grazed. Not so for cattle!

Cattle seldom die from lupine ingestion. They don't seem to eat enough for that. Early literature reports that sheep are more readily affected than cattle to the direct toxic effect of lupine (Kingsbury 1964). With cattle, on the other hand, the serious consequence is crooked calves born to cows that ate lupine during pregnancy. Incidence can be devastating--20 to 30% is not uncommon (Keeler et al. 1977). Affected calves generally have twisted limbs or spinal curvature (Shupe et al. 1967a, 1967b, 1968). Apparently from among 70 or so lupine alkaloid toxins, only anagyrine (Keeler et al. 1977) (and possibly one or two of its epimers) play a role in birth defects. Teratogenic lupine doesn't affect fetal lambs (Keeler 1984), but it does affect fetal calves when pregnant cows graze it during the 40th to 70th days of gestation (Shupe et al. 1968). Generally cows can safely eat the plants during the last two-thirds of gestation. Fortunately, anagyrine content in grazable parts of the plant varies in a striking fashion during plant growth. It is high in early vegetation growth, low during flowering, high in seeds, and low after seed drop (Keeler et al. 1976). Using toxin information can help reduce crooked calf incidence to manageable levels if cows are denied access to early growth or podded lupine between

the 40th and 70th days of gestation (Keeler et al. 1977).
Breeding date changes, turnout times, and fencing can
achieve that goal. Not all lupines contain anagyrine.
Lupinus sericeus, L. caudatus, and L. laxiflorus are three
that do and are among the worst offenders.

A third example deals with the poisonous plant Conium
maculatum. It is rather unpalatable and so is not often
heavily grazed (James et al. 1980), but occasionally
livestock deaths and poisonings are reported. The toxins
induce tremors, excessive salivation, and locomotor
disturbances (Kingsbury 1964). The main trouble arises when
pregnant cows graze the plant that same 40- to 70-day period
during which lupine-induced crooked calf disease occurs and
subsequently give birth to deformed calves (Keeler 1974;
Keeler and Balls 1978). The deformities look just like
those induced by lupine. Here are two different plants with
totally different toxins inducing what appears to be an
identical condition. The toxins and teratogens of lupine
are quinolizidine alkaloids while the toxins and teratogens
of Conium are piperidine alkaloids like coniine and γ-
coniceine (Keeler 1974; Keeler and Balls 1978).

The Conium case is also a good example of varying
potency among different animal species. In a study of the
comparative toxicity and teratogenicity of coniine in cows,
mares, and ewes, striking differences were observed (Keeler
et al. 1980). Cows were very susceptible to the toxic
action of coniine at doses as low as 3.3 mg/kg. Their
concepti were readily deformed at this dose. Doses nearly
five times as high were required to induce equally severe
toxic signs in mares; no effects were produced in their
concepti. Doses 13 times as high in ewes induced only
slight toxicity signs, and no congenital deformities were
produced in concepti. The basis for these striking species
differences between cows, mares, and ewes has not yet been
elucidated, but clearly livestock classes vary considerably
in their susceptibility.

The fourth and perhaps the classic example of
successfully avoiding poisonous plant troubles deals with
monkey face lamb disease induced by Veratrum californicum.
In this case, toxin information has been helpful in
explaining why the recommended management strategy has
worked.

Many years ago, epidemics of cyclopia and related
facial deformities occurred in newborn lambs in some parts
of Idaho (Binns et al. 1960; Binns et al. 1963).
Occasionally, incidence exceeded 25% in lambs from
individual ranches. The deformed offspring were single- or

double-globe cyclopics, usually with a proboscis above the eye, or in cases of lesser severity, the lower jaw appeared pronounced and curved up around a very much shortened upper jaw (Binns et al. 1960). Feeding trials established that Veratrum californicum was responsible for the condition (Binns et al. 1963) when it was ingested by the pregnant ewe on the 14th day of gestation (Binns et al. 1965). The main teratogenic compound in the plant is cyclopamine, a steroidal alkaloid (Keeler and Binns 1968).

From a grazing management point of view, information about the chemistry of the teratogen was unnecessary. What was needed was (a) knowledge that the plant was responsible, (b) that the insult to the conceptus occurred on the 14th day, and (c) that the plant grows in sharply defined stands in a restricted habitat. Ranchers have successfully exploited these facts to reduce incidence by preventing ewes from grazing this plant until the rams have been removed for at least 15 days.

Ranchers seldom have monkey face lamb disease now, but complaints are heard that there are low lamb crops in areas where ewes graze Veratrum at later periods of gestation. Recent research shows that Veratrum could be at fault (Keeler et al. 1986). When ewes graze Veratrum around days 28 to 30 of gestation, lambs are born with tracheal stenosis and die within two or three minutes after birth. Embryonic death occurs when ewes graze the plant around days 19 to 21. Both conditions would be counted by ranchers as lamb crop losses and each could occur at an incidence of 15 to 20%. There are other congenital defects caused by the plant that are not lethal and that occur at about days 28 to 31 (Keeler 1986). Thus, management strategy must be modified to keep ewes away from the plant until all have passed days 32 or 33 of gestation. Interestingly, the toxin cyclopamine induces monkey face lamb disease and embryonic death, but not the tracheal stenosis (R. F. Keeler, unpublished).

The fifth example relates to Delphinium spp. The toxicity of Delphinium spp. (larkspurs) is derived from the presence of diterpenoid alkaloid toxins such as lycoctonine (Keeler 1975). Poisoned animals have uneasiness, stiff gait, straddled stance, sternal recumbency, violent efforts to rise, muscular twitching, and respiratory paralysis (Kingsbury 1964; James et al. 1980). Considerable difference exists between sheep and cows in susceptibility to the toxic action of larkspurs. Kingsbury (1964) has summarized some of the reports related to these differences. In one such report, Clawson (1933) found that six times as much Delphinium barbeyi on a body weight basis was required to produce toxicity in sheep than in cattle. Olsen (1978) reported that LD_{50}s of a larkspur alkaloid extract were 175 mg of alkaloid/kg of body weight for sheep but only 43 mg/kg for cattle. The same ratio difference in susceptibility (4X) existed for ground plant material. Whatever the reason for the susceptibility differences, the information can be used to advantage in management planning by sequential or preferential grazing management. Sheep can be used as a means of larkspur control by allowing them to graze larkspur-infested areas before cattle or in preference to cattle.

The final example is one in which there is a very striking susceptibility difference to a plant toxin between sheep and cattle. It relates to the pyrrolizidine alkaloid toxins from Senecio jacobaea (tansy ragwort). Sheep and cattle that are poisoned by tansy ragwort have clinical signs suggesting coordination or vision disturbances, lack of interest in food, depression, slow gait, and hindquarter unsteadiness and dragging of hind feet just before death. Terminal signs include rectal prolapse, edema of the abomasum, appearance of blindness, recumbency, and pig-like odor (Kingsbury 1964; Johnson 1978). Common pathological changes in the liver include extensive fibrosis and bile duct proliferation. The effects are due to several but not all of the pyrrolizidine alkaloids present in the plant but only when they are metabolized to pyrroles by the ingesting animal (Mattocks 1985).

The plant grows commonly in Oregon and Washington pastures that are grazed by dairy cattle. Ranchers and dairymen have long suspected from observation that sheep are more resistant than cattle to toxicoses from tansy. Research trials in both Utah and Oregon have verified this suspected difference. Johnson (1978) showed that a cow ingesting a total amount of prebud tansy ragwort equalling 2% or more of its body weight within 20 days will likely die. Oregon workers (Anonymous 1978) found significant blood changes in cattle after six weeks on a diet that equalled 5% of its body weight during that period. They found that sheep, on the other hand, suffered no effect after ingesting two times their body weight of tansy. Thus, sheep can safely handle considerable tansy, but cows cannot. Other Senecio species grow on desert ranges of several states and may be responsible for similar problems (Johnson et al. 1985).

SUMMARY

Livestock losses may be minimized on any poisonous plant-infested range area by addressing the following questions about the poisonous plants present and their toxins.

Before Grazing Starts and Trouble Has Begun

1. What poisonous plants are present and what are their toxins?
2. Are the plants palatable and poisonous to the livestock to be grazed?
3. What are the signs of poisoning for the toxins in each plant?
4. Will management practices intensify or minimize incidence?

When Animals Die or Show Signs of Poisoning

1. What poisonous plants were grazed?
2. Did the affected animals eat those plants?
3. Are observed signs of poisoning consistent with known signs?
4. Were management mistakes made?
5. Are autopsies or blood, tissue, or plant analyses needed?
6. What immediate management changes are appropriate?

What Are the Constraints When Toxin Analyses Are Needed

1. Is a specific analysis possible and requested?
2. Is caution used in interpretation to avoid pitfalls?
3. Are the poisoning signs too general or loosely defined to be tied exclusively to a specific identified toxin in the suspect plant?

Using Existing Toxin Information to Help Solve Field Problems

1. Can tolerance to the specific toxin be increased by animal adaptation?

2.   Can tolerance differences among animal
species be exploited?
3.   Can toxin concentration changes in the plant
during the grazing season be exploited?
4.   Can variations in animal susceptibility
periods be exploited?

LITERATURE CITED

Allison, M.J. 1985. Anaerobic oxalate-degrading bacteria of the gastrointestinal tract, pp. 120-126. In: A.A. Seawright, M.P. Hegarty, L.F. James, and R.F. Keeler (eds), Plant toxicology. Proc. Aust.-USA Poisonous Plants Symp., Brisbane. Queensland Poisonous Plants Committee, Animal Res. Inst., Yeerongpilly.

Anonymous. 1978. Tansy ragwort toxin being studied by OSU scientists. Western Livestock Journal. March 6:7.

Binns, W., W.A. Anderson, and D.J. Sullivan. 1960. Cyclopian-type malformation. J. Amer. Vet. Med. Assoc. 9:515-521.

Binns, W., L.F. James, J.L. Shupe, and G. Everett. 1963. A congenital cyclopian-type malformation in lambs induced by maternal ingestion of a range plant, Veratrum californicum. Amer. J. Vet. Res. 24:1164-1175.

Binns, W., J.L. Shupe, R.F. Keeler, and L.F. James. 1965. Chronologic evaluation of teratogenicity in sheep fed Veratrum californicum. J. Amer. Vet. Med. Assoc. 147:839-842.

Clawson, A.B. 1933. Additional information concerning larkspur poisoning. Supplement to USDA Farmers Bulletin 988.

Couch, J.F. 1926. Relative toxicity of lupin alkaloids. J. Agr. Res. 32:51-67.

Crowe, M.W. 1969. Skeletal anomalies in pigs associated with tobacco. Mod. Vet. Pract. 50:54.

Fairbairn, J.W., and P. N. Suwal. 1961. The alkaloids of hemlock (Conium maculatum). II. Evidence for a rapid turnover of the major alkaloids. Phytochem. 1:38-46.

James, L.F. 1972. Oxalate toxicosis. Clin. Tox. 5:231-243.

James, L.F., R.F. Keeler, A.E. Johnson, M.C. Williams, E.H. Cronin, and J.D. Olsen. 1980. Plants poisonous to livestock in the western states. USDA Agr. Info. Bull. 415.

Johnson, A.E. 1978. Tolerance of cattle to tansy ragwort (Senecio jacobaea). Am. J. Vet. Res. 39:1542-1544.

Johnson, A.E., R.J. Molyneux, and G.B. Merrill. 1985. Chemistry of toxic range plants. Variation in pyrrolizidine alkaloid content of Senecio, Amsinkia, and Crotalaria species. J. Ag. Food Chem. 33:50-55.

Keeler, R.F. 1974. Coniine, a teratogenic principle from Conium maculatum producing congenital malformations in calves. Clin. Toxicol. 7:195-206.

Keeler, R.F. 1975. Toxins and teratogens of higher plants. Lloydia 38:56-86.

Keeler, R.F. 1978. Reducing incidence of plant-caused congenital deformities in livestock by grazing management. J. Range Mange. 31:355-360.

Keeler, R.F. 1984. Teratogenicity studies on non-food lupins in livestock and laboratory animals, p. 301. In: L. Lopez Bellido (ed), Proceedings of the II International Lupin Conference. Publicacions Agrarias, Madrid.

Keeler, R.F., and L.D. Balls. 1978. Teratogenic effects in cattle of Conium maculatum and conium alkaloids and analogs. Clin. Toxicol. 12:49-64.

Keeler, R.F., L.D. Balls, J.L. Shupe, and M.W. Crowe. 1980. The toxicity and teratogenicity of coniine in cows, ewes, and mares. Cornell Vet. 70:19-26.

Keeler, R.F., and W. Binns. 1968. Teratogenic compounds of Veratrum californicum (Durand). V. Comparison of cyclopian effects of steroidal alkaloids from the plant and structurally related compounds from other sources. Teratology 1:5-10.

Keeler, R.F., and W. Binns. 1971. Teratogenic compounds of Veratrum californicum as a function of plant part, stage, and site of growth. Phytochem. 10:1765-1770.

Keeler, R.F., and M.W. Crowe. 1985. Anabasine, a teratogen from the Nicotiana genus, pp. 324-333. In: A.A. Seawright, M.P. Hegarty, L.F. James, and R.F. Keeler (eds), Plant toxicology, Proc. Aust.-U.S.A. Poisonous Plants Symp., Brisbane. Queensland Poisonous Plants Committee, Animal Res. Inst., Yeerongpilly.

Keeler, R.F., E.H. Cronin, and J.L. Shupe. 1976. Lupin alkaloids from teratogenic and nonteratogenic lupins. IV. Concentration of total alkaloids and individual major alkaloids particularly anagyrine, the probable teratogen in crooked calf disease, as a function of plant part and stage of growth. J. Tox. Environ. Health 1:889-908.

Keeler, R.F., L.F. James, J.L. Shupe, and K.R. Van Kampen. 1977. Lupine-induced crooked calf disease and a management method to reduce incidence. J. Range Manage. 30:97-102.

362

Keeler, R.F., L.D. Stuart, and S. Young. 1986. When ewes ingest poisonous plants: the teratogenic effects. Vet. Med. 81:449-454.

Kerwin, J.F., C.P. Balant, and G.E. Ullyot. 1960. Hypotensive drugs, p. 551, In: Alfred Burger (ed), Medicinal chemistry, John Wiley & Sons, New York.

Kingsbury, J.M. 1964. Poisonous plants of the United States and Canada. Prentice-Hall, Englewood Cliffs, NJ.

Mattocks, A.R. 1985. Molecular basis of pyrrolizidine alkaloidal toxicity--recent studies, pp. 181-190. In: A.A. Seawright, M.P. Hegarty, L.F. James, and R.F. Keeler (eds), Plant toxicology, Proc. Aust.-U.S.A. Poisonous Plants Symp., Brisbane. Queensland Poisonous Plants Committee, Animal Res. Inst., Yeerongpilly.

Olsen, J.D. 1978. Larkspur toxicosis: a review of current research, p. 535. In: R.F. Keeler, K.R. Van Kampen, and L.F. James (eds), Effects of poisonous plants on livestock, Academic Press, New York.

Pelletier, S.W. 1970. Chemistry of the alkaloids. Van Nostrand Reinhold Co., New York.

Shupe, J.L., W. Binns, L.F. James, and R.F. Keeler. 1967a. Lupine, a cause of crooked calf disease. J. Amer. Vet. Med. Assoc. 151:198-203.

Shupe, J.L., W. Binns, L.F. James, and R.F. Keeler. 1968. A congenital deformity in calves induced by the maternal consumption of lupin. Aust. J. Agr. Res. 19:335-340.

Shupe, J.L., L.F. James, and W. Binns. 1967b. Observations on crooked calf disease. J. Amer. Vet. Med. Assoc. 151:191-197.

Williams, J.C., and E.H. Cronin. 1963. Effect of silvex and 2,4,5-T on alkaloid content of tall larkspur. Weeds 11:317-319.

Wink, M., and T.H. Hartmann. 1982. Diurnal fluctuation of quinolizidine alkaloid accumulation in legume plants and photomixotrophic cell suspension cultures. Z. Naturforshung 37:369-375.

26

The Importance of Poisonous Plants as Forages in the Prairies and Southwest

Charles A. Taylor, Jr., and Michael H. Ralphs

INTRODUCTION

Grazing animals are generally exposed to a wide variety of potentially poisonous plants. If grazing pressures are not too great, animals usually avoid these plants, consume them in quantities that can be safely detoxified or excreted, or select them as diet constituents when toxins are at a safe level. Since few plant species are entirely free from potentially harmful compounds, it is important to understand under what circumstances important forage plants become poisonous.

Basically, grazing animals require energy, protein, vitamins, and minerals from forage. All green plants have chemical structures that can yield these nutrients; unfortunately, many of the palatable forage plants also have other chemicals that cause sickness, death, or deviation from a normal physiological state. In general, the grazing animal consumes forage that contains a much wider range of chemical compounds than is needed for proper animal nutrition.

Forage poisoning in grazing animals usually results from three conditions: consumption of plants that are toxic in small amounts; consumption of plants considered good forage when consumed in moderate amounts, but toxic if selected in excess; or consumption of plants usually considered good forage, but consumed at a time when they are toxic (Table 26.1). Some of these plants that can potentially limit livestock production are discussed along with brief descriptions of signs of poisoning.

363

TABLE 26.1
Classification of selected plant species by their desirability as forages, toxic conditions, and animals affected.

Species	Desirable forage		Undesirable forage		Animals affected		
	Toxic when consumed in excess	Toxic only at certain periods	Toxic when forced to consume large quantities	Risk of high toxicity	Cattle	Sheep	Goats
WOODY PLANTS							
Oak (Quercus spp.)	X				X	X	X
Acorns	X	X			X	X	X
Mesquite (Prosopis glandulosa)							
Beans	X		X		X		X
Sacahuista (Nolina texana)							
Foliage	X				X	X	X
Fruit and flowers		X		X	X	X	X
Pricklypear cactus (Optunia spp.)			X				X
Guajillo (Acacia berlandieri)	X					X	X
Mescalbean (Sophora secundiflora)			X		X	X	X
Whitebrush (Aloysia gratissima)	X				Horses, mules, burros		
Hogplum (Colubrina texensis)							
Fruit	X		X			X	
GRASSES							
Kleingrass (Panicum coloratum)	X	X				X	X
Johnsongrass (Sorghum halapense)		X			X	X	X
Reed canarygrass (Phalaris arundinaceae)	X	X			X	X	?
FORBS							
Kochia (Kochia scoparia)		X			X		
Carelessweed (Amaranthus spp.)	X	X			X	X	X
Bitterweed (Hymenoxys odorata)				X	X	X	
Silverleaf nightshade (Solanum eleagnifolium)				X	X	X	X

WOODY PLANTS

Oaks

Oaks (Quercus spp.) represent a genus of approximately 500 recognized species in the Northern Hemisphere and about 250 in the New World (Correll and Johnston 1979). Poisoning of livestock from consumption of oak twigs, leaves, and acorns represents a costly loss of livestock in Texas (Reagor 1981). According to Reagor (1981), most species of oak are probably poisonous; however, liveoak (Quercus virginiana) seems to be less of a problem than shinoak (Q. pungens) or sand shinnery (Q. harvardii). Tannins are considered the causative agent and are found in higher concentrations in immature leaves and acorns (Cheeke and Shull 1985).

Signs of oak poisoning include some or all of the following: loss of appetite; constipation, often followed by diarrhea; rough hair coat; dry muzzle; excessive thirst; frequent urination; thin blood tinged with mucus coming from the nostrils; and swellings beneath the jaw and vulva (Reagor 1981). Oak contains compounds (tannic and gallic acid) which damage the kidneys, resulting in death within two to three weeks after oak consumption (Reagor 1981).

Even though livestock losses due to oak poisoning occur over many parts of the United States, the oaks provide a significant amount of forage for grazing animal production. This is especially true for sheep, goats, and deer (Kothmann 1968; Malechek 1970; Bryant et al. 1979; Rector 1983; Taylor 1983). Nutritional values for plateau oak (Q. virginiana var. fusiformis), native to the Edwards Plateau of Texas, have been reported by Huston et al. (1981). New growth of plateau oak leaves reached a peak protein content of 20% with a corresponding digestible organic matter value (DOM) of 57%. However, crude protein and in vitro digestibility values of oak leaves should be interpreted with caution (Huston et al. 1981; Nastis and Malechek 1981; Telford et al. 1983; Carter 1985). In addition to being toxic, tannins bind with plant protein, reducing digestibility and lowering the total value of the forage (McLeod 1974).

Oak mast provides a timely source of energy (DOM = 70%). Acorns usually become available in the fall, occurring during the peak breeding period for goats, deer, and sheep. Daily acorn consumption by free-ranging Angora goats has been measured as high as 75% of the diet (Taylor 1983). In general, acorn drop is gradual and extends over a period of three months or longer. This usually prevents a

large accumulation of acorns on the ground at any one time,
preventing any serious poisoning problem.

Mesquite

Livestock poisoning from mesquite (Prosopis glandulosa)
generally occurs from consumption of large amounts of the
beans over a period of a month or so. Clinical signs of
acute poisoning include loss of appetite, rapid loss of
weight, nervousness, a wild expression, and bulging eyes
(Sperry et al. 1964). These animals usually die within two
to four days of the first signs of illness. The chronic
form of the disease develops more gradually, usually after
cattle have been eating beans for two months or more. Most
of the acute poisoning occurs in pastures where large
quantities of mesquite beans accumulate.

Mesquite pods contain about 19% protein and 30%
sucrose, making them very palatable to livestock (Cheeke and
Shull 1985). Observations indicate that sheep, goats, and
deer actively select mesquite beans, regardless of current
herbaceous forage quality conditions (C.A. Taylor, personal
observation). Consumed in moderation, mesquite beans seem
to offer some benefits as a secondary or reserve forage.

Sacahuista

Sacahuista (Nolina texana) is abundant in the Edwards
Plateau and Trans-Pecos areas of Texas and extends into
southeast Arizona and northern Mexico. It usually grows on
rocky hillsides or areas characterized by shallow soil.

A member of the lily family (Liliaceae), sacahuista is
a perennial plant with grass-like leaves arising from a
thick woody base. Tall flowering stalks (0.5 to 1 m),
bearing small white flowers, rise above the foliage in April
and May. The flower stalks usually are not apparent until
the plant is in full bloom.

The flower buds, flowers, and fruits are toxic to
cattle, sheep, and goats. Sacahuista produces signs of
generalized jaundice, loss of appetite, and progressive
debilitation. Affected animals may have a yellow nasal
discharge. Ingestion of sacahuista flowers or fruit in
combination with other green plants may produce
photosensitization (Sperry et al 1964).

The leaves of sacahuista can be safely eaten by
livestock. In fact, sacahuista can be a useful reserve

forage during periods of plant shortages or dormancy (Taylor 1973, Ralphs 1983). During dormant periods, sacahuista is a standby forage, high in abundance but low in quality. Sacahuista can be eliminated from the vegetative complex if grazing pressures from cattle are too great over a sufficient period of time.

Pricklypear cactus

Many species of pricklypear (Optunia spp.) are found in the Western Hemisphere. On Texas rangelands, pricklypear has been used as livestock feed for more than a century (singeing the pear spines with a butane burner to encourage livestock use is a common practice). It is usually considered an undesirable plant, but pricklypear does have some value as a reserve or supplemental forage for livestock and wildlife during winter and drought periods; it also provides cover for game birds such as quail.

Pricklypear does not pose a threat to livestock production from plant toxins; however, the spines and seeds associated with the plant cause significant losses in sheep and goat production in the Edwards Plateau Region of Texas. Spines damage the lips, tongue, esophagus, and rumen of sheep and goats (C.W. Livingston, personal communication) but seem to cause less damage to cattle (Taylor 1973).

Consumption of large quantities of pricklypear fruit by sheep can result in either death or a significant reduction in animal production. The pricklypear seeds within the fruit are apparently nondigestible in the rumen. When sheep consume large quantities of fruit, the seeds accumulate in the rumen and abomasum and reduce or interrupt the normal functioning of the digestive tract. If impaction becomes severe, death of the animal may be prevented only by surgery.

Guajillo

Guajillo (Acacia berlandieri) is a shrub in the legume family with twice-pinnate leaves. The flattened fruit pods are four to six times as long as they are wide and have somewhat thickened margins (Sperry et al. 1964). Guajillo grows in the southern part of the Edwards Plateau of Texas and in the central and southern parts of the South Texas Plains and extends southward into Mexico.

Sheep and goats poisoned by guajillo develop a locomotor incoordination of the legs referred to as "limberleg" or "guajillo wobbles" (Sperry et al. 1964). Animals can develop poisoning after eating an exclusive diet of guajillo for nine months or longer. Losses are negligible during years of favorable rainfall, but have reached 50% in periods of extended drought (Price and Hardy 1953).

Guajillo is a valuable forage that makes a significant contribution to animal diets; however, caution should be used during extended dry periods or excessive grazing pressures.

Mescalbean

Mescalbean (Sophora secundiflora) is a small tree or shrub 90 to 240 cm tall. The leaves are once-pinnately compound, glossy green, and evergreen. The fruit is a several-seeded legume with one or two constrictions. Mescalbean grows in the Edwards Plateau and Trans-Pecos areas of Texas and extends its range into New Mexico on the west and Mexico to the south. It is found on hills, rocky ledges, and canyons.

Cattle, sheep, and goats have been poisoned by eating the leaves or seeds of mescalbeans. Cattle are very susceptible to poisoning from the leaves, but goats and sheep are more tolerant. The seeds are very poisonous but, unless crushed, apparently pass through the digestive tract of most animals without harm (Sperry et al. 1964).

Mescalbean is highly poisonous and contains the narcotic alkaloid cytisine or sophorine. An increased pulse rate and a stiffening of the hind legs results when poisoned animals are exercised. Eventually, they may fall and become comatose. Sheep usually recover after a rest period, but cattle often die. Goats were not poisoned by experimental feeding (Boughton and Hardy 1935).

Generally, mescalbean is not a very palatable plant. However, it is an evergreen, and livestock may graze it when grazing pressures are excessive or other green forage is scarce.

Whitebrush

Whitebrush (Aloysia gratissima) is a slender, many-branched shrub that can reach heights of 180 cm. The small white or bluish flowers are in open, leafy panicles of elongated spikes or spikelike racemes. Whitebrush is frequent to abundant in central, west, and south Texas and extends into New Mexico and Mexico. Horses, mules, and burros are affected by whitebrush, grazing it from 30 to 40 days before symptoms develop.

Whitebrush has some value as a forage plant. Goats, sheep, and deer browse on the leaves, and use can be very heavy during dry periods or on new sprouts following fire or mechanical treatment. Also, bees produce excellent honey from the nectar of whitebrush.

Hogplum

Hogplum (Colubrina texensis) is a shrub of the buckthorn family (Rhamnaceae). It is a much-branched spinescent shrub from 90 to 180 cm tall. The small, brownish fruit is a hard, three-celled, drupelike capsule. Hogplum is most commonly found in the Cross Timbers, Edwards Plateau, and South Texas Plains regions. It grows on rocky or gravelly slopes and along washes and arroyos. The seed and fruits are thought to contain a hepatic toxin which causes a condition similar to lechuguilla (Agave lechuguilla) poisoning in sheep. Hogplum foliage is browsed by most classes of livestock. It can be eliminated from the vegetative complex with excessive grazing pressures from sheep and goats.

GRASSES

Kleingrass

Kleingrass (Panicum coloratum) is a warm season, perennial forage grass introduced from Africa. It has gained widespread popularity and has been seeded on several hundred thousand acres of either pure or mixed stands in Texas (Huston 1978). Photosensitization (swellhead) in sheep grazing kleingrass pasture was reported in 1972 in the San Angelo area of Texas. Subsequent research at the Texas A&M University Research and Extension Center at San Angelo confirmed that the plant was responsible for swellhead. The toxic principal (a saponin) has recently been isolated and identified (B.J. Camp, personal communication). Saponins are glycosides which have profuse foaming properties (Cheeke and Shull 1985).

Research and case studies indicate that weaned lambs should not be allowed to graze kleingrass. Ewes or ewes with lambs seem to be more resistant to the problem.

Johnsongrass

Johnsongrass (Sorghum halapense) is an introduced grass that has readily adapted to the southern sections of the United Stated from Southern California to Florida. Under certain environmental conditions, Johnsongrass and the other sorghums produce cyanogenetic glycoside which yields hydrocyanic acid upon hydrolysis, which occurs during digestion (Sperry et al. 1964). The hydrocyanic acid can cause cyanide poisoning in the animal. The glycosides occur in vacuoles in plant tissue, while the enzymes are found in the cytosol. Damage to the plant from wilting, trampling, mastication, frost, drought, and bruising results in the enzymes and glycosides coming together, causing HCN (hydrogen cyanide, prussic acid) to be formed (Cheeke and Shull 1985).

Johnsongrass is considered a major forage for grazing animals in the south (Hughes et al. 1966). It is very palatable and is readily consumed by cattle, sheep, goats, and white-tailed deer. Poisoning can be avoided by not grazing during periods of regrowth following drought and by waiting ten days after frost or other causes of damage to the plant, such as cutting for hay.

Reed canarygrass

Reed canarygrass (Phalaris arundinacea) is a tall, coarse, sod-forming, cool-season perennial. It is an important improved pasture grass ranging from Canada to New Mexico and is common throughout the Midwest.

Reed canarygrass contains at least eight alkaloids, including gramine, hordenine, four derivatives of tryptamine, and two derivatives of b-carboline (Cheeke and Shull 1985). Acute phalaris poisoning results in sudden collapse and death of affected animals, caused by acute heart failure. The acute condition is characterized by hyperexcitability, incoordination, spasms, head nodding, twitching, and salivation. Chronic phalaris staggers have been reported in sheep and cattle grazing new growth. Symptoms are discussed by Cheeke and Shull (1985). This grass has cyanogenetic potential also, which is probably more important than staggers.

Other HCN producing plants include cherries and plums (Prunis spp.), velvetgrass (Holcus lanatus), mountain mahogany (Cercocarpus brevifolius), flax (Linum spp.), and sticky palefoxia (Palafoxia tripteris).

FORBS

Kochia

Kochia (Kochia scoparia), sometimes referred to as summer cypress or fireweed, is a leafy, bushy annual weed that may turn reddish in color during the fall. It is an introduced plant from Eurasia. It grows in semiarid and arid lands of the western and southwestern regions of the United States.

Kochia is successfully used as a forage plant, but can present problems during periods of drought and at seed maturation. Toxicants associated with Kochia are discussed by Cheeke and Shull (1985). It accumulates nitrates (Dickie and James 1983) and can cause photosensitization and polioencephalomalacia (Dickie and Bergman 1979).

Carelessweed

Carelessweeds (Amaranthus spp.) are vigorous, warm-season, annual forbs usually found in disturbed areas. The growth of the various species ranges from prostrate to branched upright. Most species are very palatable to both domestic and wild grazing animals.

Carelessweeds are nitrate accumulating plants (Sperry et al. 1964). Potassium nitrate levels increase with stress: nitrate levels in a green plant in a rapid stage of growth were 0.066%; plants that had stopped growing as a result of inadequate moisture had nitrate levels of 0.828%; and plants that were wilting due to inadequate moisture contained 2.874% nitrate (J.W. Dollahite and B.J. Camp, unpublished data).

Chronic nitrate toxicity has been implicated with reduced animal growth, vitamin A deficiency, abortion, infertility, goiter, and other nonspecific problems (Cheeke and Shull 1985). Acute toxicity almost always occurs when animals consume moisture-deficient plants (Sperry et al. 1964). Other potential nitrate accumulators include wheat (Triticum spp.), oats (Avena spp.), barley (Hordeum spp.), sudangrass (Sorghum spp.), and corn (Zea spp.).

Bitterweed

Bitterweed (Hymenoxys odorata) is a cool-season annual belonging to the Compositae family. Plants have numerous ascending branches with yellow flower heads and purple stems at the base. Bitterweed has a very distinct aromatic odor and a bitter taste. Germination usually occurs in late fall when soil moisture is available.

Bitterweed is considered the most serious poisonous plant problem affecting the sheep industry in Texas. Overgrazing of rangeland has increased the quantity and range of bitterweed. The plant vigorously grows on disturbed areas or where other vegetation is scarce. The heaviest infestations and the most severe losses in Texas occur in about 12 counties of the Edwards Plateau and Trans-Pecos regions (Sperry et al. 1964). Bitterweed ranges from Central Texas to California and from Kansas south into Mexico.

Bitterweed is generally unpalatable, but sheep consume the plant when forced to do so by lack of suitable forage (Calhoun 1981). The signs of acute poisoning are loss of appetite, cessation of rumination, depression, indications of abdominal pain, bloating, and green regurgitated material about the mouth and nose. Loss of weight is the most common sign of chronic bitterweed poisoning.

When bitterweed poisoning first occurs, sheep should be either moved to bitterweed-free pastures or pens or supplied extra feed. Grazing management can also play an important role in reducing losses from bitterweed poisoning. Moderate stocking rates and a combination of grazing animals and grazing systems are practices that can significantly reduce both bitterweed infestation and its associated poisoning (Merrill and Schuster 1978).

Most bitterweed problems occur when there is sufficient soil moisture in the fall for the plants to germinate and become established, followed by a dry late winter and spring which reduces growth of desirable forages. Bitterweed is consumed while in the immature growth stage, usually in the winter or early spring seasons of the year.

Silverleaf nightshade

Silverleaf nightshade (Solanum eleagnifolium) is a widely distributed, poisonous, warm-season, perennial weed found throughout the southwestern states. The leaves are oblong to linear, undulate to deeply sinuate, silvery white or stellate canescent. The flowers are violet and yellow or bluish.

Horses, cattle, sheep, goats, swine, chickens, ducks, rabbits, and humans have been poisoned by eating Solanum spp. (Sperry et al. 1964). Kingsbury (1964) has reported that considerable numbers of cattle have been lost due to the consumption of nightshade; however, grazing animals on the Texas Agricultural Experiment Station at Sonora, Texas, seem not to have been affected by silverleaf nightshade consumption (L.B. Merrill, personal communication; Ralphs 1983). The reason animals were not injured by nightshade consumption is unclear; intake may not have been at a sufficient level for any one animal to elicit a response.

Silverleaf nightshade fruit is also used in the process of making some cheeses in Mexico and Texas. Evidently this process detoxifies the poisonous principle, because ripe berries produce moderate to severe poisoning when ingested at 0.1 to 0.3% of body weight (Cheeke and Shull 1985).

SUMMARY AND CONCLUSIONS

Since the beginning of the forage livestock industry, the problem of poisonous plants has generally increased in importance. Poisonous plants have always been a component of grasslands; however, this increasing problem has been a result, in part, to excessive stocking rates and poor management. Generally, livestock do not select poisonous plants in harmful amounts unless the availability of alternate choices of forage is low. However, there are some plant species, because of their palatability, addictive properties, or some other unexplained characteristic, that are selected by grazing animals even when management and available forage are at an optimum.

Many of the plants considered toxic to livestock can serve as useful constituents of an animal's diet. They usually cause intoxication only when an animal is enticed to eat too much too fast or during certain periods of time during plant growth or for other reasons. Understanding the conditions under which intoxication may occur decreases the chance of livestock loss and also enhances the opportunity to utilize the forage value of these potential poisonous plants.

LITERATURE CITED

Boughton, I.B., and W.T. Hardy. 1935. Mescalbean (Sophora secundiflora). Tex. Agr. Exp. Sta. Bull. 519.

Bryant, F.C., M.M. Kothmann, and L.B. Merrill. 1979. Diets of sheep, Angora goats, Spanish goats, and whitetailed deer under excellent range conditions. J. Range Manage. 32: 412-417.

Calhoun, M.C. 1981. Recent research on the bitterweed (Hymenoxys odorata) poisoning problem in sheep. In: A. McGinty (ed), Proceedings of the Trans-Pecos poison plant symp. Tex. Agr. Ext. Ser. Proc.

Carter, H.C. 1985. Fecal nitrogen and phosphorus as indicators of intake and quality of Angora goat diets. M.S. Thesis, Texas A&M Univ., College Station, TX.

Cheeke, P.R., and L.R. Shull. 1985. Natural toxicants in feeds and poisonous plants. AVI Publishing Co., Inc. Westport, CN.

Correll, D.S., and M.C. Johnson. 1979. Manual of the vascular plants of Texas. Univ of Texas at Dallas. Richardson, TX.

Dickie, C.W., and J.R. Bergman. 1979. Polioencephalo-malacia and photosensitization associated with Kochia scoparia consumption in range cattle. J. Amer. Vet. Med. Assoc. 175:463-465.

Dickie, C.W., and L.F. James. 1983. Kochia scoparia poisoning in cattle. J. Amer. Vet. Med. Assoc. 183:765-768.

Hughes, D.H., M.E. Heath, and D.S. Metcalfe. 1966. Forages. Iowa State Univ. Press, Ames, IA.

Huston, J.E. 1978. Successful utilization of pastures containing a high proportion of kleingrass by sheep. In: Sheep and goat--wool and mohair. Tex. Agr. Exp. Sta. PR-3506.

Huston, J.E., B.S. Rector, L.B. Merrill, and B.S. Engdahl. 1981. Nutritional value of range plants in the Edwards Plateau of Texas. Tex. Agr. Exp. Sta. Bull. 1357.

Kingsbury, J.M. 1964. Poisonous plants of the United States and Canada. Prentice-Hall, Englewood Cliffs, NJ.

Kothmann, M.M. 1968. The botanical composition and nutrient content of the diet of sheep grazing on poor condition pasture compared to good condition pasture. Ph.D. Diss., Texas A&M Univ., College Station, TX.

McLeod, M.N. 1974. Plant tannins--their role in forage quality. Nutr. Abstr. Rev. 44:804-815.

Malechek, J.C. 1970. The botanical and nutritive composition of goat diets on lightly and heavily grazed ranges in the Edwards Plateau of Texas. Ph.D. Diss., Texas A&M Univ., College Station, TX.

Merrill, L.B., and J.L. Schuster. 1978. Grazing management practices affect livestock losses from poisonous plants. J. Range Manage. 31:351.

Nastis, A.D., and J.C. Malechek. 1981. Digestion and utilization of nutrients in oak browse by goats. J. Anim. Sci. 53:283-290.

Price, D.A., and W.T. Hardy. 1953. Guajillo poisoning of sheep. J. Amer. Vet. Med. Assoc. 122:223-225.

Ralphs, M.H. 1983. Vegetation and livestock response to increasing stocking rates in a simulated short-duration grazing system. Ph.D. Diss., Texas A&M Univ., College Station, TX.

Reagor, J.C. 1981. Overview of poison plant situation in the Trans-Pecos. In: A. McGinty (ed), Proc. of the Trans-Pecos poison plant symp. Tex. Agr. Ext. Ser. Proc.

Rector, B.S. 1983. Diet selection and voluntary forage intake by cattle, sheep, and goats grazing in different combinations. Ph.D. Diss., Texas A&M Univ., College Station, TX.

Sperry, O.E., J.W. Dollahite, G.O. Hoffman, and B.J. Camp. 1964. Texas plants poisonous to livestock. Tex. Agr. Exp. Sta. Bull. 1028.

Taylor, C.A. 1973. The botanical composition of cattle diets on a 7-pasture high-intensity low-frequency grazing system. M.S. Thesis, Texas A&M Univ., College Station, TX.

Taylor, C.A. 1983. Foraging strategies of goats as influenced by season, vegetation, and management. Ph.D. Diss., Texas A&M Univ., College Station, TX.

Telford, J.P., M.M. Kothmann, R.T. Hinnant, and K. Robinson Ngugi. 1983. Nutritive value of alfalfa and liveoak for goats. In: Sheep and goat--wool and mohair. Tex. Agr. Exp. Sta. Bull. CPR-4171.

27

The Importance of Poisonous Plants as Forages in the Intermountain Region

James E. Bowns

INTRODUCTION

Poisonous plants cause untold losses through death, physical malformations, abortions, and lowered gains of animals (Stoddart et al. 1975). Losses come not only through death and disability of livestock, but production costs are often higher where poisonous plants are a problem. These costs include plant control, fencing, supplements, and more intensive management (James and Johnson 1976, Nielsen 1978, James 1983). The presence of poisonous plants also reduces the available options for managing livestock and limits the utilization of the available forage (Cronin and Nielsen 1979).

Poisoning under range conditions often results when animals are forced, through a shortage of nontoxic plants, to consume more of the toxic species. There are exceptions, but livestock generally do not consume poisonous plants in sufficient quantities to cause problems when they are provided a choice of plants (Marsh 1909, Dwyer 1978). Most large livestock losses result from management mistakes such as confining hungry livestock on dense concentrations of poisonous plants (Ralphs and Olsen 1987). Apparently hunger lowers the smell and taste rejection thresholds, causing animals to eat plants that they would normally avoid (Arnold and Hill 1972).

Plants do not readily fall into poisonous and nonpoisonous categories. Stoddart et al. (1949) estimate that thousands of plants would be poisonous if eaten in sufficiently large quantities. Brotherson et al. (1980) compiled a list of 215 taxa of major livestock poisoning plants for the state of Utah. Some of these plants are excellent forages when eaten in moderate quantities, and

377

many definitely known to be poisonous are consumed by animals with no apparent ill effects. Plants known to be toxic are not equally poisonous at all times of the year, and different kinds of animals vary with respect to their susceptibility to poisoning from a particular plant (Stoddart et al. 1975; James 1983).

Plants containing toxins, yet commonly grazed in the Intermountain area, include gambel oak (Quercus gambelii), chokecherry Prunus virginiana var. melanocarpa), mountain mahogany (Cercocarpus mountanus), greasewood (Sarcobatus vermiculatus), halogeton (Halogeton glomeratus), tall larkspurs (Delphinium barbeyi and D. occidentale), lupine (Lupinus spp.), and locoweeds (Astragalus spp. and Oxytropis spp.). All of these taxa, with the exception of locoweeds, are rated as locally abundant climax constituents. Locoweeds are rated as present in the climax composition but not abundant (Stoddart et al. 1975).

The purpose of this paper is to discuss the use of common poisonous plants as forages. Information on this subject was obtained from scientific papers as well as personal observations and experiences of ranchers and land managers.

GAMBEL OAK

Gambel oak is the major component on more than 3.76 million ha in the western United States and is the dominant shrub at intermediate (2100 to 2400m) elevations in southwestern Utah.

Kingsbury (1964) considers oak poisoning a severe economic problem on Southwest ranges. Cattle are primarily poisoned from consuming oak leaves, buds, and acorns (Ruyle et al. 1986). Oak poisoning is generally a seasonal problem, occurring in the spring when new buds are consumed and in the fall when acorns drop. The most dangerous period lasts for about four weeks in the spring (Ruyle et al. 1986; Kingsbury 1964). Stoddart et al. (1949) suggest that freezing may increase the toxicity of young oak leaves in late spring. Kingsbury (1964) also reports poisoning of cattle, and occasionally sheep and goats, to be common in drought years.

Harper et al. (1985) consider gambel oak to have relatively low palatability to sheep and cattle, and that these animals utilize oak only after the more desirable plants have diminished. However, many livestock operators consider oak as a fair browse for livestock. Gambel oak is grazed in great quantity by livestock and mule deer without apparent harm. Kingsbury (1964) suggests that a diet of not more than 50% oak may not only be harmless, but may also contribute to the nutrition of cattle. However, Ruyle et al. (1986) suggest that since oak contains tannins, even moderate consumption of oak may reduce overall cattle productivity. This lower herd performance may manifest itself in decreased calf crops and calf weaning weights.

Forsling and Storm (1929) studied a browse range in southwestern Utah where gambel oak made up 30% of the plant composition. Oak comprised 70 to 80% of cattle diets in the late season, but no oak poisoning was experienced. The consumption of oak increased dramatically as other forage species became unavailable. Cattle made unsatisfactory gains when oak dominated the diet and lost weight during the latter part of the grazing season. Heavy oak consumption also resulted in very low calf crops.

Harper et al. (1985) consider gambel oak communities as valuable big game winter range in Utah, Arizona, and Colorado. Oak is highly valuable as forage for deer and elk and is important cover for deer. Hogs have also been grazed on oak ranges. Kingsbury (1964) reported that acorns have been used as forage by swine with no apparent adverse effects. Swine may be more resistant to oak than other animals, or not enough acorns are consumed to provoke symptoms. In the 1940s, some southern Utah oak brush ranges were grazed by hogs during years of good acorn production (Bowns 1985).

CHOKECHERRY

Chokecherry is common in valley bottoms, on north slopes of sagebrush benchlands, and on mountain ranges of the Intermountain area. It is especially common around springs and along creek banks (Stoddart et al. 1949). Chokecherry is toxic to both cattle and sheep (Stoddart et al. 1949). Hydrocyanic acid is the toxin in chokecherry, and the concentrations increase when plants are stressed from frost or drought (Stoddart et al. 1949; Stoddart et al. 1975). Cattle have been poisoned by chokecherry in southwestern Utah when the plants were frozen by a late spring frost. Cattle native to the area and familiar with the range were not affected. However, two-year-old heifers purchased from outside the local area were poisoned and died (Warren Williams, Cedar City rancher, personal communication). Toxicity also occurs when animals consume considerable quantities of leaves and twigs in a short time (James et al. 1980). When other forage is available, poisoning is unlikely. Ranchers recognize that livestock familiar with a range and its forage species are less likely to be poisoned than animals recently introduced to the area.

Chokecherry is among the most palatable of poisonous plants, and some animals prefer it to other forages (Stoddart et al. 1949; James et al. 1980). On Utah summer ranges chokecherry is a preferred forage species for sheep throughout the summer grazing period. It is common to have stands of chokecherry defoliated to the height sheep can reach. Sheep appear to prefer chokecherry to the highly desirable mountain snowberry. These high levels of utilization produce no apparent adverse effects.

MOUNTAIN MAHOGANY

True mountain mahogany has been incriminated in livestock losses because of its cyanogenetic potential (Kingsbury 1964). This shrub is an excellent browse species, but it has been considered responsible for occasional scattered loss of livestock (Kingsbury 1964). Big game and livestock avidly use the new foliage during the growing season and use the twigs in the winter (Plummer et al. 1986). Mountain mahogany is considered a premier browse for sheep, cattle, and mule deer in southern Utah. Plummer et al. (1968) make no mention of any toxicity problems associated with its consumption.

GREASEWOOD

Greasewood is a common shrub of lower elevations in the Intermountain area. It is generally confined to alkali soils on flood plains, dry washes, and gullies where soil moisture is high. It often dominates desert areas where runoff waters have accumulated (Stoddart et al. 1949).

Cattle are rarely poisoned by greasewood, but large losses of sheep have been reported (Kingsbury 1964). Heavy mortality results when hungry animals graze on almost pure stands. Trailing hungry sheep in the spring or releasing them from shearing corrals or lambing sheds into greasewood may result in high losses.

Stoddart et al. (1949) indicated that poisoning is most likely in the spring when young tender foliage is quite palatable. Local sheepmen agree. Losses as high as 10% of a herd have been reported (Bob Clark and Russell Sevy, Cedar City, Utah, ranchers, personal communication). Some ranchers recognize that there is an element of risk involved in grazing this shrub. They emphasize the fact that sheep must have a variety of forage in their diet, that they must be introduced to the plant slowly, and if necessary, be supplemented for a few days until they become accustomed to it (Bob Clark and Russell Sevy, personal communication). They also indicate that lactating ewes are more susceptible to poisoning than dry ewes, and greasewood poisoning may be more serious with pregnant ewes prior to, rather than after, lambing. Losses have also been reported in the fall when sheep eat large quantities of leaves that have fallen to the ground (James et al. 1980). The toxicity of greasewood is reported to increase as the growing season advances, reaching a maximum in August or September. Oxalates, the toxin in greasewood, accumulate in the leaves (James et al. 1980).

Oxalates are metabolized by rumen microorganisms and excreted and are not harmful if the plants are eaten over several hours or taken with other forage (Kingsbury 1964). The rumen microflora are able to adapt and thus increase their ability to detoxify the oxalates. Sheep can adapt to greasewood if they are introduced to it slowly over two or three days (James et al. 1980). Deaths occur when greasewood is eaten in large amounts in a short period of time.

Greasewood is readily eaten by sheep and is considered an excellent forage by sheepmen. Stoddart et al. (1949) consider greasewood a valuable forage plant for cattle and sheep during winter months and, therefore, do not consider eradication of greasewood as either feasible or desirable. Cattle readily graze greasewood on fall and winter ranges and will preferentially graze the woody current year's growth (M.H. Ralphs, personal communication).

HALOGETON

Halogeton, another oxalate containing plant, is also hazardous to sheep. This plant has caused severe sheep losses on desert winter ranges since it was discovered in Nevada in the 1930s. Despite its toxicity, it is considered by some ranchers to be excellent forage for cattle and good forage for sheep. Cattle often use halogeton-infested ranges without apparent ill effects, even though they can be poisoned from consuming halogeton (Cronin et al. 1978).

Cattle do, however, have locomotor disturbances following grazing on nearly pure stands of halogeton (James 1970). Livestockmen recognize that sheep can utilize halogeton as long as they have access to other forages. Losses occur when hungry animals are forced to consume halogeton because of the lack of other forage. Sheep are also able to detoxify the oxalates in halogeton if they are introduced to it slowly (James et al. 1967).

TALL LARKSPURS

Barbey larkspur (<u>Delphinium</u> <u>barbeyi</u>) and duncecap larkspur (<u>D</u>. <u>occidentale</u>) are common tall forbs on western mountain ranges. These plants are climax species and tend to increase with improving range conditions (Cronin and Nielsen 1981). They have persisted after the introduction and heavy grazing of domestic livestock. They can be the most tenacious species of pristine communities. They persist long after associated species are destroyed (Cronin et al. 1978).

Larkspurs have been recognized as both medicinal and poisonous from classical times (Kingsbury 1964). Alkaloids in larkspurs affect the respiratory system (Bailey 1978) and are highest in new growth and seeds. Toxicity decreases as the plants mature (James et al. 1980). Stems are the least preferred and least toxic portion of the plant. Leaves, flowering stem tips, and ripe seeds are the plant parts most preferred by cattle and are also the most toxic (Cronin and Nielsen 1979).

Ralphs and Olsen (1987) suggest that alkaloids are negatively correlated to palatability in some species and are considered to be bitter due to their basic nature. Larkspurs are a paradox because they are high in alkaloids yet are readily eaten under some conditions. Williams and Cronin (1963) observed that although total alkaloid content increased after herbicide treatment, utilization of sprayed plants increased. Sprayed plants became more palatable which resulted in additional cattle losses.

Cattle are very susceptible to larkspur poisoning while sheep are generally more tolerant. Sheep often prefer tall larkspurs and graze them with impunity. Sheep require four to six times as much larkspur as cattle, on a percentage basis, to produce poisoning (Kingsbury 1964, Olsen 1978).

Sheep may be used to control or reduce the availability of larkspurs. Yearling sheep were herded across larkspur patches in Montana prior to cattle grazing and strongly preferred tall larkspur to other forages. Larkspur poisoning of cattle was eliminated. It was also suggested that sheep grazing may have also reduced larkspur density and toxicity (Alexander and Taylor 1986). In Utah, Taylor (1985) reintroduced sheep to ranges currently grazed by cattle to take advantage of forbs and shrubs that have increased with cattle grazing. Specifically, sheep were used to reduce the availability of low larkspur on early summer range and to utilize tall larkspur on high elevation ranges in late summer. James et al. (1980) recommend that sheep be used to graze heavily infested larkspur areas before allowing cattle to graze.

Sheep have a distinctive pattern of grazing tall larkspur. They normally do not graze Barbey larkspur until August or September unless forced to by a shortage of other forage species. Sheep are also highly selective about the plant parts that they eat. They consistently consume only the leaf blade, leaving the petiole, stem, and flowers untouched (Bowns 1971). This delay in consuming larkspur and the selective nature of removing only leaf blades may help to account for the resistance of sheep to larkspur poisoning.

Ranchers recognize that sheep can, under some conditions, be poisoned by larkspurs. If sheep are observed to be feeding heavily on larkspur, the herders are careful to not excite the animals and to move them very slowly. Animals that become excited after heavy consumption of larkspur can die (James et al. 1980). Even though sheep can be poisoned by larkspurs, these plants are considered good forage for sheep. Many high elevation summer ranges can only be grazed by sheep because of large concentrations of larkspurs. Jones and Jones (1972) suggested that local livestock producers converted from cattle to sheep because larkspur poisoned their cattle but did not affect their sheep. Barbey larkspur persists on some of these ranges after severe long-term sheep grazing.

Sheep can be used to good advantage to reduce the availability of larkspur, but it is questionable if sheep can actually control or eliminate it. Laycock (1975) found that clipping significantly reduced plant growth and the concentration of total alkaloids in duncecap larkspur. Ellison (1954) concluded that Barbey larkspur had been reduced by grazing and was markedly less abundant on grazed ranges. However, clipping studies by Cronin (1971) suggest that tall larkspurs are resistant to grazing. He felt that there was little evidence to support the assumption that tall larkspurs can be controlled by grazing. Attempts to remove tall larkspurs by sheep would probably destroy the more valuable associated species. Cronin and Nielsen (1979) conceded that there is some evidence to indicate grazing may suppress the production of tall larkspur, but even abusive overgrazing does not eliminate it on established sites. They concluded that the management of grazing animals cannot be used to reduce the density of larkspur without detrimental effects on associated species.

LUPINE

There are approximately 100 species of lupine (Lupinus spp.) in the United States and Canada, and they are very difficult to identify (Kingsbury 1964). Not all species of lupine are known to be toxic, but they should be considered toxic until proven otherwise. Lupine is more toxic to cattle than sheep on a unit weight basis, yet literally thousands of sheep have died from eating it while few cattle deaths have been reported (James 1979).

Ingestion of a relatively large quantity of lupine over a brief period of time is generally required to cause poisoning (Kingsbury 1964). Alkaloid levels are moderate in early vegetative growth, increase to a peak prior to flowering, and decrease thereafter. Ripening seeds are very high in alkaloid concentration (Keeler et al. 1976). Losses occur when hungry sheep are given access to lupine, especially during the pod stage.

Lupine is highly nutritious and may provide an important and beneficial part of a sheep's diet. Kingsbury (1964) considers the majority of lupine species as acceptable or desirable forage under usual range conditions. The early vegetative foliage and immature seed pods are palatable to sheep (Ralphs and Olsen 1987). Sheep are especially fond of those lupine species growing under aspen (Kingsbury 1964).

Cattle are less likely to eat lupine under usual conditions (Stoddart et al. 1949). However, research has shown that congenital deformities occur in newborn calves if cows graze certain species of lupine containing the teratogen anagyrine during specific periods of gestation (Shupe et al. 1967; James et al. 1980).

Several species of lupine are found on mountain ranges in central and southern Utah. Some species are readily consumed during the summer while others are not readily eaten by sheep or cattle until after the plants freeze or dry out in the fall. These plants, however, may still be dangerous because alkaloids are not lost or detoxified on drying (Kingsbury 1964). Sheep have been poisoned when lupine has been cut and dried (James et al. 1980).

Stoddart et al. (1949) suggest that the solution to lupine poisoning is to change from sheep to cattle or to graze the plants when they are not in fruit. They also suggest that dense stands of lupine be avoided.

LOCOWEEDS

Species in the genera Astragalus and Oxytropis containing the alkaloid swainsonine (Molyneux and James 1982) comprise the locoweeds. Kingsbury (1964) reports as many as 300 species of Astragalus in North America, and Welsh (1978) lists more than 100 species in Utah alone. There is much confusion concerning the toxicity of locoweeds because of the difficulty of their identification. Some species are not toxic, and many are desirable forage species or soil binders (Kingsbury 1964).

Kingsbury (1964) considers locoweed the most widespread poisonous plant problem in the United States, and James (1979) suggests that locoweeds are the most treacherous of the poisonous plants. Signs of intoxication usually occur only after the plants have been grazed for a period of time. Locoweeds are not especially palatable to livestock but are consumed when they are green and other forage is dry (James and Johnson 1976, James 1983). However, livestock will readily consume, and are poisoned by, some species of locoweeds after they are dead and dry. James (1979) suggests that animals are more likely to eat locoweeds when other forage is scarce; therefore, poisoning may occur under conditions of drought and overgrazing. Animals often develop a craving for locoweeds once they start eating them.

Patterson (1982) considers it a fallacy to think that livestock will not eat locoweeds unless they are starved to it or that stock native to an area where loco is found will never graze it. The most apparent reason that livestock eat loco is their desire for something green and succulent. The growing season for locoweed exceeds that of most native grasses in the spring and fall.

Ralphs et al. (1985) studied the quantity and timing of white locoweed (O. sericea) consumption by cattle. Immature seed pods were the only plant part palatable to cattle and actively selected. Ralphs (1987) concluded that cattle will not select significant amounts of locoweed leaves, flowers, or mature pods when adequate forage is available.

Samuel (1981) considered prairie milkvetch (A. adsurgens) a preferred plant species, and preferential grazing was associated with high crude protein levels.

Several management strategies have been proposed which might utilize locoweeds as forage species and reduce the incidence of poisoning. James (1983) stated that any grazing system that encourages livestock to graze these plants creates a problem. Patterson (1982) in New Mexico and Everist (1981) in Australia reported that some ranchers will use very heavy stocking rates for short periods of time so that no animal will be able to eat enough loco to cause serious harm. An additional objective of high density stocking is to weaken the vigor of the locoweed plants so that competition will crowd them out. Ralphs et al. (1984) propose a grazing system that will reduce white locoweed poisoning by: (1) reducing grazing pressure to provide abundant grass and other palatable forage; (2) allowing cattle to spread out and not moving them to new pastures during the growing season; and (3) shortening the grazing season and removing the animals before intoxication becomes serious.

SUMMARY

Thousands of plants can be classified as poisonous, but only a relatively few cause serious problems. Some of the more toxic species can be good forages under some range conditions. In the Intermountain region, some of the more common plants that are valuable forage species are gambel oak, chokecherry, greasewood, tall larkspurs, and lupine.

Poisonous plants are common components of most rangeland vegetation. These plants have been and always will be present, and some are important in the climax communities. It is highly unlikely that poisonous plants can be eliminated or controlled through chemical, biological, or mechanical means. Treatment of poisoned range animals is rarely successful because of their remote locations, and death may occur soon after the appearance of symptoms. Since the control of poisonous plants or the treatment of affected animals is usually unsuccessful, it is apparent that poisonous plant problems can be alleviated only through management.

It is crucial that ranchers and land managers have the ability to identify the poisonous species. They must also know the season that these plants are most dangerous and the kind of animal most likely to be affected. Poisoning usually occurs when hungry animals are allowed to eat large quantities of toxic species over a brief period of time.

The fact that many range plants contain toxins does not preclude their value as forages. Many of these toxic species are considered good or excellent forage by livestock producers and are useful much of the time. However, under certain conditions, such as hunger, stage of growth, environmental conditions, etc., animals may consume excessive amounts of these plants too rapidly and become intoxicated. If we are aware of these conditions, we may be able to prevent the occurrence of intoxication.

LITERATURE CITED

Alexander, J.D., and J.E. Taylor. 1986. Sheep utilization as a control method on tall larkspur infested cattle range. (Abstract). 39th Annual Meeting Society for Range Manage, Kissimmee, FL.

Arnold, G.W., and J.L. Hill. 1972. Chemical factors affecting selection of food plants by ruminants, pp. 71-101. In: J.B. Harborne (ed), Phytochemical ecology. Academic Press, New York.

Bailey, E.M. 1978. Physiological responses of livestock to toxic plants. J. Range Manage. 31:343-347.

Bowns, J.E. 1971. Sheep behavior under unherded condition on mountain summer ranges. J. Range Manage. 24:105-109.

Bowns, J.E. 1985. Rehabilitation and management of Gambel oak (Quercus gambelii) dominated ranges in southwestern Utah. In: K.L. Johnson (ed), Proc. Third Utah Shrub Ecology Workshop. Utah State Univ., Logan.

Brotherson, J.D., L.A. Szyska, and W.E. Evenson. 1980. Poisonous plants of Utah. Great Basin Naturalist. 40:229-253.

Cronin, E.H. 1971. Tall larkspur: Some reasons for its continuing preeminence as a poisonous weed. J. Range Manage. 24:258-263.

Cronin, E.H., and D.B. Nielsen. 1972. Controlling tall larkspur on snowdrift areas in the subalpine zone. J. Range Manage. 25:213-216.

Cronin, E.H., and D.B. Nielsen. 1979. The ecology and control of rangeland larkspurs. Utah Agr. Exp. Sta. Bull. 499.

Cronin, E.H., and D.B. Nielsen. 1981. Larkspurs and livestock on the rangelands of western North America. Down to Earth 36:11-16.

Cronin, E.H., J.A. Young, and W.A. Laycock. 1978. The ecological niches of poisonous plants in range communities. J. Range Manage. 31:328-334.

Dwyer, D.D. 1978. Impact of poisonous plants on western U.S. grazing systems and livestock operations, pp. 13-21. In: R.F. Keeler, K.R. Van Kampen, and L.F. James (eds), Effects of poisonous plants on livestock. Academic Press, New York.

Ellison, L. 1954. Subalpine vegetation of the Wasatch Plateau, Utah. Ecol. Monog. 24:89-184.

Everist, S.L. 1981. Poisonous plants of Australia. Angus and Robertson Pub., Hong Kong.

Forsling, C.L., and E.V. Storm. 1929. The utilization of browse forage as summer range for cattle in southwestern Utah. U.S. Dept. Agr. Circ. 62.

Harper, K.T., F.J. Wagstaff, and L.M. Kunzler. 1985. Biology and management of Gambel oak vegetative type: A literature review. USDA For. Serv. Intermountain For. and Range Exp. Sta. Gen. Tech. Rep. INT-179. Ogden, UT.

James, L.F. 1970. Locomotor disturbances of cattle grazing Halogeton glomeratus. J. Amer. Vet. Med. Assoc. 156:1310-1312.

James, L.F. 1979. Management as a means of reducing sheep losses due to poisonous plants. Sheep Breeder and Sheepman, November:62-66.

James, L.F. 1983. Poisonous plants. Rangelands 5:169-170.

James, L.F., and A.E. Johnson. 1976. Some major plant toxicities of the western United States. J. Range Manage. 29:356-363.

James, L.F., R.F. Keeler, A.E. Johnson, M.C. Williams, E.H. Cronin, and J.D. Olsen. 1980. Plants poisonous to livestock in the western states. USDA Agr. Info. Bull. 415.

James, L.F., J.C. Street, and J.E. Butcher. 1967. In vitro degradation of oxalate and of cellulose by rumen ingesta from sheep fed Halogeton glomeratus. J. An. Sci. 26:1430-1444.

Jones, Y.F., and F.K. Jones. 1972. Lehi Willard Jones 1854-1947. Woodruff Printing Co., Salt Lake City, UT.

Keeler, R.F., E.H. Cronin, and J.L. Shupe. 1976. Lupin alkaloids from teratogenic and non teratogenic Lupins. IV. Concentration of total alkaloid, individual major alkaloids and the teratogen anagyrine as a function of plant part and stage of growth and their relationship to crooked calf disease. J. Toxic Environ. Health 1:899-908.

Kingsbury, J.M. 1964. Poisonous plants of the United States and Canada. Prentice-Hall Inc., Englewood Cliffs, NJ.

Laycock, W.A. 1975. Alkaloid content of duncecap larkspur after two years of clipping. J. Range Manage. 28:257-259.

Laycock, W.A. 1978. Coevolution of poisonous plants and large herbivores on rangelands. J. Range Manage. 31:335-342.

Marsh, D.C. 1909. The locoweed disease of the plains. USDA Farmers Bull. 112.

Molyneux, R.J., and L.F. James. 1982. Loco intoxication: indolizidine alkaloids of spotted locoweed (Astragalus lentiginosus). Science 216:190-191.

Nielsen, D.B. 1978. The economic impact of poisonous plants on the range livestock industry in the 17 western states. J. Range Manage. 31:325-328.

Olsen, J.D. 1978. Tall larkspur poisoning in cattle and sheep. J. Am. Vet. Med. Assoc. 173(6):762-765.

Patterson, P.E. 1982. Loco, La Yerba Mala. Rangelands 4:147-148.

Plummer, A.P., D.R. Christensen, and S.B. Monsen. 1968. Restoring big game range in Utah. Utah Div. of Fish and Game. Pub. 68-3.

Ralphs, M.H. 1987. Cattle grazing white locoweed: influence of grazing pressure and palatability associated with phenological growth stage. J. Range Manage. (In press).

Ralphs, M.H., L.F. James, D.B. Nielsen, and K.E. Panter. 1984. Management practices reduce cattle loss to locoweed on high mountain range. Rangelands 4:175-177.

Ralphs, M.H., L.F. James, and J.A. Pfister. 1986. Utilization of white locoweed (Oxytropis sericea Nutt.) by range cattle. J. Range Manage. 39:344-347.

Ralphs, M.H., and J.D. Olsen. 1987. Alkaloids and palatability of poisonous plants, pp. 70-83. USDA/FS Intmt. Res. Sta. Gen. Tech. Rep. Int-222.

Ruyle, G.B., R.L. Grumbles, M.J. Murphy, and R.C. Cline. 1986. Oak consumption by cattle in Arizona. Rangelands 8:124-126.

Samuel, M.J. 1981. Grazing of prairie milkvetch (Astragalus adsurgens Pall.) by cattle. (Abstract). 34th Annual Meeting of the Society for Range Manage., Tulsa, OK.

Shupe, J.L., W. Binns, L.F. James, and R.F. Keeler. 1967. Lupine a cause of crooked calf disease. J. Amer. Vet. Med. Assoc. 151:198-203.

390

Stoddart, L.A., A.H. Holmgren, and C.W. Cook. 1949. Important poisonous plants of Utah. Agr. Exp. Sta. Report 2. Utah State Agr. Coll., Logan.

Stoddart, L.A., A.D. Smith, and T.W. Box. 1975. Range management, 3rd ed. McGraw-Hill Book Co., NY.

Taylor, D.L. 1985. Introducing sheep for vegetation manipulation. (Abstract). 38th Annual Meeting Society for Range Manage. Salt Lake City, UT.

Welsh, S.L. 1978. Utah flora: Fabaceae (Leguminosae). Great Basin Nat. 38:225-367.

Williams, M.C., and E.H. Cronin. 1963. Effect of silvex and 2,4,5-T upon alkaloid content of tall larkspur. Weeds 11:317-319.

28

Management to Reduce Livestock Loss from Poisonous Plants

M.H. Ralphs and L.A. Sharp

INTRODUCTION

The range livestock era in the western U.S. began following the Civil War with the trail drives from Texas into the Plains States. The prairies were fully stocked by 1880 and the mountain and intermountain ranges by the early 1900s. Most ranges were overstocked until the mid 1930s, resulting in degradation of vegetation. The abundance of desirable forage species decreased while the less palatable brush, weedy, and poisonous species increased or invaded.

Much of the early range management literature blamed poisonous plant problems on poor range conditions (Marsh 1913, Marsh 1958, Stoddart and Smith 1955). Stoddart et al. (1949) stated that retrogression following misuse was the greatest single factor contributing to livestock poisoning. Poisonous plants increased and livestock were forced, through shortage of nonpoisonous forage, to eat more of the poisonous species. Examples of catastrophic losses in western U.S. are shown in Table 28.1.

The level of management on most rangelands has increased during the last 30 to 50 years with a marked improvement in range condition (Box and Malechek 1987). The incidence of widespread catastrophic losses to poisonous plants have also declined. Schuster (1978) stated that good range management is the surest and most economical means of reducing livestock loss to poisonous plants. Research has contributed to an understanding of poisonous species, their toxins, and signs of poisoning, which stockmen have utilized to manage their livestock to reduce losses to poisonous plants.

TABLE 28.1
Catastrophic livestock losses to poisonous plants.[1]

Plant	Year	Location	Livestock	Loss
Locoweed	1893	Kansas	Cattle	25,000
Deathcamas	1909	Wyoming	Sheep	500 of 1700
Deathcamas		Wyoming	Sheep	20,000 reported in 1 county
Larkspur	1913	Utah	Cattle	200 in 1 herd
Milkweed	1917	Colorado	Sheep	730 of 1,000
Greasewood	1920	Oregon	Sheep	1,000 of 1,700
Lupine	1942	Utah	Sheep	260 from 1 band
Lupine		Montana	Sheep	700 of 2,000
Lupine		Montana	Sheep	2,500
Halogeton	1945	Idaho	Sheep	1,620 in 1 band
Halogeton	1945	Idaho	Sheep	750 in 1 band
Halogeton	1945	Idaho	Sheep	250 in 1 band
Locoweed	1958	Uinta Basin	Sheep	estimate 6,000
Locoweed	1964	Uinta Basin	Sheep	1 band $45,000; 1 band $125,000
Locoweed	1964	W. Utah & Nevada	Cattle	heavy losses
Halogeton	1971	Utah	Sheep	1,200 in 1 band
Milkweed	1975	New Mexico	Cattle	200
Halogeton	1980	Utah & Wyo.	Sheep	500 in 1 band
Selenium poisoning	1981	Utah	Sheep	250 of 1,400
Senecio	1982	Oregon	Cattle	630
Deathcamas	1983	Idaho	Sheep	75 of 125
Deathcamas	1983	Idaho	Sheep	80
Deathcamas	1983	Idaho	Sheep	83
Deathcamas	1985	Idaho	Sheep	250 of 2,400
Oakbrush	1986	California	Cattle	1,700

[1] Information taken from Stoddart and Smith (1955), L.F. James (personal communications), and observation of authors.

Still, losses to poisonous plants continue to occur. Scattered incidences of catastrophic losses have occurred recently (Table 28.1). However, the aggregation of all the small losses incurred by individual ranchers across the West is of much greater magnitude. Nielsen (1987) estimates losses to livestock death and abortion in the 17 western states exceeds $230 million annually. Perhaps of greater magnitude are the insidious losses to birth defects, abortions, infertility, lowered resistance to infectious disease, and reduced production caused by sublethal doses of poisonous plants. With the present economic state of agriculture, losses to poisonous plants can no longer be accepted as a cost of grazing rangelands (Nielsen 1978). Improving efficiency is the key to survival.

PLANT-ANIMAL-ENVIRONMENTAL INTERACTIONS

Poisonous plants can no longer be considered simply a symptom of overgrazed ranges (Dwyer 1978). Some poisonous species are major components of climax plant communities. Many poisonous increaser species have declined with better management, but are still a component of plant communities and fluctuate with environmental conditions. Some poisonous introduced invader species have become naturalized. In fact, many of the important forage, pasture, and range species contain toxic compounds which are capable of poisoning livestock under some conditions (Hegarty 1981; McDonald 1981).

Toxic plant compounds have a great variety of pharmacological or inhibitory activity. Body systems can detoxify or excrete many of these harmful substances (Freeland and Janzen 1974), but poisoning occurs when the ingested toxin exceeds the body's ability to compensate. The interactive factors of the particular plant species, toxin level or combination of toxins within the plant, animal condition and propensity to consume the plant, environmental influences on both the plant and the animal, and management factors imposed on the entire system all determine whether or not an animal consumes enough of a particular species and is poisoned. These factors and interactions must be identified and understood in order to manage livestock on ranges or pastures containing toxic species if the risk of death or production losses is to be reduced.

Plant Factors

Plant species within genera and closely related genera contain similar kinds of toxins. However, the level of toxins may differ greatly among species, plant parts, and stages of growth. Species of larkspur (Delphinium) and monkshood (Aconitum) contain diterpenoid alkaloids that act as neuromuscular poisons (Benn and Jacyno 1983). Barbey larkspur (<u>Delphinium</u> <u>barbeyi</u>) is the most toxic of the tall larkspurs. Waxy leaf larkspur (<u>D</u>. <u>glaucescens</u>) is about 25% as toxic, and duncecap larkspur (<u>D</u>. <u>occidentale</u>) is only 10% as toxic (Olsen 1977). Toxicity within the same species of larkspur also varies with site (Olsen 1979).

Toxin level in larkspur species often varies between plant parts and stages of growth. Total alkaloids are highest in early vegetative growth of tall larkspur species. As plant parts differentiate, the reproductive raceme is highest in total alkaloids, followed by leaves, and then stems. Alkaloids in all parts decline as the plant matures except for a slight increase in the pods as seeds develop (Williams and Cronin 1966; Kreps 1969). Consumption of larkspur is also influenced by growth stage. The immature seed pods are relished by cattle (Cronin and Nielsen 1979), and the leaves are readily eaten after the bloom stage (Pfister and Ralphs 1988). Early management recommendations were to delay grazing larkspur-infested ranges until after flowering, when alkaloid level is low (Marsh and Clawson 1916).

Most species of <u>Lupinus</u> contain several toxic quinolizidine alkaloids that range from 0.6 to 2.4% of dry weight of the plant (Davis 1982). A number of species also contain the teratogenic alkaloid anagyrine (Keeler et al. 1976; Davis 1982). Total alkaloid concentration in tailcup lupine (<u>Lupinus</u> <u>caudatus</u>) is fairly low during early vegetative growth, increases to a peak prior to bud formation, and declines thereafter. If seeds are produced, toxin levels of pods increase (Keeler et al. 1976). The young vegetative growth of lupine and the pods are the most palatable parts.

Some plant genera, such as the <u>Astragalus</u>, contain several toxins. Miserotoxin (3-nitro-propanol or propionic acid) was found in 52% of 506 species sampled in North America, ranging in levels from 2 to 25 mg NO/g of plant material (Williams and Barneby 1977). Miserotoxin levels are low during early vegetative growth, increase until buds are formed, and then decline (Majak et al. 1976). Several species and varieties of <u>Astragalus</u> and its closely related genera, <u>Oxytropis</u>, contain the locoweed toxin swainsonine (Molyneux and James 1982). Swainsonine concentration in white locoweed (<u>Oxytropis</u> <u>sericea</u>) leaves is fairly constant throughout the growing season. Swainsonine levels are high in flowers, decline slightly in the immature pod, and increase as seeds mature (Ralphs et al. 1986). The succulent immature pod is the only locoweed plant part palatable to cattle and voluntarily selected on high mountain range (Ralphs 1987). Other Astragalus species accumulate selenium (James et al. 1981).

Four species of deathcamas (Zigadenus spp.) are toxic to both livestock and humans. The fleshy, tuberous root is most toxic, but foliar parts also contain the toxin. There is also a wide range of toxicity among species with lethal doses ranging from 4 to 71 g of plant material/kg of body weight (Marsh and Clawson 1922).

Precise species identification is important to determine the specific toxin, its level, and potential toxicity of a plant. Further research is necessary on important poisonous plants to determine variability of toxins among sites or stage of growth and to identify soil and other environmental factors that influence toxicity.

Animal Factors

Animal species differ in their grazing behavior and preference for particular plants, and in their susceptibility to specific poisonous plants. Native wildlife species appear more resistant than livestock to poisoning from native plants, presumably because of the evolutional process of selecting foods and developing preferences that would avoid harmful substances (Laycock 1978). Resistance to toxins could be from inactivation by rumen microflora, excretion, or enzymatic breakdown of toxins. Basson et al. (1984) concluded that large wild herbivores in South Africa were generally more resistant to poisonous plants than were domestic ruminants, but they were not immune. Elk and antelope have been poisoned by locoweed and waterhemlock on western U.S. rangelands (Wolf and Lance 1984, Jessup et al. 1986).

Differences in susceptibility or grazing behavior of different species or class of livestock can be utilized to properly manage poisonous plant infested ranges. Sheep are more resistant to larkspur poisoning than cattle, tolerating four times more plant material than cattle to reach similar stages of intoxication (Olsen 1978). Sheep and goats are more resistant to pyrrolizidine alkaloids [found in Senecio spp., Crotalaria spp., and hounds tongue (Cynoglossum officinale)] than are cattle and horses (Cheeke and Shull 1985). Horses are more susceptible to locoweed than cattle and sheep (Mathews 1932). Abortions caused by ponderosa pine (Pinus ponderosa) needles and broom snakeweed (Gutierrezia sarothrae) are common in cattle but have seldom been observed in sheep and goats (L.F. James, personal communication). Both plant species are toxic to the dam if fed in high concentrations. Cattle also appear to be more susceptible to oak poisoning than are sheep and goats.

Grazing behavior and management of sheep may increase the risk of poisoning by some plant species. The gregarious nature of sheep causes them to graze together and thus puts increased grazing pressure on the available forage which, in turn, may reduce selectivity. Furthermore, range sheep are kept together in large bands and moved about by a herder who controls when and where the band is to be moved to new forage (James 1980). These factors, along with their propensity to select forbs, may explain why sheep are poisoned more often by orange sneezeweed (Helenium hoopesii), bitterweed (Hymonoxys odorata), lupine, deathcamas, and halogeton (Halogeton glomeratus) than are cattle. Halogeton is actually more toxic to cattle than to sheep (James 1972), but because of their free-ranging behavior, cattle seldom eat enough to become intoxicated. Many of the large catastrophic sheep losses to halogeton were caused by trailing hungry sheep through dense patches of halogeton.

Condition of the grazing animal is often the most important factor influencing poisoning. Hungry animals in poor condition are most susceptible to poisoning (Everist 1981) because they graze less discriminately and are more likely to consume toxic doses of poisonous plants. The physiological condition of the animal may affect both intake and physiological reaction to the toxin. Lactating females require a higher intake level to meet their greater physiological requirements, thus they may graze longer and less discriminately than other animals. They may also excrete the toxin in the milk and poison the offspring from eating such plants as locoweed, white snakeroot (Eupatorium rugosum), and rayless goldenrod (Haplopappus heterophyllus) (Kingsbury 1964). Stocker cattle are less affected by high levels of selenium or teratogenic compounds that would cause low fertility or birth defects in breeding animals.

The unborn fetus is susceptible to teratogens from many plants that may not affect the dam. The damage is insidious and difficult to trace back to the causative plant. Keeler (1978) suggested several management strategies to avoid birth defects by staggering grazing so that periods of high teratogen levels in the plant did not coincide with susceptible periods of insult in the fetus.

New animals introduced to unfamiliar ranges are more susceptible to poisoning than native animals (Marsh 1916, Everist 1981). Prior grazing experience affects preferences and grazing efficiency (Arnold and Dudzinski 1978). Thus, stockmen should watch new animals closely for initial symptoms of poisoning and remove animals if they begin to consume poisonous plants.

Environment

Soil moisture and fertility influence toxin concentration in many plants. Gershenzon (1984) reviewed the literature on the effect of water and nutrient stress on toxicity. He concluded that water stress and nitrogen fertilization increased the concentration of several toxic secondary compounds such as cyanogenetic glycosides in sorghum (Sorghum bicolor), sudangrass (Sorghum sudanense), white clover (Trifolium repens), arrowgrass (Triglochin maritima), and serviceberry (Amelanchier spp); alkaloids in water hemlock, lupine, and Senecio species; and glucosinolates, phenolics, and terpenoids in other plants. He reasoned that slight water deficits retard growth. If toxin synthesis continues, or the same amount of toxin remains in a reduced amount of plant material, its concentration increases. Reduced growth rates also tend to increase levels of amino acids (due to reduced protein synthesis) that serve as precursors for many toxins. Fertilizers or high soil nitrogen increase the soluble nitrogen compounds in plants that form amino acids and other toxin precursors.

Weather changes may alter concentrations of toxins and cause normally innocuous forages to become highly toxic. Frost stress increases hydrocyanic acid contents of some forages [sorghum, sudangrass, johnson grass (Sorghum halepense), arrowgrass (Triglochin spp.), chokecherry (Prunus virginia), and mountain mahogany (Cercocarpus spp.)]. Continuous cloudy weather can increase accumulation of nitrates in plants capable of taking up toxic levels (Everist 1981). Acute poisoning from reed canarygrass (Phalaris arundinacea) is most common in foggy or cloudy weather and during early morning hours (Everist 1981). Decreasing barometric pressure prior to spring storms has been reported to reduce selective grazing and increase consumption of toxic amounts of horsebrush (Tetradymia spp.) by sheep, resulting in photosensitization on cold desert ranges (A. E. Johnson, personal communication).

Management Factors

An important factor compelling livestock to consume toxic amounts of poisonous plants is inadequate quantity and quality of desirable forage. Many of the early reports of poisoning were confused with starvation (Marsh 1909). Forage availability is under direct control of the manager, who determines season of use, range readiness, stocking rates, and length of grazing season and provides salt and mineral supplements and supplemental feed when necessary.

Many poisonous forbs and shrubs green up before grasses or other desirable forage in early spring [low larkspur, deathcamas, locoweed, oakbrush (Quercus spp.), broom snakeweed]. The early vegetative growth of these plants is usually most toxic. New green growth is appealing to livestock having been fed dry hay during the winter, and stock often consume toxic amounts of these plants. In years when these plants are abundant, stockmen should delay turnout until range readiness of desirable grasses.

Deep-rooted perennial forbs and shrubs (Senecio, sneezeweed, locoweed, lupine) remain green after grass matures in the fall. Reduced palatability or shortage of native grass may make these green poisonous forbs appealing. Livestock should be removed from these ranges when desirable forage is properly utilized.

On year-long ranges, forage availability becomes limiting during winter and early spring. Semidesert locoweed species, bitterweed, and broom snakeweed either germinate in fall or remain green over winter and start rapid growth early in spring. Dry dormant grass or shortage of feed makes this green growth appealing to livestock. Stockmen should supplement to satisfy protein requirements and keep animals in good condition.

Grazing systems may provide flexibility to manage around specific plants, or their rigidity may compound poisoning problems. Strict adherence to rest-rotation grazing has contributed to locoweed poisoning (James et al. 1969; Ralphs et al. 1984). The heavy-use pasture in the grazing sequence forced uniform utilization of all forage, including locoweed.

Short-duration grazing requires intensive management and may have the flexibility to reduce poisoning problems or to intensify them. If animals are rotated rapidly and "cream" the pasture, toxic plants are likely to be avoided. If there has been little or no regrowth on successive rotations, the lack of other desirable forage may force livestock to consume toxic plants.

The abundance of palatable poisonous plants that cause chronic poisoning can be reduced by concentrating livestock in a high-intensity, short-duration grazing system, to rapidly remove the annual growth before any one animal becomes intoxicated. This has been recommended for control of darling pea (Swainsona spp.) in Australia (Everist 1981), to remove the palatable seed pods of white locoweed (Ralphs 1987), and to remove several varieties of timber milkvetch (Astragalus miser) before poisoning occurs (James 1980).

The Merrill three-herd, four-pasture, deferred rotation grazing system (using cattle, sheep, and goats grazing in common) eliminated livestock loss to bitterweed, oakbrush, and sacahuista (Nolina texana) poisoning during a 20-year period, compared to high losses in single species, continuous grazing systems at heavy stocking rates (Merrill and Schuster 1978). Success of this common-use grazing system was attributed to reduced grazing pressure on the respective animals' preferred forage and improved range condition which increased the available forage and allowed greater diet selection.

Salt and mineral supplements are commonly recommended to reduce livestock loss to poisonous plants. They may act to reduce intake of toxic plants or to impart resistance to the toxins. A balanced salt or mineral supplement may also stimulate diet selectivity. However, few minerals have been shown to increase resistance to poisoning. On the other hand, several studies have shown that mineral supplements have no direct protective activity against tansy ragwort poisoning (Johnson 1982), lupine-induced birth defects (Keeler et al. 1977), or locoweed poisoning (James and Van Kampen 1974). The purported value of mineral supplements may be the result of better management exercised by stockmen concerned about reducing losses to poisonous plants (Cheeke and Shull 1985).

It is conceivable that alleviators may be discovered for specific poisons by identifying the mechanism of action and finding compounds that will mitigate the activity. Therapeutic treatments such as methylene blue have been discovered for nitrate poisoning (Kingsbury 1964); sodium thiosulfate and sodium nitrate for treatment of hydrocyanic acid poisoning; vitamin K for coumarin poisoning (McDonald 1981); and hydrated lime for preventing shin oak poisoning. Synthetic antioxidants that showed promise against pyrrolizidine alkaloids in laboratory animals were only marginally effective in cattle and horses (Cheeke and Shull 1985). Antidotes will probably have limited use on rangelands since animals are observed infrequently.

TABLE 28.2
General management recommendations to reduce livestock loss
to poisonous plants 1/

1. Ensure adequate palatable feed is available. Hunger lowers rejection thresholds and causes animals to eat plants they will normally avoid. Observe range readiness and proper stocking. Remove livestock when range is properly utilized.
2. Maintain range in good condition to prevent increase or invasion of noxious and poisonous weeds, and provide abundant good quality forage for grazing animals.
3. Use range when plants are least toxic. Toxin concentration is generally highest in early growth and in reproductive parts. Deferred grazing and range readiness allows desirable forage to begin growth and poisonous species to decline in toxicity.
4. Graze kind and class of animal that are least affected by particular poisonous plants.
5. Exercise caution handling animals when poisonous species are present. Avoid turning stressed and hungry animals in pastures infested with poisonous plants following trucking, trailing, or working.
6. Provide adequate water. Thirst reduces appetite. Once thirst is quenched, hungry animals regain their appetite and graze less selectively.
7. Introduce animals slowly to poisonous plants to allow rumen microorganisms to adjust to toxins. New, inexperienced animals should be watched closely.
8. Supplement livestock with salt and minerals to maintain animal health and increase food selectivity.
9. Consider vegetation manipulation if losses are severe enough to justify the cost of treatment. Selective herbicides are available for control of most poisonous plants. Type conversion is feasible if rangeland supporting dense poisonous plant infestations could be put to a better use Biological control may be helpful in keeping some exotic poisonous plants in check.

1/Literature citations:

Marsh 1913	Dwyer 1978
Marsh 1916	Schuster 1978
Huffman and Couch 1942	Krueger and Sharp 1978
Huffman et al. 1956	Merrill and Schuster 1978
James and Cronin 1974	James et al. 1980
Stoddart et al. 1975	James 1980

GENERAL MANAGEMENT RECOMMENDATIONS

Marsh (1916) made some general management recommendations to reduce poisonous plant losses. These original principles have been rephrased, expanded, and adapted to specific situations. A summary of these principles and a list of references are contained in Table 28.2. Recommendations for specific plants in local areas can be found in reference texts and State Experiment Station bulletins on poisonous plants.

There is no antidote for most plant poisonings. Reduction or elimination of livestock loss depends upon preventing livestock from consuming sufficient quantities of poisonous plants to harm them. Everist (1981) and Krueger and Sharp (1978) suggest a simple procedure for management to reduce livestock losses to poisonous plants:

1. Identify the poisonous plants on particular range or pastures.
2. Learn how they affect livestock.
3. Determine when they are most toxic and when livestock are most likely to eat them.
4. Identify conditions under which poisoning may occur.
5. Use a little common sense to devise a grazing strategy that will minimize the opportunity for poisoning.

LITERATURE CITED

Arnold, G.W., and M.L. Dudzinski. 1978. Ethology of free-ranging domestic animals. Elsevier, New York. pp. 97-124.

Basson, P.A., A.G. Norval, J.M. Hofmeyr, H. Ebedes, R. Anitra Schultz, T.S. Kellerman, and J.A. Minne. 1984. Antelopes and poisonous plants, pp. 695-701. In: 13th World Congress on Diseases of Cattle. Hoechst Pharm., Johannesburg.

Benn, M.H., and J.M. Jacyno. 1983. The toxicology and pharmacology of diterpenoid alkaloids, pp. 153-210. In: S.W. Pelletier (ed), Alkaloids--chemical and biological perspectives. John Wiley & Sons, New York.

Box, T.W., and J.C. Malechek. 1987. Grazing on the American rangelands. Proc. Western Section American Society of Animal Science 38:107-115.

Cheeke, P.R., and L.R. Shull. 1985. Natural toxicants in feeds and poisonous plants. AVI, Westport, CN. pp. 64-87.

Cronin, E. H., and D.B. Nielsen. 1979. The ecology and control of rangelands larkspurs. Utah Agr. Exp. Sta. Bull. 499.

Davis, A.M. 1982. The occurrence of anagyrine in a collection of western American lupines. J. Range Manage. 35:81-84.

Dwyer, D.D. 1978. Impact of poisonous plants on western U.S. grazing systems and livestock operations, pp. 13-22. In: R.F. Keeler, K.R. Van Kampen, and L.F. James (eds), Effects of poisonous plants on livestock. Academic Press, New York.

Everist, S.L. 1981. Poisonous plants of Australia. Angus & Robertson, Sydney. pp. 10-55.

Freeland, W.J., and D.H. Janzen. 1974. Strategies in herbivory by mammals: the role of plant secondary compounds. Amer. Natur. 108:269-289.

Gershenzon, J. 1984. Changes in the levels of plant secondary metabolites under water and nutrient stress, pp. 273-320. In: B.N. Timmermann, C. Steelink, and F.A. Loewus (eds), Recent advances in phytochemistry, phytochemical adaptations to stress. Plenum Press, New York.

Hegarty, M.P. 1981. Deleterious factors in forages affecting animal production, pp. 132-150. In: J. B. Hacker (ed), Nutritional limits to animal production from pastures. Common Agr. Bureaux, Slough Uk.

Huffman, W.T., and J.F. Couch. 1942. Plants that poison livestock, pp. 354-373. In: Keeping livestock healthy. USDA Yearbook of Agric.

Huffman, W.T., E.A. Moran, and W. Binns. 1956. Poisonous plants, pp. 118-130. In: Animal Diseases. USDA Yearbook of Agric.

James, L.F. 1972. Oxalate toxicosis. Clin. Toxico. 5:231-243.

James, L.F. 1980. Plant poisoning in livestock. Mod. Vet. Practice. Nov. 1980.

James, L.F., and E.H. Cronin. 1974. Management practices to minimize death losses of sheep grazing Halogeton-infested range. 27:424-426.

James, L.F., W.J. Hartley, and K.R. Van Kampen. 1981. Syndromes of Astragalus poisoning in livestock. J. Amer. Vet. Med. Assoc. 178:146-150.

James, L.F., R.F. Keeler, A.E. Johnson, M.C. Williams, E.H. Cronin, and J.D. Olsen. 1980. Plants poisonous to livestock in the western states. USDA/SEA Agr. Info. Bull. 415.

James, L.F., and K.R. Van Kampen. 1974. Effect of protein and mineral supplementation on potential locoweed (Astragalus spp.) poisoning in sheep. 164:1042-1043.

James, L.F., K.R. Van Kampen, and G.B. Staker. 1969. Locoweed (Astragalus lentiginosus) poisoning in cattle and horses. J. Amer. Vet. Med. Assoc. 155:525-530.

Jessup, D.A., H.J. Boermans, and N.D. Kock. 1986. Toxicosis in tule elk caused by ingestion of poison hemlock. J. Amer. Vet. Med. Assoc. 189:1173-1175.

Johnson, A.E. 1982. Failure of mineral-vitamin supplements to prevent tansy ragwort (Senecio jacobaea) toxicosis in cattle. Amer. J. Vet. Res. 43:718-723.

Keeler, R.F. 1978. Reducing incidence of plant-caused congenital deformities in livestock by grazing management. J. Range Mange. 31:355-360.

Keeler, R.F., E.H. Cronin, and J. L. Shupe. 1976. Lupin alkaloids from teratogenic and nonteratogenic lupins. IV. Concentration of total alkaloids, individual major alkaloids, plant part and stage of growth and their relationship to crooked calf disease. J. Toxicol. Environ. Health 1:899-908.

Keeler, R.F., L.F. James, J.L. Shupe, and K.R. Van Kampen. 1977. Lupine-induced crooked calf disease and a management method to reduce incidence. J. Range Manage. 30:97-102.

Kingsbury, J.M. 1964. Poisonous plants of the United States and Canada. Prentice-Hall, Englewood Cliffs, NJ.

Kreps, L. B. 1969. The alkaloids of Delphinium occidentale S. Wats. Ph.D. Diss. Utah State Univ., Logan, UT.

Krueger, W.C., and L.A. Sharp. 1978. Management approaches to reduce livestock loss from poisonous plants on rangeland. J. Range Manage. 31:347-350.

Laycock, W.A. 1978. Coevolution of poisonous plants and large herbivores on rangelands. J. Range Manage. 31:335-342.

McDonald, J.W. 1981. Detrimental substances in plants consumed by grazing ruminants, pp. 349-378. In: F.H.W. Morley (ed), World animal science, grazing animals. Elsevier, New York.

Majak, W., R.J. Williams, A.L. Van Ryswyk, and B.M. Brooke. 1976. The effect of rainfall on Columbia Milkvetch toxicity. J. Range Manage. 29:281-283.

Marsh, C. D. 1909. The locoweed disease of the plains. USDA Bur. Anim. Ind. Bull. 112.

Marsh, C.D. 1913. Stock poisoning due to scarcity of food. USDA Farmers Bull. 536.

Marsh, C.D. 1916. Prevention of losses of livestock from plant poisoning. USDA Farmers Bull. 720.

Marsh, C.D., and A.B. Clawson. 1916. Larkspur poisoning of livestock. USDA Farmers Bull. 365.

Marsh, C.D., and A.B. Clawson. 1922. The stock-poisoning death camas. USDA Farmers Bull. 1273.

Marsh, H.D. 1958. Newsoms Sheep Diseases, 2nd ed. Williams and Wilkens, Baltimore.

Mathews, F.P. 1932. Locoism in domestic animals. Texas Agr. Exp. Sta. Bull. 456.

Merrill, L.B., and J.L. Schuster. 1978. Grazing management practices affect livestock losses from poisonous plants. 31:351-354.

Molyneux, R.J., and L.F. James. 1982. Loco intoxication: indolizidine alkaloids of spotted locoweed (Astragalus lentiginosus). Science 216:190-196.

Nielsen, D.B. 1978. The economic impact of poisonous plants on the range livestock industry in the 17 western states. J. Range Manage 31:325-328.

Nielsen, D.B. 1987. Economic impact of poisonous plants on livestock. In: L.F. James, M.H. Ralphs, and D.B. Nielsen (eds), The ecology and economic impact of poisonous plants on livestock production. Westview Press, Boulder, CO.

Olsen, J.D. 1977. Toxicity of extract from three larkspur species (Delphinium barbeyi, D. glaucescens, D. occidentale) measured by rat bioassay. J. Range Manage. 30:237-238

Olsen, J.D. 1978. Tall larkspur poisoning in cattle and sheep. J. Amer. Vet. Med. Assoc. 173:762-765.

Olsen, J.D. 1979. Toxicity of larkspur (Delphinium occidentale) from two growth sites. 60th Annual Meeting of the Conference of Research Workers in Animal Disease. Chicago, IL. Nov 1979. Abstract.

Pfister, J.A., and M.H. Ralphs. 1988. Cattle grazing tall larkspur (Delphinium barbeyi) on Utah mountain rangeland. J. Range Manage. (accepted)

Ralphs, M.H. 1987. Cattle grazing white locoweed: influence of grazing pressure and palatability associated with phenological growth stage. J. Range Manage. 40:330-332.

Ralphs, M.H., L.F. James, D.B. Nielsen, and K.E. Panter. 1984. Management practices reduce cattle loss to locoweed on high mountain range. Rangelands 6:175-177.

Ralphs, M.H., L.F. James, and J.A. Pfister. 1986. Utilization of white locoweed (Oxytropis sericea Nut.) by range cattle. J. Range Manage. 39:344-347.

Schuster, J. L. 1978. Poisonous plant management problems and control measures on U.S. rangelands, pp. 23-34. In: R.F. Keeler, K.R. Van Kampen, and L.F. James (eds), Effects of poisonous plants on livestock. Academic Press, New York.

Stoddart, L.A., A.H. Holmgren, and C.W. Cook. 1949. Important poisonous plants of Utah. Utah Agr. Exp. Sta. Special Report No. 2.

Stoddart, L.A., and A.D. Smith. 1955. Range management, 2nd Ed. McGraw-Hill, New York. pp. 234-257.

Stoddart, L.A., A.D. Smith, and T.W. Box. 1975. Range management, 3rd Ed. McGraw-Hill, New York. pp. 342-355.

Williams, M.C., and R.C. Barneby. 1977. The occurrence of nitro-toxins in North America Astragalus (Fabaceae). Brittonia 29:310-326.

Williams, M.C., and E.H. Cronin. 1966. Five poisonous range weeds--when and why they are dangerous. J. Range Manage. 19:274-279.

Wolf, G.J., and W.R. Lance. 1984. Locoweed poisoning in a New Mexico elkherd. J. Range Manage. 37:59-63.

29

Toward Understanding the Behavioral Responses of Livestock to Poisonous Plants

F.D. Provenza, D.F. Balph, J.D. Olsen, D.D. Dwyer, M.H. Ralphs, and J.A. Pfister

INTRODUCTION

It is common knowledge among livestock producers that animals on rangelands are not equally vulnerable to poisonous plants, and that not all species of poisonous plants pose an equal risk. Some noxious species are sought and repeatedly eaten despite the consequences, while others are sampled and largely avoided (Cheeke and Shull 1985, Ralphs and Olsen 1987). Moreover, livestock foraging in familiar environments are apparently less likely to be poisoned than animals foraging in unfamiliar settings (Kingsbury 1964; Hughes et al. 1971; Dwyer 1978; Krueger and Sharp 1978; Behnke 1980; but see Ralphs et al. 1987). In contrast, native ungulates seldom die from eating poisonous plants (Laycock 1978; but see Fowler 1983, Wolfe and Lance 1984). Currently, there is little knowledge of the causal mechanisms behind these inequalities. Our paper seeks to fill this gap in part by exploring some fundamental processes in the relationship between livestock and poisonous plants. The specific objectives are:

1. To present an overview of the nature of phytochemical defenses and the behavioral responses of animals to noxious foods.
2. To develop working hypotheses on the mechanisms that underlie an animal's decision whether or not to eat a poisonous plant.
3. To explore implications the working hypotheses have for future research and the management of livestock on rangelands that contain poisonous plants.

407

Our assumption in pursuing these objectives is that certain phytochemicals have evolved that allow plants to counter potentially harmful biotic and abiotic interactions. This has caused animals to develop mechanisms to cope with these defenses. An understanding of how animals might cope with plant defenses is important in developing management strategies to reduce livestock losses to poisonous plants.

OVERVIEW OF PHYTOCHEMICAL DEFENSES AND HERBIVORE RESPONSES

Phytochemical Defenses

Plants often produce noxious or poisonous compounds that can neither be classified as essential for plant metabolism nor as waste products of plant metabolism (Rosenthal and Janzen 1979, Cheeke and Shull 1985). While it is difficult to prove why plants produce such compounds, a reasonable hypothesis is that they evolved as defense mechanisms. Phytochemicals are apparently of adaptive significance for plants in plant-abiotic, plant-plant, plant-pathogen, and plant-herbivore interactions (Rhoades 1979, Harborne 1982). Plant pathogens and invertebrate herbivores, rather than vertebrate herbivores, may have provided much of the selective pressure for the evolution of defensive metabolites (Kingsbury 1978, Swain 1979, Bell 1985). Presumably, if phytochemicals evolved strictly as plant defenses against herbivores, there would be selective pressure for the advertisement of their presence, for example through volatile compounds, to the herbivore to prevent loss of plant tissue. As a result of the many and varied selective pressures on the evolution of phytochemicals, the degree to which plant chemical defenses are matched to a particular herbivore will undoubtedly vary greatly (Kingsbury 1978).

Regardless of their adaptive significance for the plant, many phytochemicals are harmful when ingested by vertebrate herbivores. Hence, it is not surprising to find growing evidence that phytochemicals have a strong impact on the foraging behavior of vertebrate herbivores (Arnold and Hill 1972; Freeland and Janzen 1974; Rosenthal and Janzen 1979; Bryant and Kuropat 1980; Bryant et al. 1983, 1985, 1987; Provenza and Malechek 1983, 1984; Ralphs and Olsen 1987). The diet selection of these animals seems strongly based on avoiding plant parts high in defensive phytochemicals, even though those plant parts are often highest in nutrients and energy. Avoidance of phytochemicals by herbivores is apparently based on a response to specific compounds, not to general fractions such as terpenoids, alkaloids, resins, and tannins (Reichardt et al. 1984, 1987). Phytochemicals are numerous and diverse (Rosenthal and Janzen 1979; Reichardt et al. 1984, 1987; Cheeke and Shull 1985). While some phytochemicals (attractants and gustatory stimulants) stimulate ingestion, others (deterrents and repellents) limit or prevent ingestion, and still others (toxicants) cause malaise or death (Bate-Smith 1972, Rohan 1972, Dethier 1980). Moreover, even though some toxic compounds also act as deterrents and repellents, there are many that do not, just as there are many deterrents and repellents that are not toxic (Bate-Smith 1972, Laycock 1978, Dethier 1980). This complexity likely plays a role in determining strategies herbivores use to select a diet from among the various plant species and parts available.

Responses of Herbivores to Phytochemical Defenses

Animals might associate sensory information about harmful phytochemicals with appropriate behavior through (1) instinct (see Alcock 1984 p. 89), (2) learning as an individual, (3) learning as part of a social group, or (4) combinations of the above. The modality favored by natural selection should depend to a large extent upon temporal and spatial variability in the distribution of harmful phytochemicals. Individual learning should be favored in highly variable environments, social learning in moderately variable environments, and instinct in stable environments (Boyd and Richerson 1985, Provenza and Balph 1987). Here we outline the key characteristics of instinct, individual learning, and social learning as they relate to how animals might respond to harmful foods.

Instinct. Instinct may be favored over learning as a means herbivores use to detect and avoid harmful phytochemicals that are temporally and spatially stable,because instinctive responses would reduce the risk of error for the individual. At least two mechanisms might allow herbivores to avoid harmful plants instinctively. First, very strong stimuli of any type usually cause instinctive avoidance. Hence, plants that advertise their toxicity via strong chemical stimuli may be avoided more than plants that do not. Second, certain olfactory and gustatory receptors may have evolved to enable animals to recognize specific phytochemicals. Information from these receptors would cause the animal to avoid on first encounter any plant containing the phytochemical.

Individual learning. Conversely, natural selection should favor individual learning in environments where phytochemicals are highly variable temporally and spatially, because the error rate for individual learning will be less than for instinct or social learning. Individual animals may discover the adverse effects of ingesting harmful foods through sampling potential foods. Their aversion to harmful foods would then result from experience with the sensory, nutritional, and physiological consequences of such trial-and-error learning.

Herbivores are capable of detoxifying and eliminating many toxic plant compounds, and trial-and-error learning may even play a role here. Limitations of detoxification systems may force herbivores to consume a variety of foods to avoid overingesting toxic compounds and to ingest small amounts of novel foods.

Social learning. Finally, social learning should be favored over individual learning and instinct when temporal and spatial variability in harmful phytochemicals is moderate, because it provides for a measure of flexibility with less risk than trial-and-error learning. The pressure for this type of learning should be especially heavy for species of animals that forage in mixed-generation groups. Such cultural transmission of foraging information should be most effective when the presence of harmful phytochemicals is moderately predictable from year to year.

DEVELOPING WORKING HYPOTHESES FOR LIVESTOCK

It is probable that instinct, individual learning, and social learning are all important mechanisms used by herbivores to avoid harmful plants. In this section, we discuss empirical evidence for the existence of these mechanisms in diet selection by domestic livestock and develop working hypotheses to account for cases where livestock are poisoned from eating plants.

Instinct

The role of instinct in the recognition and avoidance of poisonous plants is problematical. Domestic livestock did not evolve in many of the areas they now occupy, and they may have been removed to some unknown degree from the selective pressure of grazing on native range. However, the two instinctive mechanisms of avoidance mentioned earlier may still operate.

First, a generalized avoidance response could be caused simply by a very strong odor or taste associated with the phytochemical. Studies clearly indicate that there is a positive correlation between plants with high concentrations of any phytochemical and avoidance by livestock and other mammals (Arnold and Hill 1972; Bate-Smith 1972; Rohan 1972; Church 1979; Robinson, 1979; Dethier 1980; Provenza and Malechek 1984; Reichardt et al. 1984, 1987; Ralphs and Olsen 1987). In addition, many naturally occurring toxins are bitter to humans, and the avoidance by livestock of plants containing bitter compounds such as alkaloids and cardiac glycosides may be instinctive (Garcia and Hankins 1975, Chapman and Blaney 1979, Dethier 1980, Brower and Fink 1985, Ralphs and Olsen 1987). This suggests that instinctive avoidance is more likely to be associated with plants that advertise their toxicity than those that do not. Moreover, if plant avoidance by livestock is based on strong stimuli, livestock are likely to be poisoned by plant toxins that are lethal in low concentrations and thus do not serve as strong stimuli.

Second, livestock could encounter and instinctively recognize in a new environment some of the same phytochemicals with which they coevolved. However, the properties of phytochemicals are dependent on their detailed structures, which are apparently quite variable among plant species and parts (Reichardt et al. 1984, 1987). Furthermore, general classes of phytochemicals, for example terpenoids, alkaloids, tannins, and resins, apparently do not contain uniformly deterrent or toxic compounds (Reichardt et al. 1984, 1987; Cheeke and Shull 1985). Hence, the likelihood may not be great that livestock coevolved with many of the harmful phytochemicals they now encounter on rangelands.

Individual Learning

It is apparent that individual animals can learn to avoid foods based on adverse gastrointestinal consequences (Braveman and Bronstein 1985, Provenza and Balph 1987). In the case of food aversion learning, the taste or odor of a food is associated with adverse gastrointestinal consequences through (1) food neophobia, (2) cue-consequence specificity, and (3) long-delay learning.

Food Neophobia. Wild rats are neophobic (Rozin 1976). On encountering a novel food or a familiar food in a novel location, rats sample the food cautiously. They gradually increase ingestion of the novel food, provided gastro-intestinal consequences are not adverse. If ingestion is followed by gastrointestinal distress, rats avoid the food in the future. In addition, rats eat only one novel food during a meal, and they associate adverse gastrointestinal consequence with the novel food, even if a familiar food contains the poison.

Limited study suggests that livestock are also neophobic with regard to food (Chapple and Lynch 1986; Thorhallsdottir et al. 1987a; Burritt and Provenza 1987). They ingest small quantities of novel foods initially and gradually increase ingestion, provided the gastrointestinal consequences are not adverse. They can learn to avoid palatable foods that contain lithium chloride (LiCl), a nonlethal gastrointestinal poison, after a single trial (Zahorik and Houpt 1977, 1981; Olsen and Ralphs 1986; Thorhallsdottir et al. 1987a; Burritt and Provenza 1987), and can be completely averted to a particular food when given intraruminal infusions of LiCl or larkspur extract (Zahorik and Houpt 1977, 1981; Olsen and Ralphs 1986). However, when LiCl is mixed with a palatable food, so that the dosage of poison depends on the amount of food ingested, livestock always eat some food (Thorhallsdottir et al. 1987a; Burritt and Provenza 1987), and appear to limit their intake just enough to avoid illness (Olsen and Ralphs 1986, Burritt and Provenza 1987). Sheep offered a familiar and a novel food associate gastrointestinal distress with, and subsequently avoid, the novel food, even if the familiar food contains the poison (Burritt and Provenza 1987). Moreover, sheep that receive LiCl after eating a meal of several familiar foods and a novel food avoid the novel food in future meals (E.A. Burritt and F.D. Provenza, unpublished data). However, when offered two novel foods, one safe and one harmful, sheep are unable to detect and avoid the harmful food (E.A. Burritt and F.D. Provenza, unpublished data).

While studies in pens suggest that neophobia may be an important strategy livestock use to detect and avoid harmful foods, the ability of livestock to cope with toxic foods on rangelands in this manner is unknown. To detect and avoid harmful foods requires cautious sampling of novel items, and eating one novel item in a foraging bout. We do not know if this strategy is used by livestock foraging on rangelands that contain a diverse array of plant species.

In addition, plant toxicity often changes with phenology (Cheeke and Shull 1985). An animal may sample a plant part that contains low concentrations of a toxic metabolite, experience no adverse effect, and subsequently regard the plant part as familiar and safe. We do not know the degree to which an increased concentration of a toxic metabolite causes livestock to perceive a familiar food as novel, thus preventing poisoning. Further, some plants contain toxins that have a cumulative effect (Cheeke and Shull 1985), and thus the animal may come to view the food as familiar and safe, even though it is cumulatively toxic. It should be remembered that it is probably particular chemical stimuli that the animals are attentive to rather than plant parts (Arnold and Hill 1972; Bate-Smith 1972; Chapman and Blaney 1979; Reichardt et al. 1984, 1987).

As an alternative strategy to neophobia, livestock may eat small quantities of a diverse array of plant species, including novel and poisonous ones. Thus, livestock might avoid overingesting any one harmful food and have the opportunity for repeated nonlethal exposure to many toxic phytochemicals. In addition, limited exposure to toxicants is often necessary to activate microbial and enzymatic detoxification systems (Hungate 1966; Freeland and Janzen 1974; Brattsten 1979; Fowler 1983; Mehansho et al. 1983; Cheeke and Shull 1985). The ability of livestock to avoid overingesting harmful phytochemicals based on this strategy will depend on the abundance of toxic relative to nontoxic plants. This strategy will be successful provided nontoxic species are relatively abundant. Should the ratio of toxic to nontoxic plant species increase, livestock may ingest relatively more harmful phytochemicals than they can detoxify and as a result be poisoned.

Cue-consequence specificity. The specific ability to form food aversions based on gastrointestinal consequences is related to cue-consequence specificity. Cue-consequence specificity is the ability of animals to differentially associate internal and external stimuli with different sensory systems within the body (Garcia et al. 1985). For example, to protect the gut from harm, mammals selectively associate gastrointestinal illness with the taste and odor of food; to protect the skin from harm, animals selectively associate exteroceptive stimuli such as noise, shock, or laceration with adverse external and internal consequences.

We believe the best hypothesis to account for causal mechanisms underlying cue-consequence specificity is based on the occurrence of neurological convergence (Garcia et al. 1985; for other hypotheses see Domjan 1985). Neurons for different defense systems, such as the gut or skin defense systems, converge at different locations within the central nervous system. Thus, neurons for gustatory, olfactory, and visceral pathways converge at common neural pathways that are distinct from other pathways in the brain.

Many poisonous plants have adverse effects on systems other than the gastrointestinal tract (Cheeke and Shull 1985). Plants that cause illnesses such as muscular paralysis, respiratory failure, or impairment of the nervous system, but do not adversely affect the gastrointestinal tract, may not be avoided, because livestock may be unable to associate the illness with the taste and odor of the plant. For example, drugs such as gallamine, which produces muscular paralysis, and naloxone, which blocks the action of endogenous opiates on pain, produce strong place avoidances but only weak taste aversions in rats (Lett 1985). Conversely, LiCl produces gastrointestinal distress and causes strong taste aversions, but only weak place avoidance in rats (Lett 1985).

Often, toxic plant metabolites also act on more than one system simultaneously (Cheeke and Shull 1985). Amphetamine produces a taste aversion and a place preference, probably by producing an emetic effect on the ventral gut system and a euphoric effect on the dorsal coping system (Garcia et al. 1985). The action of a particular plant metabolite undoubtedly depends not only on the locus of action, but also on the dose. Morphine has an emetic effect, an analgesic effect, a euphoric effect, and a stupefacient effect, all of which are dose- dependent (Garcia et al. 1985). We do not know if livestock eat or avoid plant species that contain metabolites such as amphetamine, apomorphine, and morphine that produce positively rewarding effects in addition to gastrointestinal illness (Wise et al. 1976; Reicher and Holman 1977; White et al. 1977).

Long-delay learning. Another unique characteristic of food aversion learning is the ability of individual animals to relate taste and odor of food with gastrointestinal consequences given a delay of up to 12 hours between ingestion and consequences (Rozin, 1976). This is different from other stimulus-response systems in which learning is most effective when the delay between stimulus and consequence is minimal. Long-delay learning is aptly suited to conditions encountered by the gustatory-olfactory-visceral system, because the effect of food on the gastro-intestinal tract often is not apparent for some time following ingestion.

Preliminary findings suggest that ruminant livestock can learn with a delay of at least four hours between ingestion and consequences (E.A. Burritt and F.D. Provenza, unpublished data), but that nonruminants such as horses cannot (Zahorik and Houpt 1981). Rumination may serve to reinstate food cues, which allow ruminants to associate foods and consequences better than nonruminants given a delay between food ingestion and consequences. Given the close association between cue-consequence specificity, long-delay learning, and gastrointestinal consequences in other species (Garcia et al. 1985), it is not surprising that ruminant livestock have this ability.

Social Learning

The degree to which different vertebrate species learn to eat desirable foods or avoid harmful foods through social interactions is variable. Young rats learn which foods to eat, but not which foods to avoid, through social interactions (Galef 1985). They learn which foods to avoid through cautious sampling of novel items. Conversely, birds rely heavily on visual and gustatory cues while foraging (Brower and Fink 1985) and can learn food aversions merely by watching other birds ingest a food and get sick (Mason and Reidinger 1981, 1982).

Social learning is apparently important in the foraging of young livestock (Lobato et al. 1980; Lynch et al. 1983; Green et al. 1984; Chapple and Lynch 1986; Thorhallsdottir et al. 1987b). Research suggests that lambs exposed to foods early in life with their mother consume about twice as much of those foods after weaning as do lambs exposed with an adult female not their mother; in turn, lambs exposed with an adult female consume about twice as much as lambs exposed alone (Thorhallsdottir et al. 1987b). Livestock are apparently unable to learn which foods to eat by observing different social models eat the food (Baker and Crawford 1986; Thorhallsdottir et al. 1987b). Rather, they must ingest the food to learn about its consequences.

Whether or not livestock can learn which foods to avoid through social interactions is not well established. Young livestock may learn which foods to eat from social models

and avoid new foods as a result of neophobia (Galef 1971;
Galef and Clark 1971a,b, 1972; Thorhallsdottir et al.
1987b). Limited data suggest that a lamb may be able to
learn to avoid a food by observing its mother ingest the
food and get sick (Thorhallsdottir et al. 1987b). However,
when lambs are exposed with the mother to a food which the
mother has been averted, the lambs consume limited
quantities of the food (A.G. Thorhallsdottir and F.D.
Provenza, unpublished data).

If livestock learn which foods to eat through social
interactions, they must relearn which plants to eat or avoid
through trial and error when moved from familiar to
unfamiliar environments. Such changes would likely result
in increased losses of livestock to poisonous plants. Those
individuals best able to learn which foods to eat or avoid
through trial and error would be favored. In addition,
young livestock reared on ranges in poor condition may learn
to consume poisonous plants by necessity (Merrill and
Schuster 1978), and continue consumption when placed on
ranges in better condition. Livestock eat considerable
quantities of unpalatable plants as adults if they have been
exposed to them when young (Arnold and Maller 1977, Narjisse
1981).

Species Differences

Different species of domestic and wild herbivores
undoubtedly vary in their abilities to cope with the
chemical defenses of plants. Herbivores that have evolved
with species of plants that contain particular toxic
metabolites often have detoxification mechanisms that allow
them to safely ingest such plants (Arnold and Dudzinski
1978, Basson et al. 1984).

There also may be differences in the abilities of
browsers, mixed feeders, and grazers to detoxify harmful
metabolites (Basson et al. 1984; Robbins et al. 1987).
Defenses are often associated with woody plants and,
therefore, browsers may have better developed systems for
tolerating potentially harmful phytochemicals than grazers.
For example, the larger size and greater production of
proline-rich proteins by parotid glands of browsers compared
to mixed-feeders and grazers may be adaptations that enable
browsers to live on diets high in tannins (Robbins et al.
1987).

Generally, detailed studies comparing responses of
different species of herbivores to different toxic

metabolites are lacking. Similarly, we lack information on which mechanisms enable different species of herbivores to detoxify different plant metabolites. Such information may be useful in preventing livestock losses to poisonous plants (Jones 1985).

IMPLICATIONS FOR RESEARCH AND MANAGEMENT

Poisonous plants on rangelands present a multifaceted problem for livestock producers. It is unlikely that a single management strategy will satisfactorily solve the problem. It may always be necessary to remove livestock from certain ranges during critical periods to prevent losses. We believe, however, that this overview of the relationship between herbivores and toxic plants suggests it might be possible to develop practical programs to reduce livestock losses to poisonous plants.

Among vertebrates, the rat apparently has the most sophisticated system for coping with harmful foods. However, the responses of rats to poisons are also the most studied and best understood. Limited research suggests that livestock have at least the rudiments of the same basic system. They are wary of novel foods, can learn to avoid foods that cause gastrointestinal distress, and can remember what they have learned. What we lack are many of the details necessary to develop testable management designs.

For example, it is important to know how livestock perceive the consequences of eating a particular species of poisonous plant. Changing the concentration of a phytochemical may result in consequences that vary from reward to punishment. If the toxin does not cause gastrointestinal distress, or if the gastrointestinal consequences are overridden by other rewarding effects, the odor or taste of the plant may not act as an aversive agent.

In such cases, it may be possible to train livestock to avoid eating poisonous plants (Provenza and Balph 1987). Livestock can be taught to avoid palatable foods through aversive conditioning (Zahorik and Houpt 1977, 1981; Olsen and Ralphs 1986; Thorhallsdottir et al. 1987a; Burritt and Provenza 1987), and preliminary indications are that cattle aversively conditioned to avoid fresh larkspur in pens will also avoid larkspur for at least one year in the field (M.A. Lane, unpublished data). However, livestock may have a tendency to sample small quantities of all foods, even those that are harmful, and to increase consumption when ingestion is no longer paired with gastrointestinal illness

(Thorhallsdottir et al. 1987a; Burritt and Provenza 1987). If this is the case, aversive conditioning may not be effective. Livestock that sample a palatable plant such as larkspur and do not suffer adverse gastrointestinal consequences may steadily increase ingestion.

Research on diet training must also consider how social learning affects feeding behavior of livestock (Provenza and Balph 1987). From a positive standpoint, females that have been aversively conditioned may teach their young which foods to eat and avoid, thus preventing poisoning. Conversely, livestock that have been aversively conditioned may eat poisonous plants when foraging with livestock that have not been aversively conditioned.

In addition, it is essential that we learn more about the consequences of moving livestock from familiar to unfamiliar environments. It is often stated that livestock introduced into a new environment are more susceptible to poisonous plants than those reared in the environment. However, little is known about the strategies (i.e., instinct, trial-and-error learning, social learning) that livestock use to cope with poisonous plants, or how they might tolerate ingesting poisonous plants through various detoxification systems.

Finally, it might also be instructive to learn more about how wild ungulates cope with toxins. We may find that they rely on as yet unknown avoidance or detoxification mechanisms. For example, the fact that there are poisonous plants that do not adversely affect the gastrointestinal tract should place strong selective pressure for the development of sensing systems within the body to alert the animal to such harmful food. While it may not be clear at present how the discovery of such systems in wild herbivores may help livestock, one thing is clear: we will never know what options managers have in reducing livestock losses to poisonous plants until we understand the factors that enter into the decision an herbivore makes to eat or not eat a poisonous plant.

LITERATURE CITED

Alcock, J. 1984. Animal behavior. Sinauer Associates, Sunderland.

Arnold, G.W., and M.L. Dudzinski. 1978. Ethology of free-ranging domestic animals. Elsevier, North Holland, New York.

Arnold, G.W., and J.L. Hill. 1972. Chemical factors affecting selection of food plants by ruminants, pp. 71-101. In: J.B. Harborne (ed), Phytochemical ecology. Academic Press, New York.

Arnold, G.W., and R.A. Maller. 1977. Effects of nutritional experience in early and adult life on the performance and dietary habits of sheep. Appl. Anim. Ethol. 3:5-26.

Baker, A.E.M., and B.H. Crawford. 1986. Observational learning in horses. Appl. Anim. Behav. Sci. 15:7-13.

Basson, P.A., A.G. Norval, J.M. Hofmeyr, H. Ebedes, R. Anitra Schultz, T.S. Kellerman, and J.A. Minne. 1984. Antelopes and poisonous plants, pp. 695-701. In: Thirteenth World Congress on Diseases of Cattle. Hoechest, Durban, Republic of South Africa.

Bate-Smith, E.C. 1972. Attractants and repellents in higher animals, pp. 45-56. In: J.B. Harborne (ed), Phytochemical ecology. Academic Press, New York.

Behnke, R.H. 1980. The herders of Cyrenaica: ecology, economy, and kinship among the bedouin of eastern Libya. Univ. Illinois Press, Chicago.

Bell, E.A. 1985. The biological significance of secondary metabolites in plants, pp. 50-57. In: A.A. Seawright, M.P. Hegarty, L.F. James, and R.F. Keeler (eds), Plant toxicology. Proc. Australia-U.S.A. Poisonous Plants Symp., Brisbane. Queensland Poisonous Plants Comm., Anim. Res. Inst., Yeerongpilly.

Boyd, R., and P.J. Richerson. 1985. Culture and the evolutionary process. Univ. Chicago Press, Chicago.

Brattsten, L.B. 1979. Biological defense mechanisms in herbivores against plant allelochemicals, pp. 199-270. In: G.A. Rosenthal and D.H. Janzen (eds), Herbivores: their interaction with secondary plant metabolites. Academic Press, New York.

Braveman, N.S., and P. Bronstein (eds). 1985. Experimental assessments and clinical applications of conditioned food aversions. New York Academy Sci., New York.

Brower, L.P., and L.S. Fink. 1985. A natural toxic defense system: cardenolides in butterflies versus birds, pp. 171-188. In: N.S. Braveman and P. Bronstein (eds), Experimental assessments and clinical applications of conditioned food aversions. New York Academy Sci., New York.

Bryant, J.P., F.S. Chapin III, and D.R. Klein. 1983. Carbon/nutrient balance of boreal plants in relation to vertebrate herbivory. Oikos 40:357-368.

Bryant, J.P., F.S. Chapin III, P.B. Reichardt, and T.P. Clausen. 1985. Adaptation to resource availability as a determinant of chemical defense strategies in woody plants. Rec. Adv. Phytochem. 19:219-237.

Bryant, J.P., and P. Kuropat. 1980. Selection of winter forage by sub-arctic browsing vertebrates: the role of plant chemistry. Ann. Rev. Ecol. Syst. 11:261-285.

Bryant, J.P., F.D. Provenza, and A. Gobena. 1987. Environmental controls over woody plant chemical defenses: implications for goat management, pp. 1005-1034. In: O.P. Santana, A.G. da Silva, and W.C. Foote (eds), Proc. Fourth Intnl. Conf. on Goats. Departamento de Difusao de Technologia, Brazilia, Brazil.

Burritt, E.A., and F.D. Provenza. 1987. Ability of lambs to discriminate between safe and harmful foods. (Abstract). Fortieth Annual Meeting Society for Range Management. Boise, ID.

Chapman, R.F., and W.M. Blaney. 1979. How animals perceive secondary compounds, pp. 161-198. In: G.A. Rosenthal and D.H. Janzen (eds), Herbivores: their interaction with secondary plant metabolites. Academic Press, New York.

Chapple, R.S., and J.J. Lynch. 1986. Behavioral factors modifying acceptance of supplementary foods by sheep. Res. Develop. Agr. 3:113-120.

Cheeke, P.R., and L.R. Shull. 1985. Natural toxicants in feeds and poisonous plants. AVI Publ. Co., Westport, CN.

Church, D.C. 1979. Taste, appetite and regulation of energy balance and control of food intake, pp. 281-320. In: D.C. Church (ed), Digestive physiology and nutrition of ruminants. Vol. 2, Nutrition. O & B Books, Corvallis, OR.

Dethier, V.G. 1980. Evolution of receptor sensitivity to secondary plant substances with special reference to deterrents. Amer. Nat. 115:45-66.

Domjan, M. 1985. Cue-consequence specificity and long-delay learning revisited, pp. 54-66. In: N.S. Braveman and P. Bronstein (eds), Experimental assessments and clinical applications of conditioned food aversions. New York Academy Sci., New York.

Dwyer, D.D. 1978. Impact of poisonous plants on western U.S. grazing systems and livestock operations, pp. 13-21. In: R.F. Keeler, K.R. Van Kampen, and L.F. James (eds), Effects of poisonous plants on livestock. Academic Press, New York.

Fowler, M.E. 1983. Plant poisoning in free-living wild animals: a review. J. Wildl. Dis. 19:34-43.

Freeland, W.J., and D.H. Janzen. 1974. Strategies in herbivory by mammals: the role of plant secondary compounds. Amer. Nat. 108:269-289.

Galef, B.G., Jr. 1971. Social effects in the weaning of domestic rat pups. J. Comp. Physiol. Psychol. 75:358-362.

Galef, B.G., Jr. 1985. Direct and indirect behavioral pathways to the social transmission of food avoidance, pp. 203-215. In: N.S. Braveman and P. Bronstein (eds), Experimental assessments and clinical applications of conditioned food aversions. New York Academy Sci., New York.

Galef, B.G., Jr., and M.M. Clark. 1971a. Social factors in the poison avoidance and feeding behavior of wild and domesticated rat pups. J. Comp. Physiol. Psychol. 75:341-357.

Galef, B.G., Jr., and M.M. Clark. 1971b. Parent-offspring interactions determine time and place of first ingestion of solid food by wild rat pups. Psychonom. Sci. 35:15-16.

Galef, B.G., Jr., and M.M. Clark. 1972. Mothers milk and adult presence: two factors determining initial dietary selection by weaning rats. J. Comp. Physiol. Psychol. 78:220-225.

Garcia, J., and W.G. Hankins. 1975. The evolution of bitter and the acquisition of toxiphobia, pp. 39-45. In: D.A. Denton and J.P. Coghlan (eds), Fifth International Symposium on Olfaction and Taste. Academic Press, New York.

Garcia, J., P.A. Lasiter, F. Bermudez-Rattoni, and D.A. Deems. 1985. A general theory of aversion learning, pp. 8-21. In: N.S. Braveman and P. Bronstein (eds), Experimental assessments and clinical applications of conditioned food aversions. New York Academy Sci., New York.

Green, G.C., R.L. Elwin, B.E. Mottershead, R.G. Keogh, and J.J. Lynch. 1984. Long term effects of early experience to supplementary feeding in sheep. Proc. Aust. Soc. Anim. Prod. 15:373-375.

Harborne, J.B. 1982. Introduction to ecological biochemistry. Academic Press, New York.

Hughes, J.G., D. McClatchy, and J.A. Hayward. 1971. Beef cattle on tussock country. Lincoln College Press, Canterbury, New Zealand.

Hungate, R.E. 1966. The rumen and its microbes. Academic Press, New York.

Jones, R.J. 1985. Leucaena toxicity and the ruminal degradation of mimosine, pp. 111-119. In: A.A. Seawright, M.P. Hegarty, L.F. James, and R.F. Keeler (eds), Plant toxicology. Proc. Australia - U.S.A. Poisonous Plants Symp., Brisbane. Queensland Poisonous Plants Comm., Anim. Res. Inst., Yeerongpilly.

Kingsbury, J.M. 1964. Poisonous plants of the United States and Canada. Prentice-Hall, Englewood Cliffs, NJ.

Kingsbury, J.M. 1978. Ecology of poisoning, pp. 81-91. In: R.F. Keeler, K.R. Van Kampen, and L.F. James (eds), Effects of poisonous plants on livestock. Academic Press, New York.

Krueger, W.C., and L.A. Sharp. 1978. Management approaches to reduce livestock losses from poisonous plants on rangelands. J. Range Manage. 31:347-350.

Laycock, W.A. 1978. Co-evolution of poisonous plants and large herbivores on rangelands. J. Range Manage. 31:335-342.

Lett, B.T. 1985. The pain-like effect of gallamine and naloxone differs from sickness induced by lithium chloride. Behav. Neurosci. 99:145-150.

Lobato, J.F.P., G.R. Pearce, and R.G. Beilharz. 1980. Effect of early familiarization with dietary supplements on the subsequent ingestion of molasses-urea blocks by sheep. Appl. Anim. Ethol. 6:149-161.

Lynch, J.J., R.G. Keogh, R.L. Elwin, G.C. Green, and B.E. Mottershead. 1983. Effects of early experience on the post-weaning acceptance of whole grain wheat by fine-wool merino lambs. Anim. Prod. 36:175-183.

Mason, J.R., and R.F. Reidinger. 1981. Effects of social facilitation and observational learning on feeding behavior of the red-winged blackbird (Agelaius phoeniceus). Auk 98:778-784.

Mason, J.R., and R.F. Reidinger. 1982. Observational learning of food aversions in red-winged blackbirds (Agelaius phoeniceus). Auk 99:548-554.

Mehansho, H., A. Hagerman, S. Clements, L. Butler, J. Rogler, and D.M. Carlson. 1983. Modulation of proline-rich protein biosynthesis in rat parotid glands by sorghums with high tannin levels. Proc. Natl. Acad. Sci. 80:3948-3952.

Merrill, L.B., and J.L. Schuster. 1978. Grazing management practices affect livestock losses from poisonous plants. J. Range Manage. 31:351-354.

Narjisse, H. 1981. Acceptability of big sagebrush to sheep and goats: role of monoterpenes. Ph.D. Thesis, Utah State Univ., Logan.

Olsen, J.D., and M.H. Ralphs. 1986. Feed aversion induced by intraruminal infusion with larkspur extract in cattle. Amer. J. Vet. Res. 47:1829-1833.

Provenza, F.D., and D.F. Balph. 1987. Diet learning by domestic ruminants: theory, evidence and practical implications. Appl. Anim. Behav. Sci. (in press).

Provenza, F.D., and J.C. Malechek. 1983. Tannin allocation in blackbrush. Biochem. Syst. Ecol. 11:233-238.

Provenza, F.D., and J.C. Malechek. 1984. Diet selection by domestic goats in relation to blackbrush twig chemistry. J. Appl. Ecol. 21:831-841.

Ralphs, M.H., L.V. Mickelsen, and D.L. Turner. 1987. Cattle grazing white locoweed: diet selection patterns of native and introduced cattle. J. Range Manage. 40:333-335.

Ralphs, M.H., and J.D. Olsen. 1987. Alkaloids and palatability of poisonous plants, pp. 78-83. In: F.D. Provenza, J.T. Flinders, and E.D. McArthur (eds), Proceedings--symposium on plant-herbivore interactions. USDA For. Serv. Intermtn. Res. Sta. Gen. Tech. Rep. INT-222.

Reichardt, P.B., J.P. Bryant, T.P. Clausen, and G.D. Wieland. 1984. Defense of winter-dormant Alaska paper birch against snowshoe hares. Oecologia 65:58-69.

Reichardt, P.B., T.P. Clausen, and J.P. Bryant. 1987. Plant secondary metabolites as feeding deterrents to vertebrate herbivores, pp. 37-42. In: F.D. Provenza, J.T. Flinders, and E.D. McArthur (eds), Proceedings--symposium on plant-herbivore interactions. USDA For. Serv. Intermtn. Res. Sta. Gen. Tech. Rep. INT-222.

Reicher, M.A., and E.W. Holman. 1977. Location preference and flavor aversion reinforced by amphetamine in rats. Anim. Learning Behav. 5:343-346.

Rhoades, D.F. 1979. Evolution of plant chemical defense against herbivores, pp. 3-54. In: G.A. Rosenthal and D.H. Janzen (eds), Herbivores: their interaction with secondary plant metabolites. Academic Press, New York.

Robbins, C.T., S. Mole, A.E. Hagerman, and T.A. Hanley. 1987. Role of tannins in defending plants against ruminants: reduction in dry matter digestibility? Ecology (in press).

Robinson, T. 1979. The evolutionary ecology of alkaloids, pp. 413-448. In: G.A. Rosenthal and D.H. Janzen (eds), Herbivores: their interaction with secondary plant metabolites. Academic Press, New York.

Rohan, T.A. 1972. The chemistry of flavour, pp. 57-69. In: J.B. Harborne (ed), Phytochemical ecology. Academic Press, New York.

Rosenthal, G.A., and D.H. Janzen (eds). 1979. Herbivores: their interaction with secondary plant metabolites. Academic Press, New York.

Rozin, P. 1976. The selection of food by rats, humans and other animals, pp. 21-76. In: J.S. Rosenblatt, R.A. Hinde, E. Shaw, and C. Beer (eds), Advances in the study of behavior. Academic Press, New York.

Swain, T. 1979. Tannins and lignin, pp. 657-682. In: G.A. Rosenthal and D.H. Janzen (eds), Herbivores: their interactions with secondary plant metabolites. Academic Press, New York.

Thorhallsdottir, A.G., F.D. Provenza, and D.F. Balph. 1987a. Food aversion learning in lambs with or without a mother: discrimination, novelty and persistence. Appl. Anim. Behav. Sci. (in press).

Thorhallsdottir, A.G., F.D. Provenza, and D.F. Balph. 1987b. Role of social models in the development of dietary habits in lambs. (Abstract). Fortieth Annual Meeting, Society for Range Management. Boise, ID.

White, N., L.S. Sklar, and Z. Amit. 1977. The reinforcing action of morphine and its paradoxical side effect. Psychopharmacology 52:63-66.

Wise, R.A., R.A. Yokel, and H. de Wit. 1976. Both positive reinforcement and conditioned aversion from amphetamine and apomorphine in rats. Science 191:1273-1274.

Wolfe, G.J., and W.R. Lance. 1984. Locoweed poisoning in a northern New Mexico elk herd. J. Range Manage. 37:59-63.

Zahorik, D.M., and K.A. Houpt. 1977. The concept of nutritional wisdom: applicability of laboratory learning models to large herbivores, pp. 45-67. In: L.M. Barker, M. Best, and M. Domjan (eds), Learning mechanisms in food selection. Baylor Univ. Press, Waco, TX.

Zahorik, D.M., and K.A. Houpt. 1981. Species differences in feeding strategies, food hazards, and the ability to learn food aversions, pp. 289-310. In: A.C. Kamil and T.D. Sargent (eds), Foraging behavior. Garland, New York.

List of Contributors

Bailey, E. Murl, Jr., D.V.M., Ph.D. Professor of Toxicology, Department of Physiology and Pharmacology, College of Veterinary Medicine, Texas A&M University, and The Texas Agricultural Experiment Station, The Texas A&M University System, College Station, Texas 77843-4466

Balph, David F. Professor, Department of Fisheries and Wildlife, Utah State University, Logan, Utah 84322

Bowns, James E. Professor of Range Ecology, Utah State University and Southern Utah State College, Southern Utah State College, Cedar City, Utah 84720

Calhoun, Millard C. Associate Professor, Texas Agricultural Experiment Station, 7887 North Highway 87, San Angelo, Texas 76901

Conner, J. Richard. Professor, Departments of Agricultural Economics and Range Science, Texas A&M University, College Station, Texas 77843

Cronin, Eugene H. Plant Physiologist (Retired), USDA-ARS Poisonous Plant Research Laboratory, Logan, Utah 84321

Dewey, Steven A. Extension Weed Specialist, Utah State University, Logan, Utah 84322

Dwyer, Don D. Professor Emeritus, Department of Range Science, Utah State University, Logan, Utah 84322

425

Gartner, F. Robert. Professor, Department of Animal and Range Sciences, and Director, South Dakota State University West River Agricultural Research and Extension Center, 801 San Francisco Street, Rapid City, South Dakota 57701

Godfrey, E. Bruce. Associate Professor, Department of Economics, Utah State University, Logan, Utah 84322

Gordon, Hal W. Department of Agricultural Economics and Agricultural Business, New Mexico State University, Las Cruces, New Mexico 88003

Harper, Kimball T. Professor, Department of Botany and Range Science, Brigham Young University, Provo, Utah 84602

Hatch, S. L. Associate Professor, Department of Range Science, Texas A&M University, College Station, Texas 77843

James, Lynn F. Research Leader, USDA-ARS Poisonous Plant Research Laboratory, Logan, Utah 84321

Johnson, A. Earl. Animal Physiologist, USDA-ARS Poisonous Plant Research Laboratory, Logan, Utah 84321

Johnson, Frederic D. Professor, Department of Forest Resources, University of Idaho, Moscow, Idaho 83843

Keeler, Richard F. Research Chemist, USDA-ARS Poisonous Plant Research Laboratory, Logan, Utah 84321

Lacey, John R. Extension Range Management Specialist, Montana State University, Bozeman, Montana 59715

Laycock, William A. Chairman, Range Management Division, University of Wyoming, Laramie, Wyoming 82071

Lorenz, Russell J. Professor of Range Science, North Dakota State University, Mandan, North Dakota 58554

McDaniel, Kirk C. Associate Professor, Department of Animal and Range Sciences, New Mexico State University, Las Cruces, New Mexico 88003

McGinty, Allan. Texas A&M University, Fort Stockton, Texas 79735

Mayland, H. F. Soil Scientist, USDA-ARS Soil and Water Management Research Unit, Route 1, Box 186, Kimberly, Idaho 83341

Morgan, Penelope. Assistant Professor, Department of Forest Resources, University of Idaho, Moscow, Idaho 83843

Nielsen, Darwin B. Professor, Department of Economics, Utah State University, Logan, Utah 84322

Ogden, Phil R. Division of Range Resources, University of Arizona, Tucson, Arizona 85721

Olsen, John D. Veterinary Medical Officer, USDA-ARS Poisonous Plant Research Laboratory, Logan, Utah 84321

Panter, Kip E. Research Animal Scientist, USDA-ARS Poisonous Plant Research Laboratory, Logan, Utah 84321

Pfister, James A. Range Scientist, USDA-ARS Poisonous Plant Research Laboratory, Logan, Utah 84321

Provenza, Frederick D. Associate Professor, Department of Range Science, Utah State University, Logan, Utah 84322

Ralphs, Michael H. Range Scientist, USDA-ARS Poisonous Plant Research Laboratory, Logan, Utah 84321

Reagor, J. C. Head, Diagnostic Toxicolgy, Texas Veterinary Medical Diagnostic Laboratory, College Station, Texas 77843

Rector, B. S. State Extension Range Specialist, Texas Agricultural Extension Service, College Station, Texas 77843

Rimbey, Neil R. Range Economist, University of Idaho-- Caldwell Research and Extension Service, Caldwell, Idaho 83605

Rittenhouse, L. R. Professor, Range Science Department, Colorado State University, Fort Collins, Colorado 80523

Rowe, L. D. Veterinary Medical Officer, USDA-ARS Southern Plains Area Veterinary Toxicology & Entomology Laboratory, College Station, Texas 77843

Ruyle, G. B. Range Management Specialist, School of Renewable Natural Resources, University of Arizona, Tucson, Arizona 85721

Schuster, Joseph L. Professor and Head, Department of Range Science, Texas A&M University, College Station, Texas 77843

Sharp, Lee A. Professor, Forestry Range and Wildlife Department, University of Idaho, Moscow, Idaho 83843

Sharrow, Steven H. Professor, Department of Rangeland Resources, Oregon State University, Corvallis, Oregon 97331

Short, Robert E. Research Physiologist, USDA-ARS Ft. Keogh Livestock and Range Research Station, Miles City, Montana 59301

Sosebee, Ron E. Professor, Department of Range and Wildlife Management, Texas Tech University, Lubbock, Texas 79409

Taylor, Charles A., Jr. Research Station Superintendent, Texas A&M University Agricultural Experiment Station, P.O. Box 918, Sonora, Texas 76950

Torell, L. Allen. Department of Agricultural Economics and Agricultural Business, New Mexico State University, Las Cruces, New Mexico 88003

Ueckert, Darrell N. Range Research Scientist, Texas Agricultural Experiment Station, 7887 North Highway 87, San Angelo, Texas 76901

Welsh, Stanley L. Department of Botany and Range Sciences, Brigham Young University, Provo, Utah 84602

Williams, M. Coburn. Plant Physiologist, USDA-ARS Poisonous Plant Research Laboratory, Logan, Utah 84321

Young, James A. Research Leader and Range Scientist, USDA-ARS, 920 Valley Road, Reno, Nevada 89512

Printed and bound by CPI Group (UK) Ltd, Croydon, CR0 4YY

23/10/2024

01778232-0008